International Conference on Many Particle Spectroscopy of Atoms, Molecules, Clusters and Surfaces 2010

Journal of Physics: Conference Series Volume 288

Sendai, Japan
4 – 7 September 2010

ISBN: 978-1-61782-713-6
ISSN: 1742-6588

Printed from e-media with permission by:

Curran Associates, Inc.
57 Morehouse Lane
Red Hook, NY 12571

Some format issues inherent in the e-media version may also appear in this print version.

Printed by Curran Associates, Inc. (2011)

Additional copies of this publication are available from:

Curran Associates, Inc.
57 Morehouse Lane
Red Hook, NY 12571 USA
Phone: 845-758-0400
Fax: 845-758-2634
Email: curran@proceedings.com
Web: www.proceedings.com

TABLE OF CONTENTS

Electron-impact Ionization of Oriented Molecules Using the Time-dependent Close-coupling Approach 1
J Colgan, M S Pindzola Doi

Close-coupling Calculations of 64.6 eV e-He Ionization 10
I Bray, D V Fursa, A S Kadyrov, A T Stelbovics, J Colgan, M S Pindzola

Electron Impact Double Ionization of Rare Gases: a Kinematical Analysis of the Non-first Order Effects at Intermediate Impact Energy 16
A Lahmam-Bennani, E M Staicu Casagrande, C Li

Electron Impact Double Ionization of Single Oriented Water Molecules: Evidence of the Role of the Two-step Mechanism 22
C Dal Cappello, I Kada, A Mansouri, C Champion

Coherent Electron Emission for Dissociative Ionization of Molecular Hydrogen by Electron Impact 34
O A Fojón, C R Stia, R D Rivarola

(e,2e) Experiments on C_{60} 40
P Bolognesi, L Praviça, J Berakdar, Y Pavlyukh, R Camilloni, D Cvejanovic, L Avaldi Doi:

Dynamical (e, 2e) Studies Using Tetrahydrofuran as a DNA Analogue 49
C J Colyer, S M Bellm, G F Hanne, O Al-Hagan, D Madison, C G Ning, B Lohmann

Calculation of (e, 2e) Triple Differential Cross Sections of Mg in Coplanar Geometry 55
G Purohit, Vinod Patidar, K K Sud

Theory of Laser-assisted Electron Momentum Spectroscopy: Beyond the Volkov Wave Born Approximation 63
Konstantin A Kouzakov, Yuri V Popov, Masahiko Takahashi

Ionization of Liquid Water by Fast Electron Impact: Multiple Differential Cross Sections for the $1B_1$ Orbital 68
O A Fojón, M L De Sanctis, R Vuilleumier, C R Stia, M -F Politis

Differential Positron and Electron Ionization Studies: How a Plus or Minus Sign Makes a Difference 75
R D Dubois

Surface-sensitive Measurements of Excitonic Processes in Rare Gas Clusters by Electron Energy Loss Spectroscopy 81
Hayato Kubotera, Tatsuya Sekitsuka, Satoshi Jinno, Takayuki Tachibana, Takato Hirayama

Double Differential Cross Sections and Generalized Oscillator Strength Distributions of Ammonia 87
Karin Yamamoto, Keisuke Nogami, Yuta Hino, Yasuhiro Sakai

Elastic Electron Scattering from the DNA Bases: Cytosine and Thymine 94
C J Colyer, S M Bellm, F Blanco, G Garcia, B Lohmann

Spin-orbit Effects in the (e,2e) Scattering from a W(110) Surface and Thin Gold Layer 99
S N Samarin, O M Artamonov, A D Sergeant, L Pravica, D Cvejanovic, P Wilkie, P Guagliardo, A A Suvorov, J F Williams

Auger Electron Spectra of Hydrogenated Si(111)-1×1 Surface Obtained from Si-L_{23}VV Auger Electron Si-2p Photoelectron Coincidence Measurements 107
Takahiko Yamazaki, Shogo Hashimoto, Takuhiro Kakiuchi, Kazuhiko Mase, Masatoshi Tanaka

Elastic Scattering and Impact Ionization by Returning Electrons Induced in a Strong Laser Field 112
Junliang Xu, Yaqiu Liang, Zhangjin Chen, C D Lin

Infrared Laser Assisted Photoabsorption of Ne Atoms 128
X M Tong, N Toshima

Ultra-intense X-ray Induced Non-linear Processes in Molecular Nitrogen 133
L Fang, M Hoener, N Berrah

Application of a VMI Spectrometer to Near-threshold Photoionization with Synchrotron Radiation 142
P O'Keeffe, P Bolognesi, A Mihelic, R Richter, A Moise, E Ovcharenko, G C King, L Avaldi

Dissociative Photoionization of Perfluorocyclobutane and *cis*-1,1,2,2,3,4-hexafluorocyclobutane 149
K Okada, T Nakashima, M Sakai, A Suemitsu, C Huang, H Yagi, H Katayanagi, K Mitsuke, K Tabayashi

Site- and Geometry-specific CH...O Interaction in Small Acetaldehyde Clusters Studied with Core-electron Excitation Spectroscopy in the Carbon K-edge Region 156
K Tabayashi, Y Tsutsumi, M Chohda, O Takahashi, Y Tamenori, I Higuchi, I H Suzuki, S Nagaoka, T Gejo, K Honma

Dissociation and Recapture Dynamics in H_2O Following O 1*s* Inner-shell Excitation 167
T Gejo, M Oura, M Kuniwake, K Honma, J R Harries

Theoretical Spectroscopy of O 1s and N 1s Excited States of N_2O 174
M Ehara, T Horikawa, R Fukuda, H Nakatsuji, T Tanaka, M Hoshino, H Tanaka, K Ueda

Interference and Non-franck-condon Effects in Ionization of H_2 Molecules by Photon Impact 185
O A Fojón, J Fernández, A Palacios, F Martín
Coherence and Partial Localization in Electron Emission from Asymmetric Diatomic Molecules 197
C A Tachino, M E Galassi, F Martín, R D Rivarola
Multiple Ionization and Fragmentation of Polyatomic Molecules by Fast Heavy Ions 207
A Itoh, T Mizuno, T Majima
Dynamic Correlation in the Electron Angular Distribution in Ionization of Helium by Ion Impact 216
J M Monti, O A Fojón, J Hanssen, R D Rivarola
Author Index

PREFACE

The 17th edition of the International Conference on Many Particle Spectroscopy of Atoms, Molecules, Clusters, and Surfaces (MPS) was held in Sendai, Japan, from September 4–7, 2010. It was the first time that a meeting of this series of biennial conferences was hosted in a non-European country. The conference was attended by 110 researchers (90 regular participants and 20 students) from 15 different countries around the world.

The themes that the conference covered can be divided into three broad areas: lepton impact, photon impact and heavy-particle impact. A total of 43 oral presentations – including 2 plenary talks, 29 progress reports and 12 hot topics – and 87 poster presentations were held during the course of the program. Rapid progress both in experimental and theoretical techniques has led to discussions across a broad range of currently hot topics, such as many-body dynamics and electron correlation effects in excitation processes, as well as in single and multiple ionization processes for various kinds of targets including atoms, molecules, clusters, solid state and even biological systems. A snapshot of the present status of many particle spectroscopy is given in this proceedings.

The chairs of the conference gratefully acknowledge the financial support from the Morino Foundation for Molecular Science, Iwatani Naoji Foundation, Sendai Tourism and Convention Bureau, and Intelligent Cosmos Academic Foundation. They are indebted to the Institute of Multidisciplinary Research for Advanced Materials, Tohoku University, for co-hosting the conference, as well as to the international advisory board members for their extremely helpful suggestions to make the program attractive. The united effort of the local organizing committee, involving N Watanabe (Secretary), M Okunishi (Treasurer), H Fukuzawa, M Yamazaki, Y Kino, and N Kishimoto, is also gratefully acknowledged. Without the help of these institutions and individuals it would have been impossible to organize the conference. Finally, the chairs would like to express their thanks to all the participants for contributing to lively and fruitful discussions throughout the conference.

Masahiko Takahashi and Kiyoshi Ueda

International Advisory Board

All papers published in this volume of *Journal of Physics: Conference Series* have been peer reviewed through processes administered by the proceedings Editors. Reviews were conducted by expert referees to the professional and scientific standards expected of a proceedings journal published by IOP Publishing.

Int. Conf. on Many Particle Spectroscopy of Atoms, Molecules, Clusters and Surfaces IOP Publishing
Journal of Physics: Conference Series **288** (2011) 012001 doi:10.1088/1742-6596/288/1/012001

Electron-impact ionization of oriented molecules using the time-dependent close-coupling approach

J. Colgan[1] **and M. S. Pindzola**[2]

[1]Theoretical Division, Los Alamos National Laboratory, Los Alamos, NM 87545, USA
[2]Department of Physics, Auburn University, Auburn, AL 36849, USA

E-mail: jcolgan@lanl.gov

Abstract. An overview is given on recent progress on computing triple differential cross sections for electron-impact ionization of the hydrogen molecule using a time-dependent close-coupling approach. Our calculations, when averaged over all molecular orientations, are generally in very good agreement with (e,2e) measurements made on H_2, where the molecular orientation is unknown, for a range of incident energies and outgoing electron angles and energies. In this paper, we present TDCS for ionization of H_2 at specific molecular orientations. It is hoped that this study will help stimulate future measurements of TDCS from oriented H_2 at medium impact energies.

1. Introduction

Ionization of small molecules has come under increasing scrutiny in recent years. This work has built on a major success of atomic collision physics in the last 20 years - the development of increasingly sophisticated techniques to measure the triple differential cross sections (TDCS) arising from the ionization of small atoms, particularly hydrogen and helium [1, 2, 3, 4]. These methods have been refined alongside the development of advanced close-coupling techniques which compute all possible cross sections for single ionization of hydrogen and helium with few approximations. Experiment and theory are in excellent agreement for the differential cross sections for electron-impact ionization of atomic hydrogen [5, 6, 7]. For helium, several techniques are in very good, although not perfect, agreement with most measurements and with each other [8, 9, 10, 11].

It is thus natural to consider other few-electron targets of increasing complexity, such as molecular hydrogen. H_2 offers major challenges, in particular to theory. Since the spherical symmetry, which is so advantageous in atomic collisions, has been broken, the standard partial-wave expansions of the electronic wavefunctions can no longer be readily made. In addition, the extra vibrational and rotational degrees of freedom unique to molecules can complicate the interpretation of measurements made on non-cooled molecules. The TDCS from single ionization of H_2 have been measured by several groups, over a range of incident energies and outgoing electron energies and angles [12, 13, 14, 15]. Almost all of these measurements have been made on molecules where the molecular orientation is unknown, apart from a few very recent studies made at higher impact electron energies [16, 17]. Recent theoretical developments have produced calculations which are, for the most part, in good agreement with these measurements [18, 19, 20, 21], when the calculations are averaged over all possible orientations.

It is of considerable interest to study the TDCS arising from ionization of oriented H_2. Although several recent studies have highlighted the differences between the TDCS from H_2 and He (both two-electron systems) [18, 22], one might expect further differences to be found when considering oriented molecules. Studies of the analogous process of double photoionization of oriented H_2 finds strong molecular effects, particularly with regard to the dependence of the TDCS on the molecular orientation [23].

In this report, we present a theoretical study of the TDCS from electron-impact single ionization of oriented H_2 molecules, for a range of moderate impact electron energies, and several different outgoing electron geometries. Differences between the TDCS from oriented molecules and randomly-oriented molecules, as well as He, are highlighted. It is hoped that such studies may further stimulate measurements of such quantities at moderate impact electron energies.

2. The time-dependent close-coupling method

Our approach uses the time-dependent close-coupling (TDCC) technique [24], used previously to obtain total cross sections and TDCS for electron-impact ionization of H_2 [19, 20, 25], where details of the TDCC approach to molecular ionization can be found. Briefly, an expansion of the total electronic wavefunction in products of four-dimensional radial angular functions and rotational functions is used to express the Schrödinger equation in terms of a set of time-dependent coupled partial differential equations. The non-ionized electron of H_2 is frozen and its interaction with the two outgoing electrons is represented through direct and local exchange potential terms. The interaction between the two outgoing electrons is treated in full. The time-dependent equations are propagated for each M (projection of the angular momentum onto the z-axis), L (the incident angular momentum of the incident electron, represented by a Gaussian wavepacket), and S (total spin angular momentum), for a given incident electron energy and at the equilibrium internuclear separation of H_2. The TDCS can be expressed as

$$\begin{aligned} \frac{d^3\sigma}{dE_1 d\Omega_1 d\Omega_2} &= \frac{\pi}{4k^2}\frac{1}{k_1 k_2}\sum_S (2S+1) \\ &\times \int dk_1 \int dk_2 \delta\Big(\alpha - \tan^{-1}\frac{k_2}{k_1}\Big)|\mathcal{M}|^2 , \end{aligned} \tag{1}$$

where k is the incident electron's momentum and k_1 and k_2 are the outgoing electron momenta (ejected into solid angles $\Omega_{1,2}$), with α the hyperspherical angle between k_1 and k_2. For diatomic molecules, where the z-axis is defined along the internuclear direction, and the incoming electron beam is oriented at angles (θ_k, ϕ_k) with respect to the z-axis,

$$\begin{aligned} \mathcal{M} &= \sum_L \sum_{M=-L}^{+L} i^L Y^*_{LM}(\theta_k, \phi_k) \sum_{l_1,l_2} \sum_{m_1 m_2} (-i)^{l_1+l_2} e^{i(\sigma_{l_1}+\sigma_{l_2})} \\ &\times P^{LMS}_{l_1 m_1 l_2 m_2}(k_1, k_2, T) Y_{l_1 m_1}(\hat{k_1}) Y_{l_2 m_2}(\hat{k_2}) \delta_{m_1+m_2, M} , \end{aligned} \tag{2}$$

where $P^{LMS}_{l_1 m_1 l_2 m_2}(k_1, k_2, T)$ is the wavefunction which results after projection of the four-dimensional radial and angular wavefunctions onto products of H_2^+ continuum states after propagation to a suitable time T. In Eq. (2), $Y_{lm}(\hat{k})$ is a spherical harmonic, and σ_l is the Coulomb phase shift. Our TDCS expression defined by Eqs. (1,2) is given in the molecular frame. To compare with experiment, a transformation must be made into the Laboratory frame. In the measurements shown here, the molecular orientation is unknown, and so the TDCS is computed for all possible (θ_N, ϕ_N) (these are the angles the molecule makes with the z-axis in the Lab frame, where in the Lab frame the z-axis is defined by the electron beam direction), which are then averaged to compare with the measurements. TDCS for specific orientations of

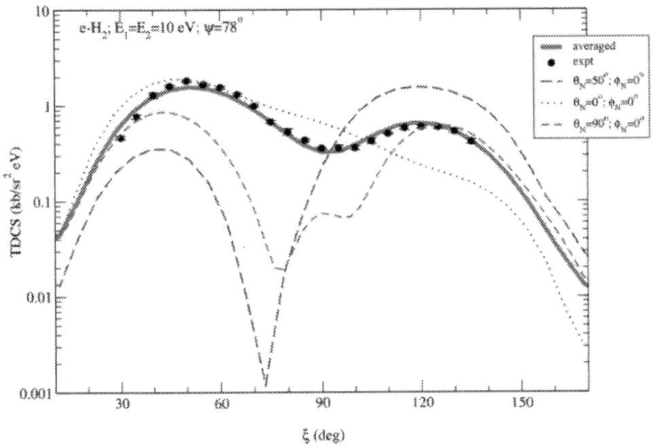

Figure 1. TDCS for H_2 for E_1=E_2=10 eV and for $\psi = 78°$. The measurements previously presented in [18] are compared with the TDCC calculations (solid red line) averaged over all molecular orientations. The dashed purple line indicates a TDCC calculation for a molecular orientation of $\theta_N = 50°$, $\phi_N = 0°$. The dotted blue line and short-dashed green line indicate TDCC calculations for molecular orientations $\theta_N = 0°$, $\phi_N = 0°$ and $\theta_N = 90°$, $\phi_N = 0°$, respectively.

the molecule are also presented. All calculations are made for a fixed internuclear separation of 1.4 a.u., the equilibrium separation of H_2, which implies that we employ the Franck-Condon approximation. This approximation should be quite suitable for the cases considered here, where the electrons are emitted with several eV, and so move much faster than the heavier nuclei.

3. Results and discussion

In a previous study [21], we presented TDCS from oriented H_2 for equal energy sharing outgoing electrons, in which it was demonstrated that a deep minimum is present in the TDCS for a specific molecular orientation and a specific "gun" angle $\psi = 67.5°$ (where the gun angle is the angle between the incident electron beam and the detection plane in the Manchester experimental set-up [2]). This deep minimum was first shown for TDCS from helium, and has recently been attributed to the presence of vortices in the outgoing electronic wavefunction [26]. The minimum is found only for highly symmetric configurations. For example, the two outgoing electrons must have equal energy, and both electrons must make the same angle (ξ) with respect to the axis defined by the incident electron beam. This orientation implies that only singlet amplitudes contribute to the TDCS, in both the He and H_2 cases. Our previous study showed that the minimum is not observed in the measurements made with molecules in which the orientation is unknown. In figure 1 we show that a similar minimum is also present for another gun angle value of $\psi = 78°$, where in this case, a deep minimum is found for the molecular orientation $\theta_N = 50°$, $\phi_N = 0°$. When the molecule is aligned along the incident beam direction ($\theta_N = 0°$, $\phi_N = 0°$), or perpendicular to it ($\theta_N = 90°$, $\phi_N = 0°$), no deep minimum is observed.

Our previous analysis [21] found that the minimum occurred in the TDCS for He as a result of deep destructive interference between the various contributing partial waves which are included in the TDCC calculations. In figures 2 and 3 we show that a similar interference effect causes the minimum in the TDCS for H_2. Figure 2 shows the TDCS for H_2 at the specific orientation of $\theta_N = 50°$, $\phi_N = 0°$, where the number of partial waves included in the calculation have been truncated, as indicated. Figure 3 shows the contribution from the interference terms which arise in the coherent sum over L in equation (2). In these figures, the contributions from each M

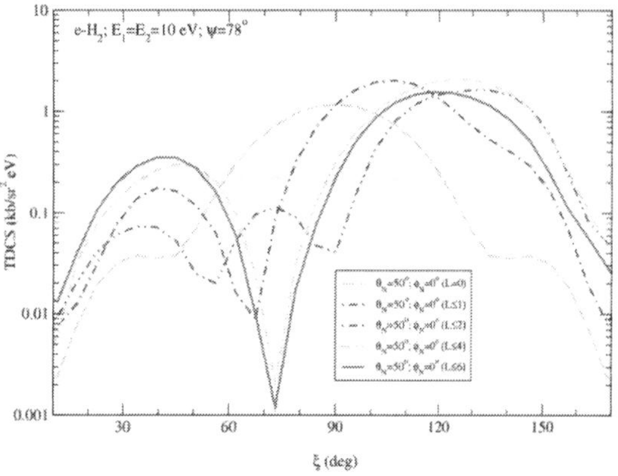

Figure 2. TDCC calculations for the TDCS of H_2 for the molecular orientation $\theta_N = 50°$, $\phi_N = 0°$. The various dashed lines show calculations which include fewer partial wave contributions, as indicated.

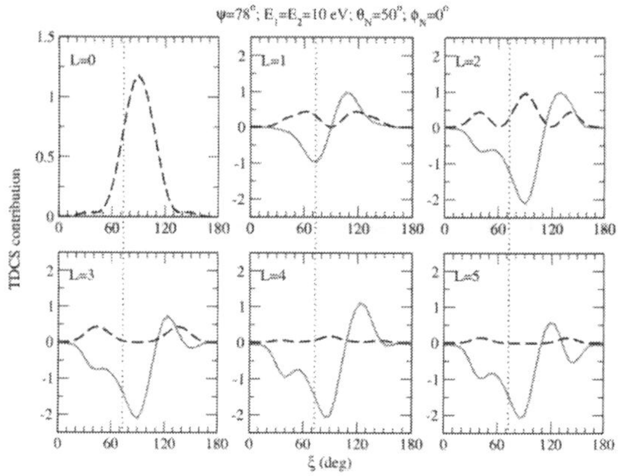

Figure 3. Contributions from individual partial waves as indicated (dashed black lines). The solid red lines signify the contribution from the interference (cross) terms inherent in the coherent sum in equation (2). The vertical dotted lines indicate the position of the minimum in the TDCS in figure 1.

amplitude corresponding to a given L have been coherently summed. Figures 2 and 3 (for H_2) are qualitatively very similar to the same analysis applied to He, as shown in figures 3 and 4 of [21].

Figure 3 clearly demonstrates that it is the interference between the partial waves which causes the minimum in the TDCS. The cross terms (with respect to L) in equation 2 destructively contribute to the TDCS in the region where the minima is observed in the complete TDCC calculations. It would be of great interest to explore whether the presence of a vortex in the outgoing electronic wavefunction in this case could also be observed, as has been postulated in the helium case [26].

In figure 4 we also show that a similar deep minimum may be observed at a lower incident electron energy of 25.4 eV. Again, no indication of a minimum is found in the TDCC calculations averaged over all molecular orientations (which are in excellent agreement with measurement). However, a deep minimum is found when the molecule is oriented at $\theta_N = 40°$, $\phi_N = 0°$ with respect to the z-axis.

These results show, perhaps counter-intuitively, that the TDCS for H_2 is more similar to the TDCS for He for the *oriented* case than for the *unoriented* case. This implies that molecular effects are *not* enhanced when we examine the TDCS from a specific orientation. However, this appearance is probably misleading. The similarity between the TDCS for He and for oriented H_2 for these highly symmetric cases is more likely due to the dominance of electron-electron correlation effects over the influence of the different potentials (atomic or molecular) in which the outgoing electrons move. In the following sections we find other cases in which the TDCS for oriented molecules are quite different to those from He.

In figure 5 we present TDCS for perpendicular plane cases. These are geometries where the experimental gun angle is perpendicular to the detection plane of the two outgoing electrons [18]. A previous study [18] highlighted the differences between He and H_2 in this configuration: the TDCS for He is found to be a dominant central ($\xi = 90°$) peak with smaller side peaks; whereas

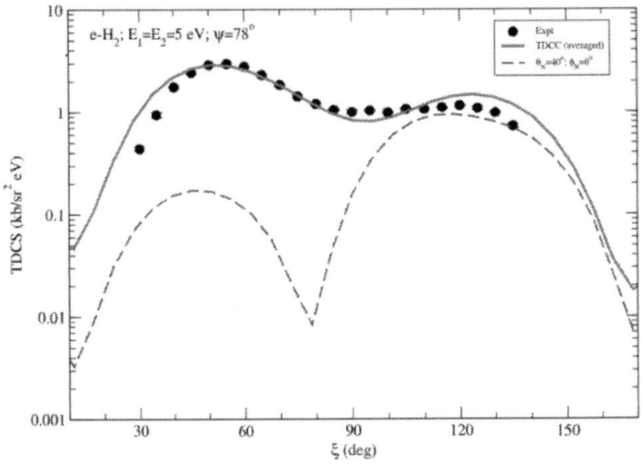

Figure 4. TDCS for H_2 for E_1=E_2=5 eV and for $\psi = 78°$. The measurements of [27] are compared with the TDCC calculations (solid red line) averaged over all orientations. The dashed purple line indicates a TDCC calculation for a molecular orientation of $\theta_N = 40°$, $\phi_N = 0°$.

the TDCS for H_2 is found to be two equal peaks at either side of a shallow minimum at $\xi = 90°$. This difference was ascribed to the positioning of the nuclei in the two systems. For helium, the large central peak resulted from one of the scattered electrons being very close to the nucleus, so that is elastically backscatters at 180° from the nucleus (corresponding to $\xi = 90°$, since the angle between the outgoing electrons is 2ξ). For H_2, the electron-electron collision occurs between the nuclei, where the net attractive force cancels, resulting in little backscattering and so no peak at 180°. A more recent study [27] has examined the change in the H_2 TDCS in the perpendicular plane as the outgoing electron energies decrease. It was found that the central minimum feature eventually becomes a weak maximum for cases where the two electrons share 2 eV, due to the dominance of post-collision interaction between the slowly moving outgoing electrons.

In the upper panels of figure 5 we show H_2 TDCS measurements in the perpendicular plane for two sets of outgoing electron energies as indicated. The TDCC calculations are in excellent agreement with experiment and clearly show the central minimum found previously at these energies [18, 27]. In the middle and lower panels we present TDCC calculations for the same geometries where now the molecular orientation is fixed at $\theta_N = 0°$, $\phi_N = 0°$ and $\theta_N = 90°$, $\phi_N = 0°$, respectively. Some interesting features are evident.

For cases where the molecule is aligned along the z-axis (i.e. along the incoming electron beam), the central minimum in the TDCS disappears. For the E_1=E_2=10 eV case (left-hand-side), the TDCS is flat in the central region, and at the lower energies a small maximum is found. The two side peaks (near $\xi = 60°$ and 120°) remain. This can be understood since the two outgoing electrons have little probability of colliding between the nuclei, since the initially bound electron is constrained by the fixed molecule position. The electron-electron collision is most likely to occur in a region where the attractive force from the nuclei is substantial, allowing greater backscatter and so a larger cross section in the $\xi = 90°$ region.

For cases where the molecule is perpendicular to the incoming beam ($\theta_N = 90°$, $\phi_N = 0°$, lower panels), the TDCS is quite different from the averaged case. For 10 eV outgoing electrons, the dip in the TDCS is very pronounced, but for the 5 eV outgoing electron case, this dip has turned into a maximum in the central region. At even lower energies (not shown), we find that this central peak completely dominates the cross section, resulting in a TDCS which is quite He-like. This trend appears to indicate that post-collision interaction (PCI) begins to influence

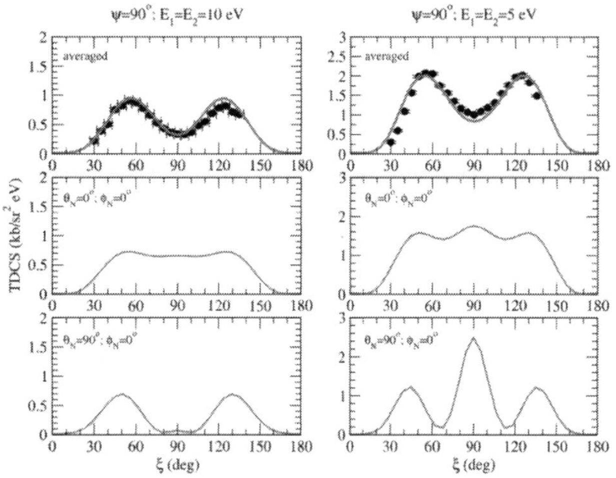

Figure 5. TDCS for H_2 for $\psi = 90°$. The left panels show TDCS for E_1=E_2=10 eV and the right-hand panels show TDCS for E_1=E_2=5 eV. The measurements of [18,27] are compared with the TDCC calculations (solid red line) averaged over all molecular orientations in the upper panels. The center and lower panels show TDCS for two molecular orientations as indicated.

the TDCS shape for the 5 eV outgoing electrons, and as the electron energy decreases, PCI dominates the shape of the TDCS. The electron-electron interaction pushes the electrons apart so that their preferred escape direction is back-to-back. This also suggests that the Wannier region may extend somewhat further out in energy for oriented molecules than was previously found for the randomly oriented molecule case [27].

In figure 6 we turn to coplanar examples (i.e. where the incident and detected elecrons all lie in the same plane). In this case, as previously noted [19, 20], the TDCC calculations for $E_1 = E_2 = 10$ eV averaged over all geometries are in only marginal agreement with the magnitude of the measurements. At the lower energies of $E_1 = E_2 = 5$ eV the agreement is somewhat better, although certainly not perfect. The oriented TDCS for these geometries are

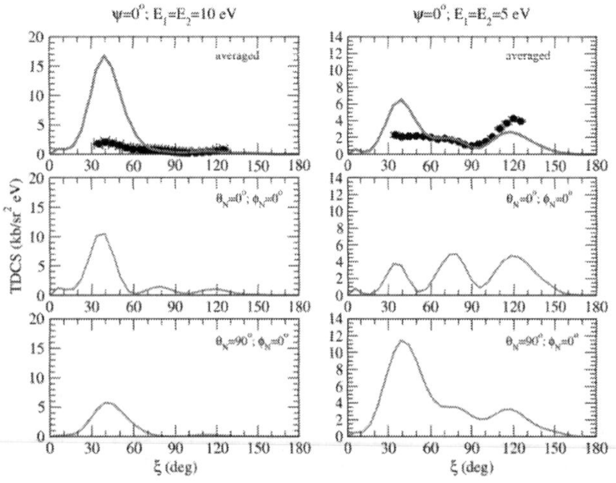

Figure 6. Same as figure 5, except for $\psi = 0°$.

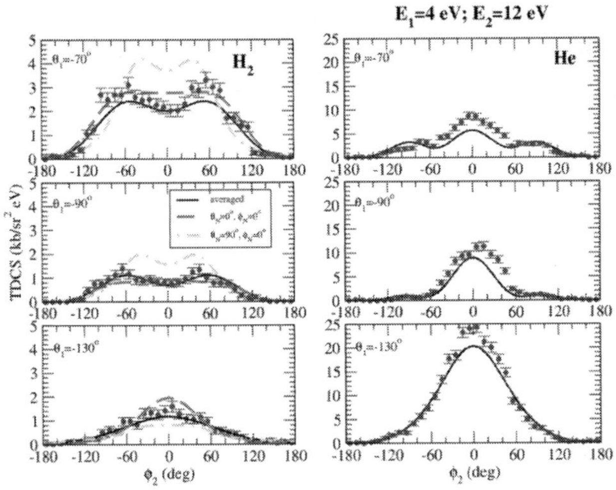

Figure 7. TDCS for H_2 (left-hand-side) and He (right-hand-side). TDCC calculations averaged over all orientations (solid red lines) are compared with experiment [22]. The electrons are emitted in the perpendicular plane with the energies indicated and the TDCS is shown as a function of the azimuthal angle of the second electron for fixed first electron angles as indicated. The left panels for H_2 also show the TDCS for two fixed orientations of the molecule as indicated.

again quite different from the averaged cases. For orientations along the z-axis, a third peak is found in the TDCS between the usual binary and recoil peaks (at around $\xi = 40°$ and $\xi = 120°$, respectively). This extra peak is quite prominent in the lower energy case. The extra feature observed for this geometry is again in the region where the electrons escape in a back-to-back configuration, and as in the previous example illustrated in figure 5, this back-to-back escape mode seems to be favoured more for lower energies. For molecules oriented perpendicular to the incoming beam, but still within the detection plane (lower panels), the TDCS is quite similar to the averaged case.

Finally, in figures 7 and 8, we examine oriented TDCS for H_2 and compare our calculations to recent measurements made on randomly oriented H_2 and on He [22]. Again, the TDCC calculations averaged over all orientations are in good agreement with the absolute measurements, in shape as well as in magnitude. Our calculations for He are also in very good agreement with the measured values, and also in good agreement with convergent close-coupling calculations [22]. The TDCS are shown for fixed energy sharing and 3 different θ_1 values (in figure 7) and for 3 different energy sharings at a fixed θ_1 value of $-90°$ (figure 8). These measurements and calculations are made in the perpendicular plane, where in these cases this signifies that the cross sections are measured for a fixed θ_1 and as a function of ϕ_2, with θ_2 fixed at $-90°$ [22] (i.e. perpendicular to the incident electron direction). The TDCS for H_2 oriented in the perpendicular plane (dot-dashed green lines) are somewhat similar to the averaged results, although the oriented TDCS are generally larger than the averaged results. The results for H_2 oriented along the incident electron direction ($\theta_N = 0°$, $\phi_N = 0°$) are flatter than the averaged results, with little evidence of a minimum in the TDCS near $\phi_2 = 0°$. In all cases, the averaged H_2 TDCS, as well as the oriented H_2 TDCS, are quite different in shape compared with the corresponding cross sections for He. The comparisons of the averaged H_2 TDCS with He are discussed in more detail elsewhere [22].

4. Conclusions

We have presented a range of TDCS for the electron-impact ionization of oriented H_2 molecules. We have identified a number of geometries where interesting effects are found, with substantial differences between the oriented molecule TDCS and the TDCS arising from unoriented H_2. We also compare and contrast our calculated TDCS with those obtained for He at similar geometries and kinematics.

It would be highly desirable to obtain measurements of the TDCS from oriented H_2 to check our predictions. We realize that this is a formidable experimental challenge. The only measurements made to date which report differential cross sections for H_2 have been obtained at quite large impact energies [16, 17], which require prohibitively large close-coupling calculations due to the need to include a large number of partial waves in the wavefunction expansions to ensure convergence. Moreover, the orientation of the molecule at the time of ionization is obtained by selecting those ionization events which lead to dissociation, so that one of the ionic fragments may be captured, and its momenta recorded. Unfortunately, to model such an experiment, it may be sometimes necessary to compute the electron-impact simultaneous ionization and excitation of H_2, so that the residual ion is left in a (dissociative) excited state. This implies a complete three-electron multi-center calculation, which is beyond the scope of current computational capabilities.

An alternative possibility is that the usual (e,2e) experiment may be performed on molecules which have been laser aligned [28]. Although this will most likely dramatically reduce the count rate compared to experiments in which no alignment is used, this approach appears to be the most promising in attempts to measure the TDCS from oriented molecules at low energies, and which are left in the residual ion ground state.

Acknowledgments
The Los Alamos National Laboratory is operated by Los Alamos National Security, LLC for the National Nuclear Security Administration of the U.S. Department of Energy under Contract No. DE-AC5206NA25396. A portion of this work was performed through DOE and NSF grants to Auburn University. Computational work was carried out at NERSC, in Oakland, CA, and through a LANL Institutional Computing Resources award.

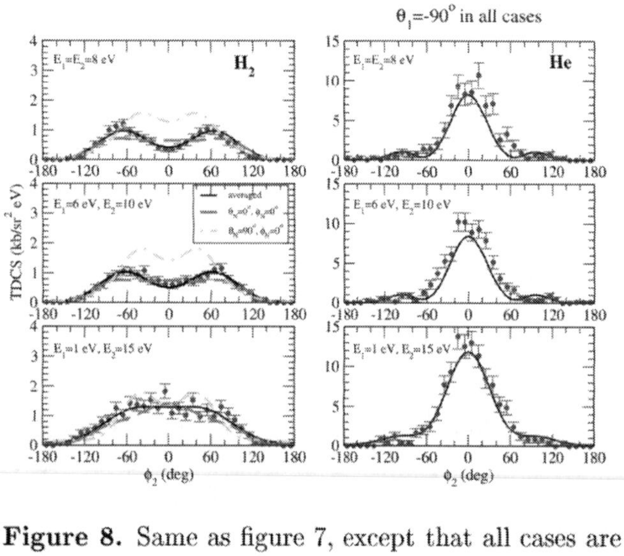

Figure 8. Same as figure 7, except that all cases are for a fixed electron angle (-90°), and for various outgoing electron energy sharings, as indicated.

References

[1] Erhardt H, Jung K, Knoth G and Schlemmer P 1986 *Z. Phys. D* **1** 3
[2] Murray A J, Woolf M B J and Read F H 1992 *J. Phys. B: At. Mol. Opt. Phys.* **25** 3021
[3] Rioual S, Röder J, Rouvellou B, Erhardt H, Pochat A, Bray I and Fursa D V 1998 *J. Phys. B: At. Mol. Opt. Phys.* **31** 3117
[4] Dürr M, Dimopoulou C, Najjari B, Dorn A and Ullrich J 2006 *Phys. Rev. Letts.* **96** 243202
[5] Rescigno T N, Baertschy M, Isaacs W A and McCurdy C W 1999 *Science* **286** 2474
[6] Bray I 2002 *Phys. Rev. Letts.* **89** 273201
[7] Colgan J and Pindzola M S 2006 *Phys. Rev. A* **74** 012713
[8] Stelbovics A T, Bray I, Fursa D V and Bartschat K 2005 *Phys. Rev. A* **71** 052716
[9] Colgan J, Pindzola M S, Childers G and Khakoo M 2006 *Phys. Rev. A* **73** 042710
[10] Colgan J, Foster M, Pindzola M S, Bray I, Stelbovics A T and Fursa D V 2009 *J. Phys. B: At. Mol. Opt. Phys.* **42** 145002
[11] Bray I, Fursa D V, Kadyrov A S and Stelbovics A T 2010 *Phys. Rev. A* **81** 062704
[12] Murray A J 2005 *J. Phys. B: At. Mol. Opt. Phys.* **38** 1999
[13] Takahashi M, Watanabe N, Udagawa Y and Eland J H D 2005 *Phys. Rev. Letts.* **94** 213202
[14] Milne-Brownlie D S, Foster M, Gao B, Lohmann B and Madison D H 2006 *Phys. Rev. Letts.* **96** 233201
[15] Staicu-Casagrande E M, Naja A, Mezdari F, Lahmam-Bennani A, Bolognesi P, Joulakian B, Chuluunbaatar O, Al-Hagan O, Madison D H, Fursa D V and Bray I 2008 *J. Phys. B: At. Mol. Opt. Phys.* **41** 025204
[16] Bellm S, Lower J, Weigold E and Mueller D W 2010 *Phys. Rev. Letts.* **104** 023202
[17] Senftelben A, Pflüger T, Ren X, Al-Hagan O, Najjari B, Madison D, Dorn A and Ullrich J 2010 *J. Phys. B: At. Mol. Opt. Phys.* **43** 081002
[18] Al-Hagan O, Kaiser C, Madison D H and Murray A J 2008 *Nature Physics* **5** 59
[19] Colgan J, Pindzola M S, Robicheaux F, Kaiser C, Murray A J and Madison D H 2008 *Phys. Rev. Letts* **101** 233201
[20] Colgan J, Al-Hagan O, Madison D H, Kaiser C, Murray A J and Pindzola M S 2009 *Phys. Rev. A* **79** 052704
[21] Colgan J, Al-Hagan O, Madison D H, Murray A J and Pindzola M S 2009 *J. Phys. B: At. Mol. Opt. Phys.* **42** 171001
[22] Ren X, Senftelben A, Pflüger T, Dorn A, Colgan J, Pindzola M S, Al-Hagan O, Madison D, Bray I, Fursa D V and Ullrich J 2010 *Phys. Rev. A* **82** 032712
[23] Reddish T J, Colgan J, Bolognesi P, Avaldi L, Gisselbrecht M, Lavollée M, Pindzola M S and Huetz A 2008 Phys. Rev. Letts. **100** 193001
[24] Pindzola M S *et al* 2007 *J. Phys. B: At. Mol. Opt. Phys.* **40** R39
[25] Pindzola M S, Robicheaux F J and Colgan J 2006 *Phys. Rev. A* **73** 052706
[26] Macek J H, Sternberg J B, Ovchinnikov S Y and Briggs J S 2010 *Phys. Rev. Letts* **104** 033201
[27] Al-Hagan O, Madison D H, Murray A J, Kaiser C and Colgan J 2010 *Phys. Rev. A* **81** 030701
[28] Murray A J 2010 *J. Phys.: Conference Series* **212** 012016

Journal of Physics: Conference Series **288** (2011) 012002 doi:10.1088/1742-6596/288/1/012002

Close-coupling calculations of 64.6 eV e-He ionization

I Bray[1], D V Fursa[1], A S Kadyrov[1], A T Stelbovics[1], J Colgan[2] and M S Pindzola[3]

[1]Centre for Antimatter-Matter Studies, Curtin University, GPO Box U1987 Perth, Western Australia 6845, Australia
[2]Theoretical Division, Los Alamos National Laboratory, NM 87545, USA
[3]Department of Physics, Auburn University, Auburn, AL 36849, USA

Abstract. The convergent close-coupling and the time-dependent close-coupling methods are applied to the calculation of 64.6 eV electron-impact ionization of the ground state of helium resulting in two 20 eV outgoing electrons. The results of the calculations are compared with measured fully differential cross sections in various geometries. For the generally large in-plane geometries there is good agreement between the two theories and experiment. For the out-of-plane case the cross sections are generally smaller and some differences between the two calculations are evident, as well as experiment.

1. Introduction

The problem of calculating fully differential electron-impact ionization cross sections has seen enormous progress in the last decade. The goal of yielding accurate cross sections irrespective of geometry or kinematics can only be attained by non-perturbative approaches that aim to numerically solve the full few-body problem. The exterior complex scaling method of Rescigno *et al* [1, 2] was first to show that this was indeed possible for the e-H ionization problem. This was followed soon after by the convergent close coupling (CCC) method [3], and the time-dependent close-coupling approach [4]. It is our view that these techniques are able to fully solve the e-H three-body problem by providing results of sufficiently high accuracy for comparison with experiment.

The e-He problem is a four-body problem and may be expected to be substantially more difficult than the e-H system. Due to the tight binding of the He^+ electron, the He discrete spectrum consists of only one-electron excitations. Hence, the e-He interactions are dominated by one-electron transitions. This allows the reduction of the e-He four-body problem to essentially a three-body problem, not much more difficult than the e-H system. The CCC approach typically treats e-He in the frozen-core approximation, that keeps one electron in the ground state of He^+. This can be relaxed, if necessary, to use a multi-configuration treatment of He, but this has been found only necessary for more complex targets such as Mg. Sufficiently good agreement with e-He experiments of the frozen-core model is a reflection of the quality of the approximation. It has to be noted that the e-He system also has ionization plus excitation, and double ionization processes. These are much smaller in magnitude than the one-electron processes, and remain a formidable challenge to theory, see Pindzola *et al* [5] for example.

Electron scattering on helium is one of the most experimentally studied collision problems in atomic physics. The determination of (often) absolute fully differential cross sections for

Journal of Physics: Conference Series **288** (2011) 012002 doi:10.1088/1742-6596/288/1/012002

one-electron ionization by, in particular, the Kaiserslautern and Manchester groups over two decades [6]-[17] has given an excellent foundation for the testing of theory. So much so that we wonder whether the e-He single ionization problem can also be regarded as solved in the same way as the e-H system. The purpose of this article is to briefly address this issue.

2. Theory

We shall present the results of two non-perturbative approaches to the problem of calculating one-electron ionization of helium by electron impact. The specific problem we consider is 64.6 eV electron-impact ionization of helium with two 20 eV outgoing electrons, leaving the He^+ ion in the ground state.

2.1. Convergent close-coupling (CCC) method

The CCC method for the general e-He collision problem [18] expands the total wavefunction using states obtained by diagonalizing the target Hamiltonian in a two-electron Laguerre basis. In the frozen-core approximation one of the electrons is restricted to be the 1s orbital of He^+. This generally suffices for scattering from the He ground state, but a multi-configuration treatment may be required when scattering from the helium metastable states or more complex two-electron targets, see Refs. [19, 20] for example. The expansion states have negative- and positive-energy states relative to the ground state energy of He^+ (-54.4eV). With increasing basis size the negative-energy states yield an increasing number of He discrete eigenstates, and the positive-energy states provide an increasingly dense discretization of the He one-electron continuum. Solution of the resulting close-coupling equations is performed in momentum space yielding scattering amplitudes for all open states. Ionization is associated with excitation of the positive-energy states.

The extension of the CCC method to generate the ionization scattering amplitudes has proved to be remarkably simple [21], though it took some time to fully appreciate the consequences. Briefly, the scattering amplitude for a positive energy state is multiplied by the overlap of the state and the corresponding true continuum state of the same energy. The origin of this treatment has only been fully understood following the derivation of the expression necessary to calculate true ionization scattering amplitudes [22, 23]. Furthermore, as the availability of the computational resources continued to grow it was realized that the underlying scattering amplitudes exhibited step-function behaviour, with the step being at the equal-energy sharing point [24]. At that point the ionization amplitudes can always be obtained ab initio from coherent combinations of the underlying amplitudes for both hydrogen [25] and helium [26]. It is the latter formulation that we use here.

2.2. Time-dependent close-coupling (TDCC) method

The TDCC method as applied to the electron-impact single ionization of helium has previously been described in some detail [27, 28]. Briefly, an expansion of the total electronic wavefunction for the two outgoing electrons in products of radial functions and coupled spherical harmonics is used to express the Schrödinger equation in terms of a set of time-dependent coupled partial differential equations.. The non-ionized electron of helium is frozen and its interaction with the two outgoing electrons is represented through direct and local exchange potential terms. The interaction between the two outgoing electrons is treated in full. The time-dependent equations are propagated for each LS (total orbital angular and spin angular momentum, respectively), for sufficient L so that the resulting triple differential cross sections are completely converged. In the equal-energy calculations presented here, it was found that the TDCS was well converged with around 10 L contributions, although for asymmetric energy sharings, more terms are usually needed.

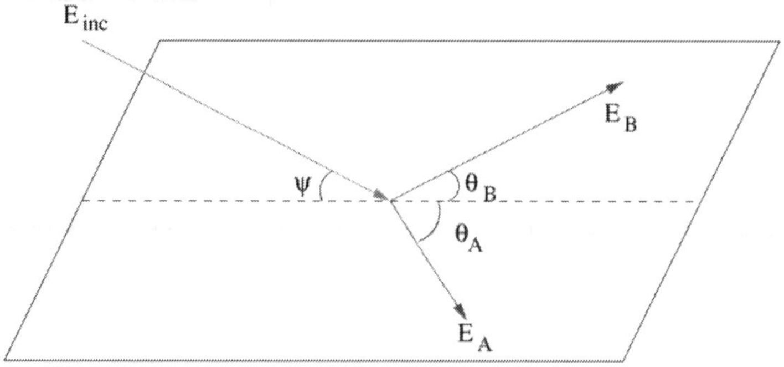

Figure 1. Orientation of detectors in an (e,2e) experiment, for coplanar ($\psi = 0°$), and out-of-plane geometries.

3. Results

A convenient way to parametrize the (e,2e) geometries is given in figure 1. The angle of the incident beam to the scattering plane is ψ. For example, in the coplanar case $\psi = 0°$, and in the perpendicular-plane case $\psi = 90°$. We use negative and positive angles to indicate detectors

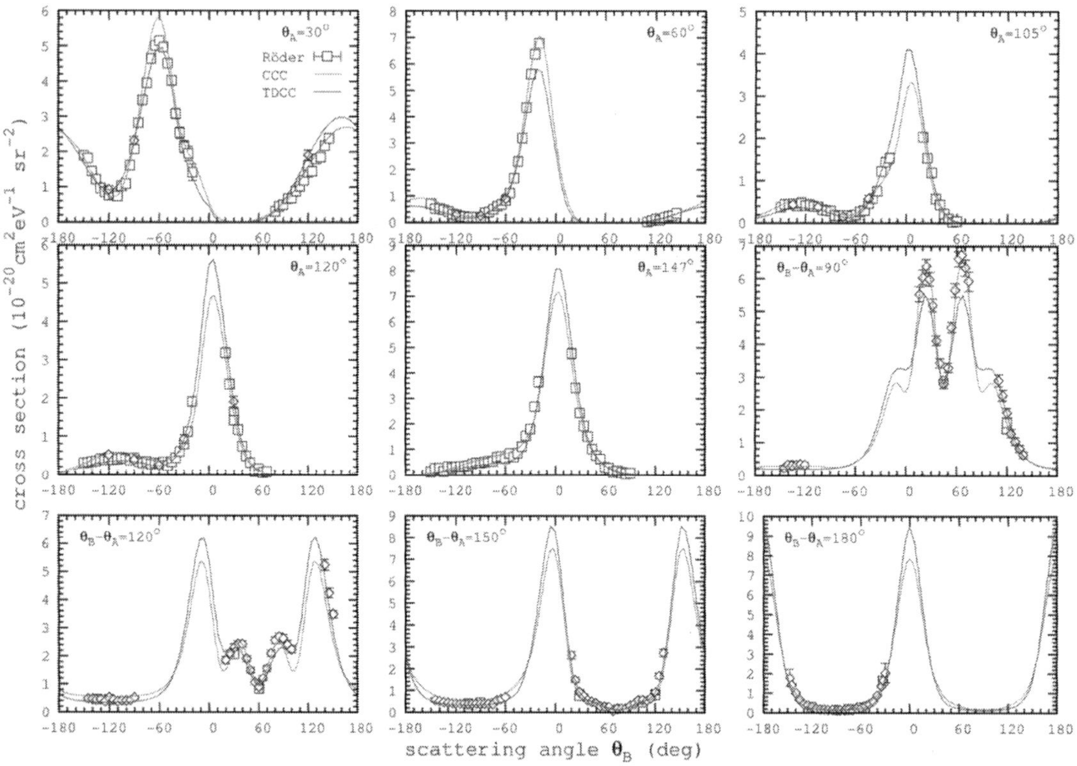

Figure 2. (Colour online) Fully differential cross sections for 64.6 eV electron-impact coplanar ionization of the ground state of helium with two 20 eV outgoing electrons. The data of Röder [29] have absolute uncertainty of 25% and have been multiplied by 1.2 for overall best visual fit to the theory.

on opposite sides to the incident beam.

In figure 2 we consider the coplanar case measured by Röder, and first presented by Bray *et al* [29]. The measurements are given with statistical uncertainty of around 5% and an absolute uncertainty of 25%. To enable best visual comparison between theory and experiment the latter have been uniformly multiplied by 1.2, which is well within the experimental uncertainty. Two types of geometries are being considered. One has a detector fixed at θ_A and the other is rotated in the plane from $-180°$ to $180°$, where possible. Such geometries show the typical behaviour of a dominant binary peak and and a smaller recoil peak. The other geometry has the angular separation of the two detectors $\theta_B - \theta_A$ fixed, which are then rotated together. These geometries readily show how electron-electron repulsion drops the cross sections as this angle is reduced.

We see that the agreement between the results of the two theories and experiment is remarkable for all cases considered. Seeing such a spectacular agreement it would be tempting to suggest that the e-He problem may well be solved.

In figure 3 we present the out-of-plane cross sections measured by Murray and Read [15]. The geometries are such that the incident electron is at the angle ψ to the plane containing the two detectors at symmetric angles on the opposite side of the incident beam. The $\psi = 0°$ case is the so-called coplanar doubly symmetric geometry, has also been measured by Röder [29]. One interesting aspect of these geometries is that for all values of $0° \leq \psi \leq 90°$ the $\theta_B = -\theta_A = 90°$ point is common. This seems an ideal point of normalization since the data of Murray and Read [15] were not independently normalized.

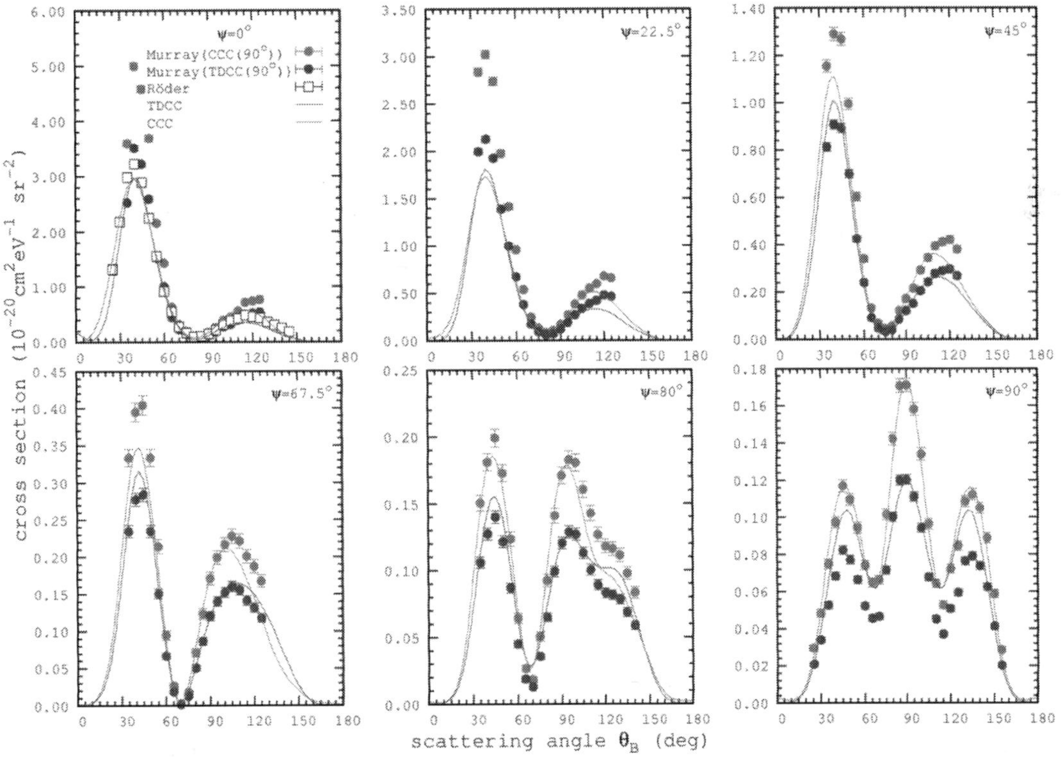

Figure 3. (Colour online) Fully differential cross sections for 64.6 eV electron-impact out-of-plane ionization of the ground state of helium with two 20 eV outgoing electrons. The data of Murray and Read [15] are presented twice: once normalized to the CCC(90°) point and again to the TDCC(90°) point. The data of Röder [29] is as for figure 2.

Though the $\theta_B = 90°$ point is common, its value is near the minimum of the $\psi = 0°$ geometry, and yet a maximum for the perpendicular-plane $\psi = 90°$ geometry. To make things more complicated, the two theories differ substantially in their predictions for the cross section at this point. Accordingly we have chosen to present the data normalized in two ways: once to the CCC(90°) point and also to the TDCC(90°) point.

Starting at the $\psi = 0°$ case we see the largest cross sections, and the normalization to TDCC(90°) yields best agreement with the data of Röder [29]. The theories are in good agreement with each other and the experimental data. The same holds for $\psi = 22.5°$ and $\psi = 45°$ cases. However, as ψ increases further we see that the discrepancy between the CCC and TDCC theory begins to grow with the shape of the data being unable to differentiate as to which theory may be the more accurate. The exception to that is the $\psi = 90°$ case where the CCC result yields remarkable shape-agreement with the data. This may be a coincidence as normalization here leads to a major discrepancy for the largest cross sections when $\psi = 0°$.

4. Conclusions
We have presented the results of the CCC and TDCC calculations for 64.6 eV electron-impact single ionization of the ground state of helium yielding two 20 eV outgoing electrons. The difference between the results of the two calculations are only evident when the cross sections are particularly small. Existing measurements are in sufficiently good agreement with the theory so as to be unable to differentiate between them. In order for that to be possible, new measurements are required. For example, obtaining the ratio of the cross sections at the 90° point of figure 3 and say the 45° point of the $\psi = 0°$ geometry would be very helpful. We are hopeful that the measurements currently underway will be able to differentiate between the CCC and TDCC theories for the presently considered equal energy sharing case, and also for asymmetric energy sharing [30].

Acknowledgments
This work was supported by the Australian Research Council. We are grateful for access to the Australian National Computing Infrastructure Facility and its Western Australian node iVEC. The Los Alamos National Laboratory is operated by Los Alamos National Security, LLC for the National Nuclear Security Administration of the U.S. Department of Energy under Contract No. DE-AC5206NA25396.

References
[1] Rescigno T N, Baertschy M, Isaacs W A and McCurdy C W 1999 *Science* **286** 2474–2479
[2] Baertschy M, Rescigno T N and McCurdy C W 2001 *Phys. Rev. A* **64** 022709
[3] Bray I 2002 *Phys. Rev. Lett.* **89** 273201
[4] Colgan J and Pindzola M S 2006 *Phys. Rev. A* **74** 012713
[5] Pindzola M S, Robicheaux F and Colgan J 2008 *J. Phys. B* **41** 235202
[6] Jung K, Müller-Fiedler R, Schlemmer P, Ehrhardt H and Klar H 1985 *J. Phys. B* **18** 2955–2966
[7] Ehrhardt H, Jung K, Knoth G and Schlemmer P 1986 *Z. Phys. D* **1** 3–32
[8] Rösel T, Dupre C, Röder J, Duguet A, Jung K, Lahmam-Bennani A and Ehrhardt H 1991 *J. Phys. B* **24** 3059–3067
[9] Schlemmer P, Srivastava M K, Rösel T and Ehrhardt H 1991 *J. Phys. B* **24** 2719–2736
[10] Rösel T, Röder J, Frost L, Jung K, Ehrhardt H, Jones S and Madison D H 1992 *Phys. Rev. A* **46** 2539–2552
[11] Röder J, Ehrhardt H, Bray I and Fursa D V 1997 *J. Phys. B* **30** 1309–1322
[12] Röder J, Ehrhardt H, Pan C, Starace A F, Bray I and Fursa D V 1998 *J. Phys. B* **31** L525–L531
[13] Hammond P, Read F H, Cvejanovic S and King G C 1985 *J. Phys. B* **18** L141–L148
[14] Murray A J, Read F H and Bowring N J 1993 *Journal de Physique* **3** 51–58
[15] Murray A J and Read F H 1993 *J. Phys. B* **26** L359–L365
[16] Murray A J and Read F H 1992 *J. Phys. B* **25** L579–L583
[17] Murray A J, Woolf M B J and Read F H 1992 *J. Phys. B* **25** 3021–3036
[18] Fursa D V and Bray I 1995 *Phys. Rev. A* **52** 1279–1298

[19] Fursa D V and Bray I 2003 *J. Phys. B* **36** 1663
[20] Bartschat K, Zatsarinny O, Bray I, Fursa D V and Stelbovics A T 2004 *J. Phys. B* **37** 2617–2639
[21] Bray I and Fursa D V 1996 *Phys. Rev. A* **54** 2991–3004
[22] Kadyrov A S, Bray I, Mukhamedzhanov A M and Stelbovics A T 2008 *Phys. Rev. Lett.* **101** 230405
[23] Kadyrov A S, Bray I, Mukhamedzhanov A M and Stelbovics A T 2009 *Ann. Phys.* **324** 1516–1546
[24] Bray I 1997 *Phys. Rev. Lett.* **78** 4721–4724
[25] Stelbovics A T 1999 *Phys. Rev. Lett.* **83** 1570–1573
[26] Stelbovics A T, Bray I, Fursa D V and Bartschat K 2005 *Phys. Rev. A* **71** 052716(13)
[27] Colgan J, Pindzola M S, Childers G and Khakoo M A 2006 *Phys. Rev. A* **73** 042710
[28] Colgan J, Foster M, Pindzola M S, Bray I, Stelbovics A T and Fursa D V 2009 *J. Phys. B* **42** 145002 (6pp)
[29] Bray I, Fursa D V, Röder J and Ehrhardt H 1997 *J. Phys. B* **30** L101–L108
[30] Murray A 2010 private communication

Journal of Physics: Conference Series **288** (2011) 012003 doi:10.1088/1742-6596/288/1/012003

Electron impact double ionization of rare gases: A kinematical analysis of the non-first order effects at intermediate impact energy

A Lahmam-Bennani[1], E M Staicu Casagrande and C Li
Université Paris-Sud 11 and CNRS, Institut des Sciences Moléculaires d'Orsay (ISMO), UMR 3363, Bât. 351, 91405 Orsay cedex, France

E-mail: azzedine.bennani@u-psud.fr

Abstract. The (e,3-1e) four-fold differential cross sections (4DCS) are measured for the electron impact double ionization of rare gas atoms (He, Ne, Ar) in coplanar asymmetric geometry for a wide range of ejected electron energies (5 to 144 eV) and at an incident energy of about 600 - 700 eV. Though we only show here a sample of the results, *all* experimental angular distributions of the 4DCS are characterized by large angular shifts of the forward and backward lobes with respect to the momentum transfer direction or its opposite, respectively. This validates our previously published results [Lahmam-Bennani *et al* 2002 *J. Phys. B: At. Mol. Opt. Phys.* **35** L59] which were questioned by Götz *et al* [2003 *J. Phys. B: At. Mol. Opt. Phys.* **36** L77]. A qualitative, kinematical analysis is given which allows relating these shifts and the observed structures (or sub-lobes) in the cross section distributions to the second order, 'two-step 2' double ionization mechanism, which is shown to predominate over the first-order 'shake-off' and 'two-step 1' mechanisms under the present kinematics.

1. Introduction

In an electron impact double ionization (DI) experiment, an incident electron (indexed 0) scatters off an atomic or molecular target provoking the emission of two electrons from the target. Though indistinguishable, these electrons are indexed for convenience "a" for the fast-scattered one and "b" and "c" for the slower-ejected ones. A kinematically complete experiment is realized when these electrons are energy and angle selected and are detected in *triple coincidence* to insure that they are issued from the same ionizing event. However, these so-called (e,3e) experiments [1,2] are very difficult to perform, because the corresponding cross sections are inherently very low due to their multiply (five-fold) differential character. Therefore, it is also of interest to consider (e,3-1e) experiments [3,4] where, with respect to the (e,3e) case, one electron (say "c") is not detected, hence only necessitating a *double coincidence* experiment similar to an (e,2e) single ionization (SI) but with energetics corresponding to DI. Such experiments were clearly demonstrated [4] to provide a very sensitive mean to identify the various mechanisms responsible for electron impact DI and to gauge their relative importance.

A few years ago, there was an article published by our group reporting (e,3-1e) experiments for DI of He at about 600 eV impact energy [5]. In this paper, we claimed that evidence is given of the great importance of second- or higher-order effects in the projectile-target interaction. These effects mostly

[1] To whom any correspondence should be addressed.

showed up in the measured angular distributions as large shifts of the forward and backward lobes with respect to the momentum transfer direction **+K**, or its opposite, **-K**. It is well known [6] that these shifts cannot be reproduced by first-order models, for example the 'first Born – 3 Coulomb waves' (FB-3C) model described in [5] which is peaking, as expected, in the ± **K** directions.

A year later, in 2003, Götz *et al* [7] published a tentative theoretical description of our results, where the four-particle continuum was described in different approximations leading up to a 6C wave function which takes all two-body interactions into account as well as the interactions between each outgoing electron and the nucleus, and hence goes beyond the second Born approximation. However, the authors could not reproduce the strong asymmetries seen in the measured data, so that they came to implicitly question the experiments by writing : <<we suggest that it is very important to establish the validity or otherwise of the asymmetry in the experimental data>>.

Given this questioning, we decided to re-examine the problem by taking advantage of the extended capabilities and sensitivity of our spectrometer gained by including to it new developments [8]. Our aim is to produce more extensive experimental data which would validate or invalidate the observations we made in our earlier work, and against which the theoretical predictions can be tested. The experimental set-up used for these investigations has been fully described elsewhere [8]. It essentially features multi-angle analysis and double/triple coincidence detection of the two/three final state electrons, based on three toroidal analysers coupled with three position sensitive detectors (PSD). The key point is that the angular distribution of the scattered or ejected electrons in the collision plane is preserved upon arrival on the PSD's.

2. Results and discussion

Since we are aiming to discuss the results in terms of DI mechanisms, it is convenient to start this section with a brief description of the different DI processes usually invoked in the literature. *Direct* electron impact DI may proceed mostly via three routes [9]. (i) In the shake-off mechanism (SO), the incident electron interacts only once with a single target electron which is ejected without further interaction with other target electrons. Subsequent ejection of a second electron occurs, due to the sudden change in the distribution of potential seen by this electron. (ii) In the two-step mechanism termed TS1, the projectile electron interacts with one target electron which subsequently collides with a second electron leading to ejection of the pair. (iii) In the two-step mechanism termed TS2, the projectile electron successively collides with two target electrons, resulting in their double ejection. Both SO and TS1 involve only one single interaction of the projectile with the target, hence their denomination as first order or first Born mechanisms (in the Born series). In contrast, TS2 involves two successive projectile - target interactions and hence is a second order or second Born mechanism. The most obvious signature of the presence of these mechanisms is that first order mechanisms yield ejected electron(s) angular distributions which are symmetrical with respect to the momentum transfer vector direction; whereas this symmetry is broken in the case of second order, TS2 mechanism, see [6]. Another high-order mechanism was recently invoked by M Schulz *et al* [10] in the context of p + He collisions. The authors labelled it Two-Step 1–Elastic mechanism (TS1-El). It can be viewed as follows: first, a TS1 process leads to the ejection of both target electrons, and then the projectile is elastically scattered from one of the two continuum electrons (hence the alternative label '3steps – 2 interactions' 3S2). As in TS2, the symmetry about **K** direction is broken because of this double interaction of the projectile with the target. In the following, we will concentrate only on the contribution of TS2.

Let us now discuss the results of (e,3-1e) experiments. Several targets were investigated (He, Ne, Ar, H_2, N_2), under various kinematics. We used a fixed scattered electron energy and angle, 500 eV and -6 degrees respectively. (Throughout this work, positive angles are counted clockwise starting from the incident direction). In contrast, the energies of the ejected 'b'-electron (E_b) which is detected and that of the 'c'-electron (E_c) which is undetected were varied over a large range, from 5 to 144 eV. Only a selection of these results will be shown here (the full results will be published elsewhere), as the observations and conclusions are more or less the same for all cases. But before discussing the DI data, it is important to mention that the experimental procedure and the good response of the spectrometer for measuring coincidence angular distributions were validated as follows. Very briefly, the (e,2e) triple differential cross section (TDCS) for SI of the He 1s orbital is measured at the energies E_a and E_b of the current DI experiments. The so-obtained experimental TDCS-distributions are compared with the calculated convergent close coupling (CCC) results. The CCC model is commonly known to produce very accurate TDCS results for electron impact SI of He when the residual ion is left in its ground state. Consequently, the very good agreement obtained between experiments and CCC calculations clearly validates our measurement procedure.

With this, we can now look with confidence to our DI results. Some of the He data are displayed in figure 1 for two energy sharings $(E_b:E_c)$ = (37:5) and (74:5) eV. They are compared with the predictions of the FB-3C model where the final state is described by the product of three Coulomb waves. All the experimental as well as theoretical distributions of the 4DCS display the well-known two-lobes structure: a forward lobe pointing roughly in the momentum transfer direction, **+K**, and a backward lobe pointing in the opposite direction, **-K**, (these two directions are indicated by the vertical dashed lines). Also, the calculated backward-to-forward maximum intensity ratio is more or less in agreement with the measured one at (37:5) eV, but not so at (74:5) eV where the theory predicts quite a small backward lobe. However, the most striking observations from figure 1 are the breaking of symmetry about ±**K** directions, the large angular shifts (~30° to 60°) of the measured lobes from these directions, and the existence of structures or sub-lobes in these lobes. For all energy sharings considered in this work, the symmetry breaking and the large angular shifts are very much alike those we reported earlier in Ref. [5] which were obtained at $(E_b : E_c)$ = (51:10) eV using an older version of our spectrometer. This close resemblance between the 'old' results and the extended new data definitely confirms the validity of our previous results, and should put a final end to the question left open by Götz *et al* [7] about this validity.

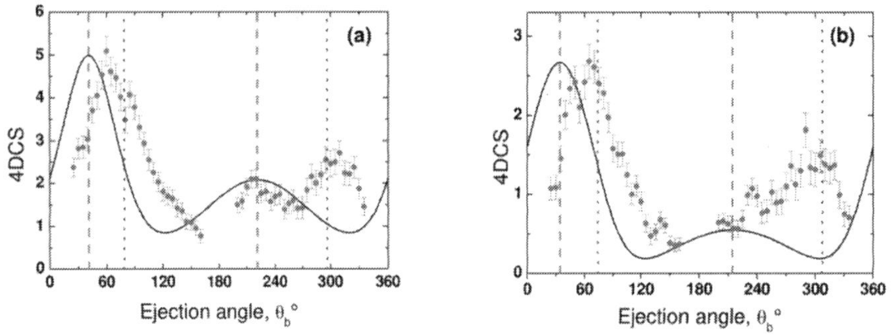

Figure 1. Relative (e,3-1e) 4DCS measured for He (full circles). E_a = 500 eV, θ_a = -6° and $(E_b : E_c)$ = (37:5) eV (left panel) and (74:5) eV (right panel). Full line: results from the first-order FB-3C model. Experiment is arbitrarily normalized to theory at the maximum of the forward lobe. The dashed lines indicate the direction of the momentum transfer and its opposite, ±**K**. The dotted lines indicate the most probable direction of ejection predicted by our TS2 interpretation.

In terms of DI mechanisms, the features observed in figure 1 (i.e. breaking of symmetry, large angular shifts and existence of structures in the measured lobes) cannot be present in any first-order mechanism, namely the SO and TS1 which involve only one single interaction of the projectile with the target. Hence, they are clear signatures of non-first order processes such as TS2 or TS1-EI which involve two or more successive target - projectile interactions. In the absence of detailed second- or higher-order calculations which could be compared with our data (such calculations are very difficult to perform), it was proposed in [11] a qualitative interpretation of the origin of the observed structures, as well as an estimate of the angular positions of the sub-lobes, based on kinematical arguments for TS2 built on our knowledge of SI (e,2e) processes. For the sake of brevity, the discussion is briefly summarized in the following and is illustrated using the data of Fig. 1(a) at $(E_b : E_c) = (37 : 5)$ eV and $E_0 = 621$ eV. However, the ideas and the resulting conclusions hold for all the data and targets considered in this work.

TS2 is a two-step process: in the first step, the slowest c-electron (with energy 5 eV in figure 1) is ejected in an (e,2e)-like process where the corresponding scattered electron (call it 'a*') has an energy E_a* imposed by the energy balance, $E_a* = E_0 - E_c$ - IP = 591 eV (IP is the He SI potential) and appears with the highest probability at the Compton scattering angle, $\pm\ \theta_a$*, corresponding to the Bethe ridge condition [12] : $E_0 - E_a* = E_0 \sin^2(\theta_a*)$. This corresponds to $\theta_a* \sim \pm 12°$. The ± sign stands for the fact that the intermediate, scattered a*-electron may appear on both sides of the incident beam direction. The associated c-electron in this (e,2e) process is most likely ejected in the corresponding momentum transfer direction, that is $\theta_K* \sim \pm 62°$ with respect to the incident direction. Note that neither the a*-nor the c-electrons are detected in the (e,3-1e) experiments.

In the second step of TS2, the a*-electron plays the role of the incident projectile in a new (e,2e) ionization of the target, resulting in the pair of electrons effectively detected, that is the a-scattered one with ($E_a = 500$ eV, $\theta_a = -6°$) and the fast b-ejected one with $E_b = 37$ eV. Here, two scenarios may occur, depending whether the intermediate a*-electron is scattered under $\theta_a* \sim -12°$ or under + 12°. We show in the following that these 2 scenarios might be responsible for the forward and backward sub-lobes observed in the experimental data.

Scenario 1 corresponds to $\theta_a* \sim -12°$. In the corresponding (e,2e) event, the a-electron is detected at $\theta_a = -6°$, and the b-electron is essentially ejected in the corresponding θ_K -direction, that is $\theta_K \sim -48°$ from the a*-incident direction (at the high energies involved here, the recoil intensity appearing about **–K** direction is quite small). In other words, the b-electron is mostly ejected at $\theta_b \sim -60°$, or alternatively at $\theta_b \sim (360° - 60°) = 300°$ from the primary incident beam direction. Scenario 2 corresponds to $\theta_a* \sim +12°$. In a similar way, we find that the b-electron is mostly ejected at $\theta_b \sim +78°$ from the primary incident beam direction.

In the experimental data (figure 1(a)), the backward and the forward intensity distributions are peaking at about 305° and 85°. On the one hand, these values are quite far from the **±K** directions, (41° and 221°, respectively) where the first-order SO and TS1 contributions should be at their maximum. But on the other hand, these values are in excellent agreement with the values 300° and 78° derived from scenarios 1 and 2 above. We thus conclude that the positions of the backward and the forward lobes in the measured data are essentially determined by the TS2 contribution, according to these two scenarios. In contrast, SO and TS1 contributions are of course also present, but their intensity appears to be appreciably smaller than that of the TS2.

This discussion and the resulting conclusions may be likewise repeated for the other data we obtained either for He at different energies $(E_b : E_c)$, or for other targets. Two more examples are given respectively in figure 1(b) for DI of He and in figure 2 for Ne: the splitting in two sub-lobes of both the forward and backward intensity distributions is also clearly seen, and the position of the large angle sub-lobes nicely agrees with the prediction of our qualitative two-step-2 interpretation indicated by the vertical dotted lines.

In summary, all our data show that for all targets and energies investigated, and under the present kinematics, the "non-first order mechanisms such as TS2" dominates over the first-order ones, SO and TS1, and are mostly responsible for the structures and angular positions of the measured lobes. The argument is very nicely supported by very recent second-Born calculations by Dal Cappello *et al* [13] displayed in figure 3 for the He case at (37:5) eV. As noticed above, the first Born-3C prediction does not reproduce the observed shifts and structures of the cross section distribution, whereas the 2nd Born treatment (dashed curve) very clearly constitutes a considerable improvement, both for the position of the forward lobe which is correctly reproduced and for the structure of the backward lobe which is qualitatively well predicted.

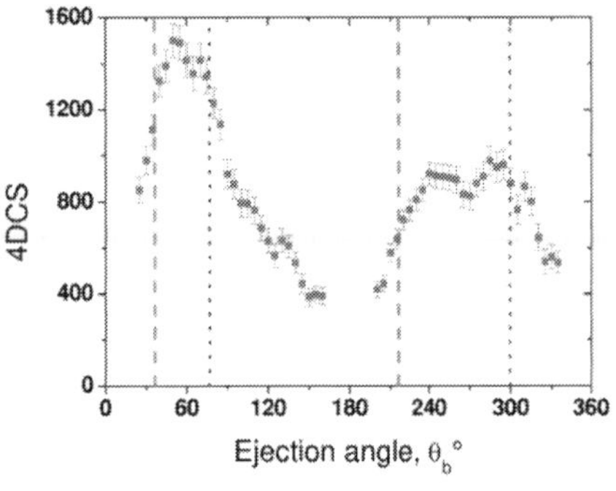

Figure 2. Relative (e,3-1e) 4DCS measured for Ne (full circles). E_a = 500 eV, θ_a = -6° and (E_b : E_c) = (72:12) eV. The dashed lines indicate the direction of the momentum transfer and its opposite, ±**K**. The dotted lines indicate the most probable direction of ejection predicted by our TS2 interpretation.

Similarly, TS2 results by M Schulz and co-workers (not shown here, and to be published elsewhere) are also in quite good agreement with experiments.

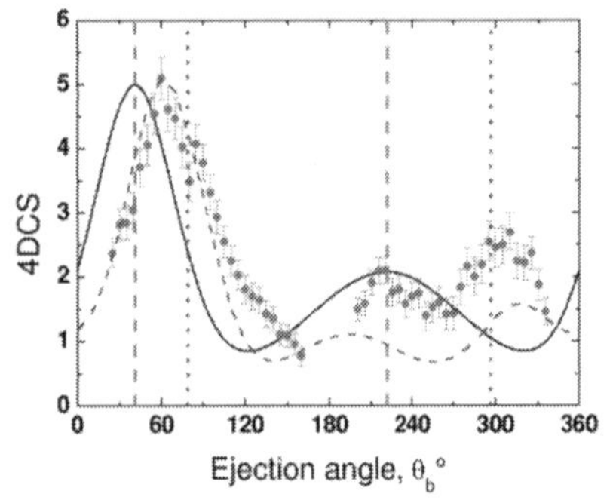

Figure 3. Same as in figure 1(a). The experimental data are in addition compared here to 2nd-Born results [13], shown as a dashed (blue) line.

3. Conclusion

In conclusion, we have presented in this contribution a sample of our new (e,3-1e) experiments for DI of a series of atoms and molecules, at an incident energy of about 600-700 eV and different energy sharings (E_b:E_c) among the two ejected electrons. All our data display large shifts of the forward and backward lobes in the ejected electron angular distributions, thus confirming and validating our earlier observations [5].

We have discussed a qualitative analysis of the data based on kinematical arguments which shows that under the present kinematics the non-first order mechanisms such as TS2 dominate over the first order ones (SO and TS1), and are mostly responsible for the structures or sub-lobes observed in the measured distributions. Other mechanisms such as TS1-El may also contribute, but the present kinematical analysis cannot be straightforwardly applied to these mechanisms.

Acknowledgment
C Li wishes to acknowledge a PhD grant from the China Scholarship Council.

References

[1] Lahmam-Bennani A, Dupré C and Duguet A 1989 *Phys. Rev. Lett.* **63** 1582
[2] Taouil I, Lahmam-Bennani A, Duguet A and Avaldi L1998 *Phys. Rev. Lett.* **81** 4600
[3] Dupré C, Lahmam-Bennani A and Duguet A 1991 *Meas. Sci. Technol.* **2** 327
[4] El Marji B, Duguet A, Lahmam-Bennani A, Lecas M and Wellenstein H F 1995 *J. Phys. B: At. Mol. Opt. Phys.* **28** L733
[5] Lahmam-Bennani A, Duguet A and Roussin S 2002 *J. Phys. B: At. Mol. Opt. Phys.* **35** L59
[6] Dal Cappello C and LeRouzo H 1991 *Phys. Rev.* A **43** 1395
[7] Götz J R, Walter M and Briggs J S 2003 *J. Phys. B: At. Mol. Opt. Phys.* **36** L77

[8] Catoire F, Staicu Casagrande E M, Lahmam-Bennani A, Duguet A, Naja A, Ren X G, Lohmann B and Avaldi L 2007 *Rev. Sci. Instrum.* **78** 013108
[9] Carlson T A and Krause M O 1965 *Phys. Rev.* **140** 1057
[10] Schulz M, Ciappina M F, Kirchner T, Fischer D, Moshammer R and Ullrich J 2009 *Phys. Rev.* A **79**, 042708
[11] Lahmam-Bennani A, Staicu Casagrande E M, Naja A, Dal Cappello C and Bolognesi P 2010 *J. Phys. B: At. Mol. Opt. Phys.* **43** 105201
[12] Inokuti M 1971 *Rev. Mod. Phys.* **43** 297
[13] Dal Cappello C 2010 private communication

Electron impact double ionization of single oriented water molecules: evidence of the role of the two-step mechanism

C Dal Cappello[1], I Kada[2], A Mansouri[2] and C Champion[1]

[1] Université Paul Verlaine-Metz, LPMC, Institut de Physique, 1 bd Arago, 57078 Metz Cedex 3, France

[2] Université Ferhat Abbas, LPQSD, Département de physique, 19000 Sétif, Algérie

E-mail : cappello@univ-metz.fr

Abstract. Double ionization of single oriented water molecules is here investigated within a theoretical approach based on the second Born approximation. The initial wave function describing the two active electrons is taken from a single-centre description previously used with success for describing the single ionization process whereas the final state wave function describing the two ejected electrons is the approximate BBK wave function. Secondary electron angular distributions for an incident energy close to 600 eV are then reported for particular kinematical conditions and compared to their first Born homologous. Strong similarities are observed in terms of maxima localization as well as identification of the main mechanisms involved in the double ionization. On the contrary, for particular kinematical conditions we demonstrate that the first-order treatment is unable to explain the observations contrary to the second-order approximation which points out pure TS2 contributions.

1. Introduction

Very recently Lahmam-Benanni *et al.* [1] have shown that the second Born approximation was needed for describing the experimental observations on double ionization of helium by 600-700 eV electron impact. In this kind of experiments - called (*e*,3-1*e*) [1-2] - only the scattered electron and one of the bound ejected electrons are detected in coincidence contrary to the so-called (*e*,3*e*) experiments [3-4] where the three "final" electrons are detected in coincidence both in direction and in energy, these latter being issued from the same ionizing event. However, due to their low cross sections, the (*e*,3*e*) experiments remain up to now limited to noble gases [5-7] and helium atom, this latter being the ideal target for theoretical studies [4, 8-11].

In fact, similarly to the well-known (*e*,2*e*) experiments for single ionization process, the (*e*,3-1*e*) experiments permit the identification of the mechanisms of the double ionizing process, namely, the shake-off (SO), the two-step 1 (TS1) and the two-step 2 (TS2) [12]. Note that contrary to the SO process which consists in a single ionization of one of the target electrons - the other one being simply ejected due to the relaxation of the target - the TS1 and TS2 mechanisms imply two electrons: in the first one, the incident electron interacts with one target electron which is ejected and then collides another target electron whereas in the second one the incident electron successively ejects two target electrons itself. This last mechanism clearly involves two interactions between the incident electron and the target and then needs the use of *the second Born approximation* whose signature is an absence of symmetry around the momentum transfer [13].

In the present work, we treat the double ionization within the second Born approximation by using a 2CWG model previously used with success [14] for describing the big shift of the binary peak observed by Lahmam-Bennani *et al.* [1]. In this model, the two ejected electrons are described in terms of Coulomb waves together with a Gamow factor to take into account the repulsion between them whereas plane waves are used for treating the incident and the scattered electrons. The Gamow factor was first used by Brauner *et al.* [15] in their study on the ionization of atomic hydrogen. In fact, in its original version, the Gamow factor was seen as a normalization constant of the third Coulomb wave function (BBK model). A successful agreement in reproducing a great number of (*e*,2*e*) cross section measurements was then reported by many authors (see for example Röder *et al.* [16]) although this latter was unable to reproduce the Wannier law, what was corrected by Ward and Macek [17] who proposed in 1994 a "revised" Gamow factor (almost identical to the original Gamow factor in particular in shape). It was then shown that introducing the Gamow factor into the cross section calculations allowed reproducing the dominant angular behaviour of the differential cross sections although such a multiplicative factor annihilates the overall normalization of the theoretical differential cross sections. Under these conditions, our model appears as a valid approximation for predicting new experimental results and then exploring the possible mechanisms.

Despite its importance in many fields like astrophysics, plasma physics and especially in radiobiology where it is considered as the privileged molecule for modelling the biological medium, water has been only rarely approached in (*e*,3*e*) studies. Let us nevertheless cite the recent work of Champion *et al.* [18] where it has been clearly shown that the target orientation might play a major role in the secondary electron angular distributions for isolated water molecules impacted by 1 keV electrons. In their study, the authors have described the double ionization process within the 1st Born approximation. Kheifets [19] and El Mkhanter and Dal Cappello [20] have shown that the second Born approximation was especially important at 0.5 keV. In the present work we consider an incident energy close to 600 eV and use the second Born approximation in order to know if the TS2 mechanism can be identified when considering the orientation of the molecule.
Atomic units are used throughout unless otherwise stated.

2. Theory

In the second Born approximation, the fivefold differential cross section (FDCS) is given by

$$\sigma^{(5)}(\alpha,\beta,\gamma)=\frac{d^5\sigma(\alpha,\beta,\gamma)}{d\Omega_a d\Omega_b d\Omega_s dE_a dE_b}=\frac{k_s k_a k_b}{k_i}\left|f_{B1}+f_{B2}\right|^2, \quad (1)$$

where $d\Omega_s$, $d\Omega_a$ and $d\Omega_b$ denote the elements of solid angles for the scattered and the ejected electrons a and b, respectively, whereas the energy intervals of the ejected electrons are represented by dE_a and dE_b. The momenta of the incident, scattered, and ejected electrons are denoted by $\vec{k}_i$, $\vec{k}_S$, $\vec{k}_a$ and $\vec{k}_b$, respectively. Furthermore, note that the FDCS depends here on the Euler angles (α,β,γ).

The first Born term f_{B1} is written as

$$f_{B1}=-\frac{1}{2\pi}\left\langle \exp(i\vec{k}_s.\vec{r}_0)\Psi_f(\vec{k}_a,\vec{k}_b,\vec{r}_1,..\vec{r}_{10})\right|V\left|\exp(i\vec{k}_i.\vec{r}_0)\Phi_i(\vec{r}_1,\vec{r}_2,..\vec{r}_{10})\right\rangle, \quad (2)$$

where $\Phi_i(\vec{r}_1,..\vec{r}_{10})$ is the wave function of the initial state of the water molecule and $\Psi_f(\vec{k}_a,\vec{k}_b,\vec{r}_1,..\vec{r}_{10})$ is the wave function for the double continuum state of the water molecule. The potential V represents the Coulomb interaction between the incoming electron and the target and is given by

$$V = -\frac{8}{r_0} - \frac{1}{\left|\vec{r}_0 - \vec{R}_1\right|} - \frac{1}{\left|\vec{r}_0 - \vec{R}_2\right|} + \sum_{i=1}^{10} \frac{1}{r_{0i}}, \tag{3}$$

with $R_1 = R_2 = R_{OH} = 1.814$ a.u. while $\vec{r}_i$ is the position vector of the ith bound electron of the target with respect to the centre of the oxygen nucleus and r_0 denotes the coordinate of the incident particle with $\vec{r}_{0i} = \vec{r}_0 - \vec{r}_i$. Furthermore, note that we have used in this work the well-known frozen-core approximation in order to reduce the 10-electron target problem to a *two-active* electron target [21] *i.e.* those which will be ejected after the collision, what is coherent with the fact that we are here essentially interested by the ejection of electrons from valence shells. In these conditions, it is reasonable to assume that the remaining electrons of the doubly charged ion core are unaffected by the ionization process itself [21-22].

The initial wave function describing the two active electrons is $\left|\Phi_i(\vec{r}_1, \vec{r}_2)\right\rangle$ and is taken from the single centre description reported by Moccia [23] whereas the final state wave function describing the two ejected electrons is the approximate BBK wave function (often called 2CWG model) [24] such as

$$\Psi_f(\vec{k}_a, \vec{r}_1, \vec{k}_b, \vec{r}_2) = \frac{1}{\sqrt{2}}\left[\Psi_C^-(\vec{k}_a, \vec{r}_1)\Psi_C^-(\vec{k}_b, \vec{r}_2) + \Psi_C^-(\vec{k}_a, \vec{r}_2)\Psi_C^-(\vec{k}_b, \vec{r}_1)\right].\varphi(\left|\vec{k}_a - \vec{k}_b\right|) \tag{4}$$

and

$\Psi_C^-(\vec{k}, \vec{r}_1) = \frac{1}{(2\pi)^{3/2}} \exp(i\vec{k}.\vec{r}_1)\Gamma(1 - i\alpha)\exp(-\frac{\pi}{2}\alpha)\,{}_1F_1\ (i\alpha, 1, -i(\vec{k}.\vec{r}_1 + kr_1))$, with $\alpha = -Z/k_e$, $Z = 2$. The Gamow factor is $\varphi(\left|\vec{k}_a - \vec{k}_b\right|) = \exp(-\frac{\pi\chi_{ab}}{2})\Gamma(1 - i\chi_{ab})$ and $\chi_{ab} = \frac{1}{\left|\vec{k}_a - \vec{k}_b\right|}$.

The second Born term f_{B2} is given by

$$\begin{aligned} f_{B2} &= \frac{1}{8\pi^4}\sum_n \int \frac{d\vec{q}}{q^2 - k_n^2 - i\varepsilon} \\ &\left\langle \exp(i\vec{k}_s.\vec{r}_0)\Psi_f(\vec{k}_a, \vec{r}_1, \vec{k}_b, \vec{r}_2)\right| V \left|\exp(i\vec{q}.\vec{r}_0)\Phi_n(\vec{r}_1, \vec{r}_2)\right\rangle \\ &\left\langle \exp(i\vec{q}.\vec{r}_0)\Phi_n(\vec{r}_1, \vec{r}_2)\right| V \left|\exp(i\vec{k}_i.\vec{r}_0)\Phi_i(\vec{r}_1, \vec{r}_2)\right\rangle, \end{aligned} \tag{5}$$

where the summation over n means that we take into account all the contributions of the n discrete states and those of the continuum states of the water molecule. Eq.5 means that the incident electron collides two times with the target and corresponds to the well-known two-step 2 mechanism (TS2) [12]. Then, by performing the integration over $\vec{r}_0$ in Eq.(5) and by applying the closure approximation, we get [25]

$$\begin{aligned} \bar{f}_{B2} &= \frac{2}{\pi^2}\int \frac{d\vec{q}}{q^2 - p^2 - i\varepsilon}\frac{1}{K_i^2 K_f^2} \\ &\left\langle \Psi_f(\vec{k}_a, \vec{r}_1, \vec{k}_b, \vec{r}_2)\right| \exp(i\vec{K}_f.\vec{r}_1) + \exp(i\vec{K}_f.\vec{r}_2) - 2 \Big| \\ &\Big| \exp(i\vec{K}_i.\vec{r}_1) + \exp(i\vec{K}_i.\vec{r}_2) - 2 \left|\Phi_i(\vec{r}_1, \vec{r}_2)\right\rangle, \end{aligned} \tag{6}$$

where $\vec{K}_i = \vec{k}_i - \vec{q}$ and $\vec{K}_f = \vec{q} - \vec{k}_s$, $\vec{K} = \vec{K}_i + \vec{K}_f = \vec{k}_i - \vec{k}_s$ being the momentum transfer. Here we have

Journal of Physics: Conference Series **288** (2011) 012004 doi:10.1088/1742-6596/288/1/012004

$$\frac{p^2}{2} = \frac{k_i^2}{2} - \overline{w} , \qquad (7)$$

where $\overline{w}$ refers to the average excitation energy.

Finally, note that the integrals on $d\vec{q}$ must be performed numerically with a great care.

3. Results and discussion

We here investigate the double ionization of the four molecular states of the water molecule hereafter referred as $1b_1$, $3a_1$, $1b_2$ and $2a_1$. The incident and ejected electron energies have been selected in order to have a scattering energy of 500 eV, what corresponds to the experimental conditions reported by Lahmam-Bennani *et al.* [1]. Furthermore, we only consider the target orientation given by $\alpha = \beta = \gamma = 0$ and report in the following (*e*, 3*e*) FDCS as a function of the ejected angles θ_a and θ_b in a coplanar geometry ($\varphi_s = \varphi_a = \varphi_b = 0$). Note that this kinematics is similar to that studied by Champion *et al.* in Ref. [18].

In Figure 1, fivefold differential cross sections (FDCS) for the double ionization of the water molecular orbital $1b_2$ oriented in the direction $\alpha = \beta = \gamma = 0$ are reported as a function of the ejected angles θ_a and θ_b for the kinematical conditions above detailed ($E_a = E_b = 10$ eV and $E_s = 500$ eV). For the first time we observe the signature of the TS2 mechanism which here appears without any interference effects due to other double ionization mechanisms in particular the SO process. Thus, contrary to Champion *et al.* who reported negligible cross sections within the 1st Born approximation for the $1b_2$ molecular orbital in this particular target orientation ($\alpha = \beta = \gamma = 0$) [18] we here show that the (*e*,3-1*e*) FDCS are not null. For these particular conditions, it is clear that the double ionization process has to be described as a double collision between an incoming electron and two target ones. However, note that two successive collisions with a scattering angle (θ_{s1} and θ_{s2}, respectively) different from 0 are needed such as the sum $\theta_{s1} + \theta_{s2} = \theta_s = 0$. Furthermore, note that the positions of the maxima here obtained, namely, ($\theta_a \approx 90^0$, $\theta_b \approx 270^0$) and ($\theta_a \approx 270^0$, $\theta_b \approx 90^0$) are easily interpretable. Indeed, in a fast collision between an incident electron and a molecular target, a pure elastic collision occurs during which a target electron is ejected along a θ_a direction defined by $\theta_{s1} + \theta_a \approx 90^0$ (with $\theta_{s1} \cong 0$) whereas a second target electron is ejected during a second elastic collision for which $\theta_{s2} + \theta_b \approx 270^0$.

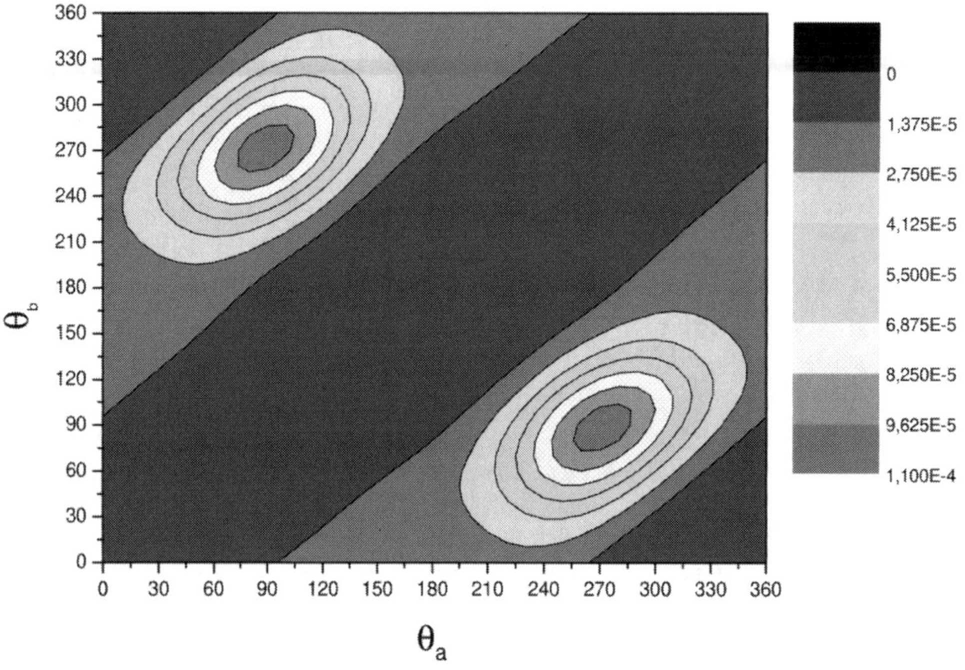

Figure 1. Fivefold differential cross sections (FDCS) for the double ionization of the water molecular orbital *1b*$_2$ oriented in the direction $\alpha = \beta = \gamma = 0$ for $\theta_s = 0$ and $\varphi_s = \varphi_a = \varphi_b = 0$ as a function of the ejected angles θ_a and θ_b relative to the incident electron. The ejected energies are $E_a = E_b = 10$ eV and $E_s = 500$ eV. The FDCS have been calculated within the second Born approximation.

In Figs.2a and 2b we successively report the FDCS for the molecular state $2a_1$ within the first and the second Born approximations. As previously reported by Champion *et al.* [18] for an incident energy of 1 keV, Fig.2a clearly shows that the TS1 mechanism is the main mechanism involved in the double ionization process when the first Born approximation is considered. In this case, the maxima are located at $|\theta_a - \theta_b| = 140^0$: the incoming electron collides one target electron which is ejected along the direction of the momentum transfer $\vec{K}$, this latter colliding - in a second step - another molecular electron during a *not-pure* collision [10] leading to a $|\theta_a - \theta_b|$ angle greater than 90° due to the electronic repulsion. On the other hand, when the collision is treated within the second Born approximation, the observations differ (see Fig.2b). Indeed, the maxima are in this case given by $|\theta_a - \theta_b| = 180^0$ what means that the SO mechanism becomes now the main mechanism of the double ionization. The signature of the SO mechanism is characterized by a first electron ejected along the momentum transfer leading to an electronic rearrangement of the target and to the ejection of a second electron in the opposite direction. We nevertheless observe a secondary (smaller) maximum for

$|\theta_a - \theta_b| = 140^0$. In fact, in this particular case we may assess that the TS2 and TS1 mechanisms destructively interfere together.

Figure 3a reports the first Born FDCS for the molecular state $1b_1$. Like previously, the observations are similar to those reported by Champion *et al.* [18] for an incident energy of 1 keV. We here observe four hills: a first one extending from θ_a=15^0 to θ_a=135^0 (while θ_b varying from θ_b=215^0 to θ_b=345^0) and a second one (with a bigger maximum) extending from θ_a=75^0 to θ_a=150^0 (while θ_b varying from θ_b=200^0 to θ_b=280^0), the two other hills being obtained by symmetry. These two groups of maxima verify $|\theta_a - \theta_b| \approx 120^0$ for the biggest and $|\theta_a - \theta_b| \approx 240^0$ for the smallest, what indicates that the TS1 is the main mechanism, as already observed by Kada *et al.* [26] for the double ionization of a non-oriented $1b_1$ molecular state.

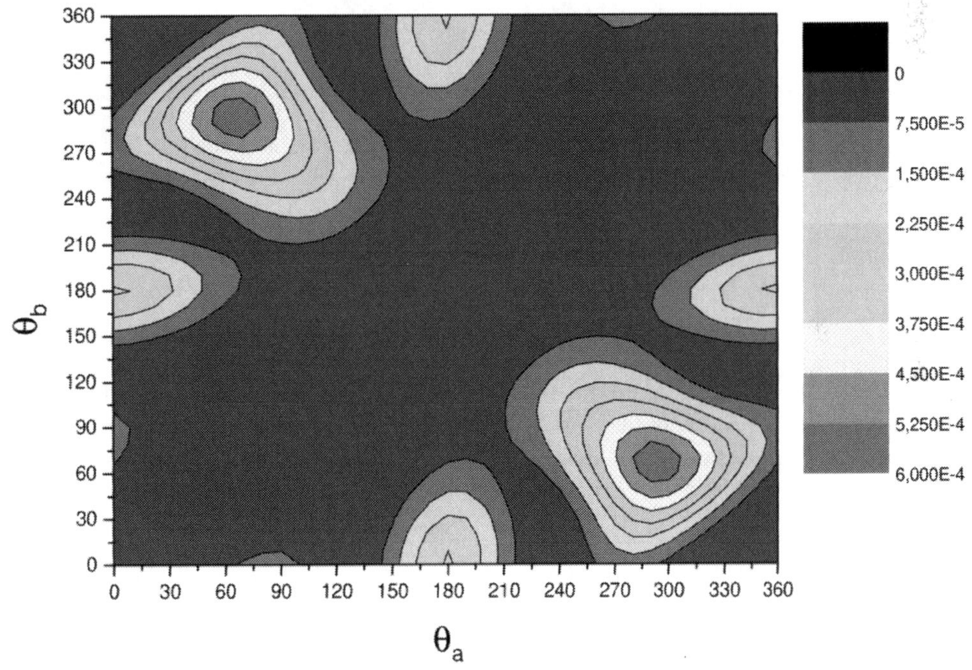

Figure 2a. Fivefold differential cross sections for the double ionization of the water molecular orbital $2a_1$ oriented in the direction $\alpha = \beta = \gamma = 0$ for $\theta_s = 0$ and $\varphi_s = \varphi_a = \varphi_b = 0$ as a function of the ejected angles θ_a and θ_b relative to the incident electron. The ejected energies are $E_a = E_b = 10$ eV and $E_s = 500$ eV. The FDCS have been calculated within the first Born approximation.

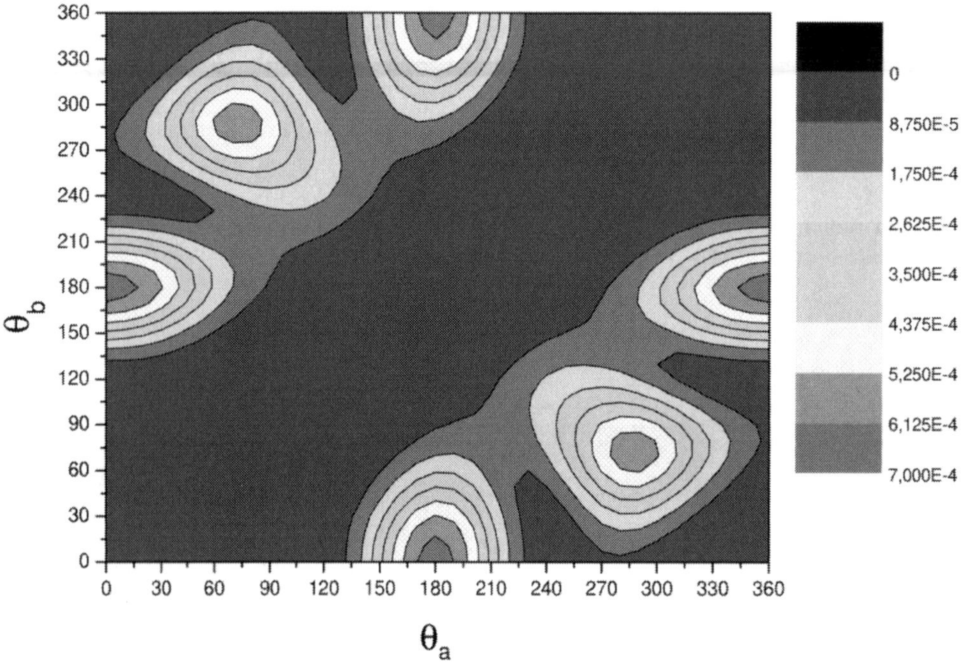

Figure 2b. Same as in Fig.2a by using the second Born approximation

When the second Born approximation is considered (see Fig.3b) the amplitude of the forward scattering increases while that of the backscattering decreases. Thus, the TS2 mechanism plays here no important role except for the magnitudes of the maximum.

The FDCS of the molecular state $3a_1$ are reported in Figs. 4a and 4b within the 1st and 2nd Born predictions, respectively. In both cases we observe maxima given by $|\theta_a - \theta_b| \approx 180^0$ what corresponds to the signature of the SO mechanism with nevertheless a slight decreasing of the corresponding magnitudes when the second Born approximation is used.

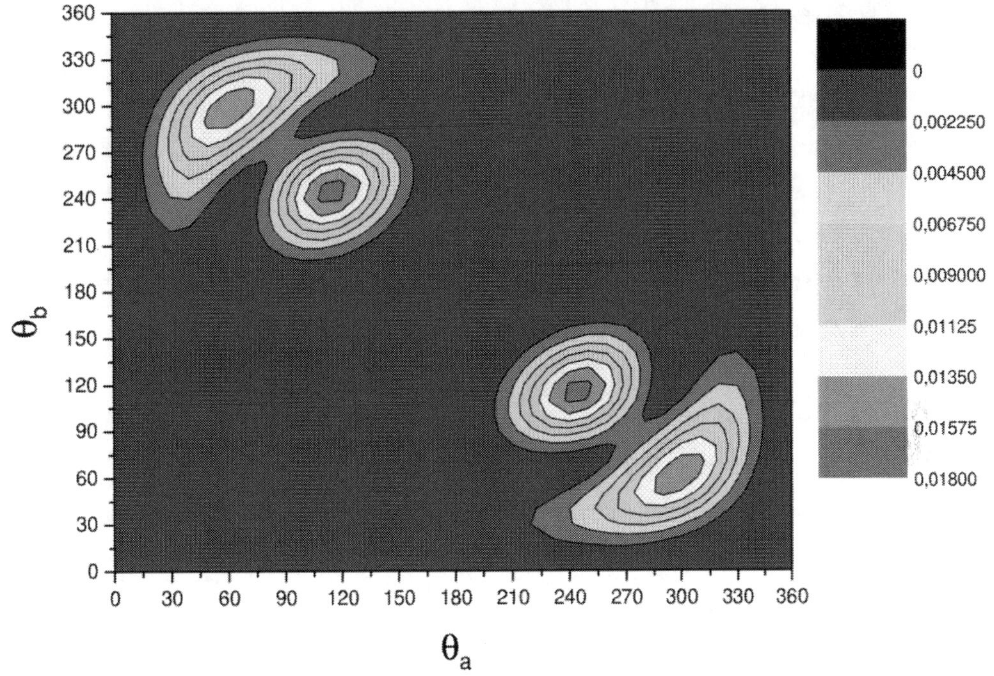

Figure 3a. Fivefold differential cross sections for the double ionization of the water molecular orbital *1b₁* oriented in the direction $\alpha = \beta = \gamma = 0$ for $\theta_s = 0$ and $\varphi_s = \varphi_a = \varphi_b = 0$ as a function of the ejected angles θ_a and θ_b relative to the incident electron. The ejected energies are $E_a = E_b = 10$ eV and $E_s = 500$ eV. The FDCS have been calculated within the first Born approximation.

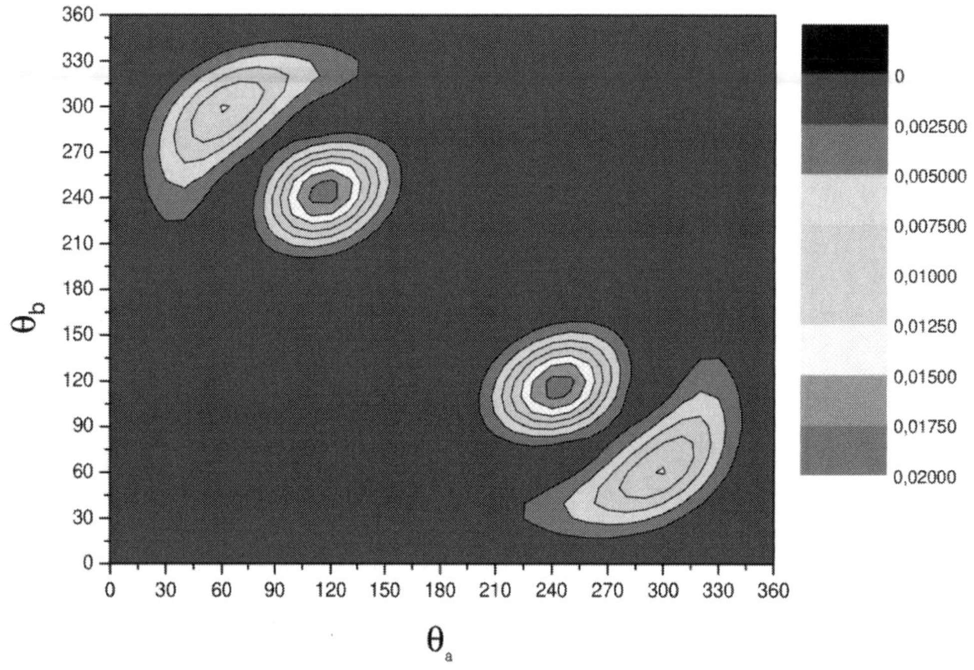

Figure 3b. Same as in Fig.3a by using the second Born approximation.

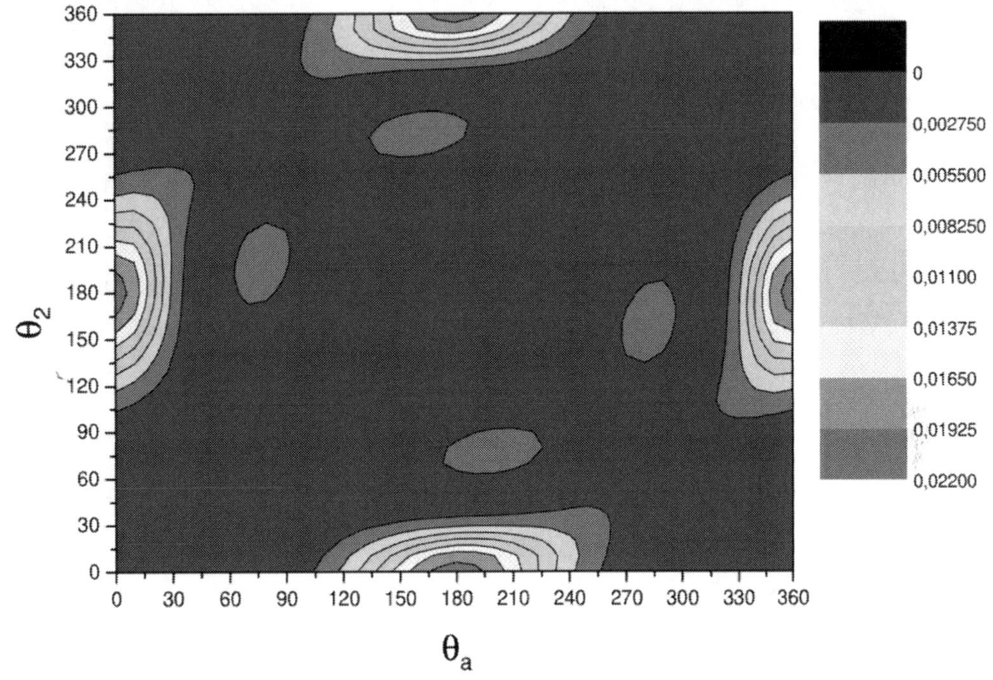

Figure 4a. Fivefold differential cross sections for the double ionization of the water molecular orbital *3a₁* oriented in the direction $\alpha = \beta = \gamma = 0$ for $\theta_s = 0$ and $\varphi_s = \varphi_a = \varphi_b = 0$ as a function of the ejected angles θ_a and θ_b relative to the incident electron. The ejected energies are $E_a = E_b = 10$ eV and $E_s = 500$ eV. The FDCS have been calculated within the first Born approximation.

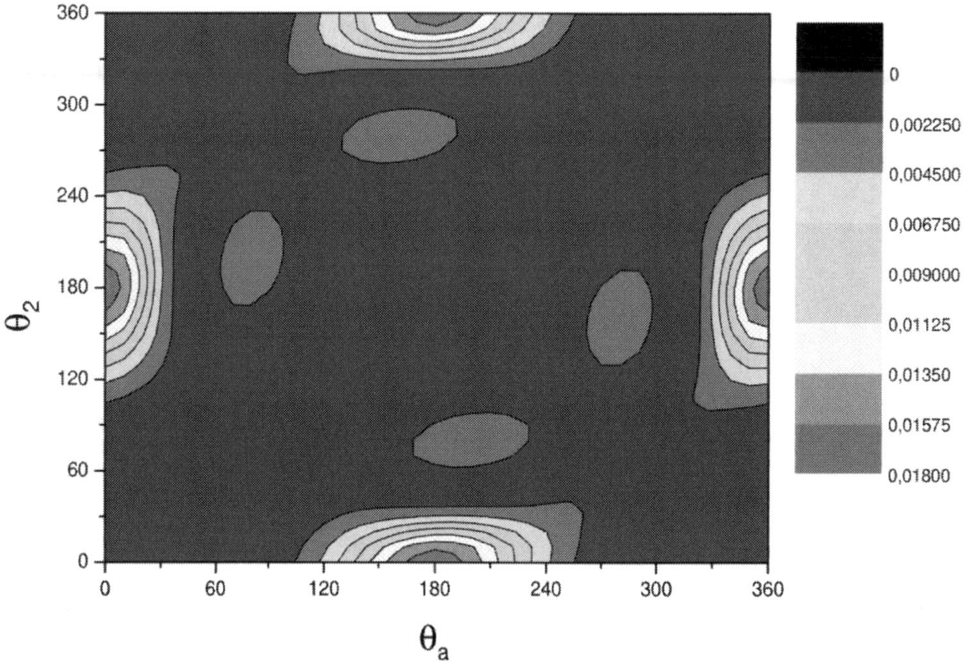

Figure 4b. Same as in Fig.4a by using the second Born approximation.

4. Conclusion

In this work, we have carried out a theoretical analysis of the double ionization of single oriented water molecules within the second Born approximation. Fivefold differential cross sections were then reported and compared to their 1st Born homologous recently provided by Champion *et al.* [18]. Strong similarities were then reported in terms of maxima localization as well as identification of the main mechanisms involved in the double ionization whereas for particular kinematical conditions, it has been clearly shown that the first-order treatment was unable to explain the observations contrary to the second-order approximation which highlighted pure TS2 contributions.

Acknowledgments
We would like to thank the PMMS (Pôle Messin de Modélisation et de Simulation) for computer time.

References

[1] Lahmam-Bennani A, Staicu-Casagrande E M, Naja A, Dal Cappello C and Bolognesi P 2010 *J. Phys. B: At. Mol. Opt. Phys.* **43** 105201
[2] El Marji B, Duguet A, Lahmam-Bennani A, Lecas M and Wellenstein H F 1995 *J. Phys. B: At. Mol. Opt. Phys.* **28** L733
[3] Lahmam-Bennani A, Dupré C and Duguet A 1989 *Phys. Rev. Lett.* **63** 1582
[4] Taouil I, Lahmam-Bennani A, Duguet A and Avaldi L 1998 *Phys. Rev. Lett.* **81** 4600
[5] El Marji B, Schröter C, Duguet A, Lahmam-Bennani A, Lecas M and Spielberger L 1997 *J. Phys. B: At. Mol. Opt. Phys.* **30** 3677

[6] Schröter C, El Marji B, Lahmam-Bennani A, Duguet A, Lecas M and Spielberger L 1998 *J. Phys. B: At. Mol. Opt. Phys.* **31** 131
[7] Jha L K, Santosh K, Roy O P and Kumar P 2008 *Phys. Scr.* **77** 015304
[8] Dorn A, Moshammer R, Schröter C D, Zouros T J M, Schmitt W, Kollmus H, Mann R and Ullrich J 1999 *Phys. Rev. Lett.* **82** 2496
[9] Dorn A, Kheifets A, Schröter C D, Najjari B, Höhr C, Moshammer R and Ullrich J 2001 *Phys. Rev. Lett.* **86** 3755
[10] Dorn A, Kheifets A, Schröter C D, Najjari B, Höhr C, Moshammer R and Ullrich J 2002 *Phys. Rev. A* **65** 032709
[11] Dürr M, Dorn A, Ullrich J, Cao S P, Kheifets A, Götz J R and Briggs J S 2007 *Phys. Rev. Lett.* **98** 193201
[12] Carlson T A and Krause M O 1965 *Phys.Rev.* **140** 1057
[13] Dal Cappello C and Le Rouzo H 1991 *Phys.Rev. A* **43** 1395
[14] Dal Cappello C, Haddadou A, Menas F and Roy A C submitted to *J. Phys. B: At. Mol. Opt. Phys*
[15] Brauner M, Briggs J S and Klar H 1989 *J. Phys. B: At. Mol. Opt. Phys* **22** 2265
[16] Röder J, Rasch J, Jung K, Whelan C T, Ehrhardt H, Allan R J and Walters H R J 1996 *Phys. Rev. A* **53** 225
[17] Ward S J and Macek J H 1994 *Phys. Rev. A* **49** 1049
[18] Champion C, Oubaziz D, Aouchiche H, Popov Yu V and Dal Cappello C 2010 *Phys. Rev. A* **81** 032704
[19] Kheifets A 2004 *Phys. Rev. A* **69** 032712
[20] El Mkhanter R and Dal Cappello C 1998 *J. Phys. B: At. Mol. Opt. Phys.* **31** 301
[21] Hda H, Dal Cappello C and Langlois J 1994 *Z. Phys. D* **29** 25
[22] Cooper J W and Van Boeyen R W 2004 *J. Phys. B: At. Mol. Opt. Phys.* **37** L77
[23] Moccia R 1964 *J. Chem. Phys. A* **40** 2186
[24] Dal Cappello C, Joulakian B and Langlois J 1993 *J. Physique* **3** 125
[25] Grin M, Dal Cappello C, El Mkhanter R and Rasch J 2000 *J. Phys. B: At. Mol. Opt. Phys.* **33** 131
[26] Kada I, Mansouri A, Dal Cappello C, Hervieux P A and Roy A C 2009 *J. Phys. B: At. Mol. Opt. Phys.* **42** 025201

Coherent electron emission for dissociative ionization of molecular hydrogen by electron impact

O. A. Fojón, C. R. Stia and R. D. Rivarola

Instituto de Física Rosario, CONICET and Universidad Nacional de Rosario, Av. Pellegrini 250, 2000 Rosario, Argentina.

E-mail: ofojon@fceia.unr.edu.ar

Abstract. The single ionization of hydrogen molecules is studied theoretically as a function of the molecular alignment. Within the framework of the two-effective center model, multiple differential cross sections as a function of both electron momenta in the final channel of the reaction, and the internuclear orientation, are computed for both non-dissociative and dissociative final H_2^+ states. Preliminary results show that the interference pattern arising from the two-center character of the molecular target changes strongly with the final state of the residual molecular ion.

1. Introduction

In this work, the single ionization of H_2 molecules is studied theoretically as a function of the molecular alignment. The interest is focused here on the interference effects arising from the two-center nature of the molecular target. In a previous work [1], molecular transition amplitudes obtained within the framework of a two-effective center (TEC) approximation [2] were shown to be described as an *effective* one-center contribution modulated by an interference factor similar to the one appearing in Young-type experiments with light. Oscillatory patterns that depend markedly on the molecular orientation were predicted. Also, it was found that interference phenomena may be evident even in the integrated cross sections. It is worth noting that the ionization process was considered there as a pure vertical electronic transition from the ground electronic state ($^1\Sigma_g^+$) of H_2 target to the non-dissociative ground electronic state ($^2\Sigma_g^+$) of the residual H_2^+ [1]. This channel gives the dominant contribution for the electron ionization reaction. The TEC model was successfully used to explain many experimental evidences of interference patterns (see for instance Ref. [3] and references therein). It is worthy to mention that molecules were in general randomly oriented in those experiments.

Only recently, Senftleben and co-workers [4] presented experimental 5-fold differential cross sections (5DCS's) in coplanar geometry of electron impact ionization of hydrogen molecules as a function of the molecular orientation. Indeed, dissociative ionization of the molecular target was considered in order to define the alignment of the internuclear axis. Under those conditions, it was found that experimental cross sections corresponding to molecules oriented along the momentum transfer direction are larger than the ones for molecular alignment perpendicular to the momentum transfer. Moreover, they showed that calculated triply differential cross sections of ionization of He targets multiplied by the interference factor predicted by Ref. [1] to the case of final non-dissociative state, exhibit the opposite behavior.

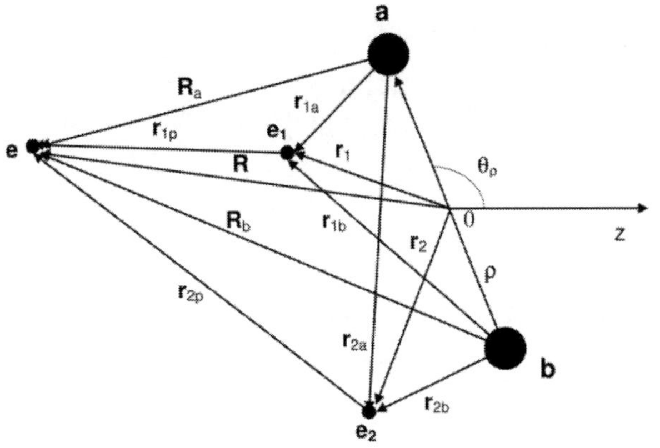

Figure 1. Coordinates used in the text.

In the present work, single ionization of hydrogen molecules by the impact of fast electrons is studied for both non-dissociative *gerade* and dissociative *ungerade* final H_2^+ states. 5DCS's in coplanar geometry as a function of the scattered and the ejected electron momenta, and of the internuclear orientation are computed by using the first-order TEC approximation. Interference effects due to the coherent electron emission from the molecular target are explored. Particular attention is paid to the influence of the final state of the residual H_2^+ molecular ion.

Atomic units will be used except where otherwise stated.

2. Theory

The reaction of interest is the electron-impact ionization of H_2 molecular targets, namely,

$$e^- + H_2(^1\Sigma_g^+) \rightarrow 2e^- + H_2^+(^2\Sigma_{g,u}^{\pm}) \tag{1}$$

The ionization process may be considered as a pure electronic transition by applying closure relations over all possible final rotational and vibrational states of the residual target [5, 6]. At the energies involved in this work, the molecule can be considered as frozen during all the reaction. As a consequence, only vertical transitions from the ground electronic state of the target to the specific final electronic state of the H_2^+ molecule are assumed. Also, exchange effects in the collision dynamics are disregarded in the present calculations.

The 5DCS in coplanar geometry is given then by the following expression [1]:

$$\sigma^{(5)} = \frac{d^5\sigma}{d\Omega_\rho d\Omega_e d\Omega_s d(k_e^2/2)} \cong 2(2\pi)^4 \frac{k_e k_s}{k_i} \, |t_{fi}(\boldsymbol{\rho}_0)|^2 \tag{2}$$

where the emitted electron is ejected with momentum $\mathbf{k}_e$ into the differential solid angle Ω_e with respect to the incidence direction, and the projectile is scattered with momentum $\mathbf{k}_s$ into the solid angle Ω_s. The incidence direction is defined by the wave vector $\mathbf{k}_i$ of the incident electron. In equation (2), $\boldsymbol{\rho}_0$ denotes the equilibrium internuclear vector of the molecular target and Ω_ρ, the corresponding solid angle.

The *prior* version of the transition matrix element reads

$$t_{fi}(\boldsymbol{\rho}_0) = \langle \Psi_f^-(\boldsymbol{\rho}_0, \mathbf{R}, \mathbf{r}_1, \mathbf{r}_2) | \, \frac{1}{r_{1p}} + \frac{1}{r_{2p}} - \frac{Z_T}{R_a} - \frac{Z_T}{R_b} \, | \Psi_i^+(\boldsymbol{\rho}_0, \mathbf{R}, \mathbf{r}_1, \mathbf{r}_2) \rangle \tag{3}$$

where Ψ_i^+ is the non-perturbed electronic wavefunction in the initial channel and Ψ_f^- is the final wavefunction, both satisfying the proper asymptotic boundary conditions. $Z_T = 1$ represents the charge of the molecular nuclei. Coordinates are illustrated in Fig. 1.

The initial wavefunction Ψ_i^+ is represented as

$$\Psi_i^+ = \frac{e^{i\mathbf{k}_i.\mathbf{R}}}{(2\pi)^{3/2}} \Phi_i(\rho_0, \mathbf{r}_1, \mathbf{r}_2) \tag{4}$$

where Φ_i is the initial molecular bound state described here by a simple Heitler-London type wavefunction [7],

$$\Phi_i(\rho_0, \mathbf{r}_1, \mathbf{r}_2) = N_{HL}(\rho_0)\ \{e^{-\beta r_{1a}}\, e^{-\beta r_{2b}} + e^{-\beta r_{1b}}\, e^{-\beta r_{2a}}\} \tag{5}$$

with $\beta = 1.166$, $\rho_0 = 1.406$ and N_{HL} the normalization function.

The final-state wavefunction is approximated as,

$$\Psi_f^- \cong \frac{e^{i\mathbf{k}_s.\mathbf{R}}}{(2\pi)^{3/2}} \Phi_{g,u}(\rho_0, \mathbf{r}_2)\, \xi_c(\mathbf{k}_e, \mathbf{r}_1) \tag{6}$$

where Φ_g (Φ_u) represents the wavefunction corresponding to the *gerade* ${}^2\Sigma_g^+$ (*ungerade* ${}^2\Sigma_u^-$) state of the residual H_2^+ molecular ion. In the calculations, the wavefunctions $\Phi_{g,u}$ are represented by a simple linear combination of atomic orbitals

$$\Phi_{g,u}(\rho_0, \mathbf{r}_2) = N_{g,u}(\rho_0)\ \{e^{-\alpha r_{2a}} \pm e^{-\alpha r_{2b}}\} \tag{7}$$

with $N_{g,u}(\rho_0)$ the corresponding normalization factor and $\alpha = 1.3918$ the variational parameter.

In the first-order TEC model [2], the final continuum wavefunction ξ_c is taken as

$$\xi_c(\mathbf{k}_e, \mathbf{r}_1) = \frac{e^{i\mathbf{k}_e.\mathbf{r}_1}}{(2\pi)^{3/2}}\ C(\mathbf{k}_e, \mathbf{r}_{1j}, \gamma_e) \tag{8}$$

where j = a or b indicates the center from which the target electron is ionized, $\gamma_e = -Z_T/k_e$ is the corresponding Sommerfeld parameter, and the Coulomb factor $C(\mathbf{k}, \mathbf{r}, \gamma)$ is given by

$$C(\mathbf{k}, \mathbf{r}, \gamma) = \Gamma(1 - i\gamma)\ e^{-\pi\gamma/2}\ {}_1F_1[i\gamma; 1; -i(kr + \mathbf{k}.\mathbf{r})] \tag{9}$$

In preceding equations, it is assumed that electron 1 is ionized from the center labelled j while the *passive* electron is supposed to screen completely the charge of the other nucleus from which ionization is not produced. This is the main assumption made in the TEC approximation. The exchange with electron 2 is taken into account by means of the factor 2 included in equation (2).

Using equations (4)-(9), the transition matrix element t_{fi} of equation (3) can be written as a sum of direct and indirect terms as follows,

$$t_{fi}(\boldsymbol{\rho}_0) = \frac{1}{(2\pi)^{9/2}}\, N\, N_{g,u}(\rho_0)\, N_{HL}(\rho_0) \sum_{\substack{j,l \\ j\neq l}} (t_{j,l}^{dir} + t_{j,l}^{ind}) \qquad j, l = a, b \tag{10}$$

where

$$t_{j,l}^{dir} = \Big\langle e^{i\mathbf{k}_e.\mathbf{r}_1}\ {}_1F_1[i\gamma_e; 1; -i(k_e r_{1j} + \mathbf{k}_e.\mathbf{r}_{1j})]\ (e^{-\alpha r_{2j}} \pm e^{-\alpha r_{2l}}) \mid \left(\frac{1}{r_{1p}} - \frac{1}{R_j}\right) \mid e^{i\mathbf{K}.\mathbf{R}}\ e^{-\beta r_{1j}}\ e^{-\beta r_{2l}} \Big\rangle \tag{11}$$

and

$$t_{j,l}^{ind} = \Big\langle e^{i\mathbf{k}_e.\mathbf{r}_1}\ {}_1F_1[i\gamma_e; 1; -i(k_e r_{1j} + \mathbf{k}_e.\mathbf{r}_{1j})]\ (e^{-\alpha r_{2j}} \pm e^{-\alpha r_{2l}})\ |\left(\frac{1}{r_{2p}} - \frac{1}{R_l}\right)|\ e^{i\mathbf{K}.\mathbf{R}}\ e^{-\beta r_{1j}}\ e^{-\beta r_{2l}}\Big\rangle \tag{12}$$

Here, $\mathbf{K} = \mathbf{k}_i - \mathbf{k}_s$ is the momentum transfer and $N = \Gamma(1-i\gamma_e)\ e^{-\pi\gamma_e/2}$. The direct and indirect terms can be interpreted as follows [8]. The direct terms describe ionization of electron 1 from center j = a (or b) by means of the interaction of the projectile with this electron and with center j. The indirect terms consider also that the electron 1 is ionized from center j but now through the interaction of the projectile with the *passive* electron and with the other center labelled $l \neq j$. As stated in Ref. [8], electrons are shared by both nuclei in the molecule but matrix elements allow this interpretation.

In the case in which the contribution of the indirect terms (12) may be neglected, the 5DCS given by equation (2) reduces to the approximate expression [1],

$$\sigma_{g,u}^{(5)} \simeq 2\ [1 \pm \cos(\boldsymbol{\chi}.\boldsymbol{\rho}_0)]\ \sigma_{g,u}^{(3)} \tag{13}$$

which depends on the symmetry of the H_2^+ final state, $\boldsymbol{\chi} = \mathbf{k}_e - \mathbf{K}$ and the internuclear vector $\boldsymbol{\rho}_0$. In equation (13), $\sigma_{g,u}^{(3)}$ represents a one-center differential cross section corresponding to *effective* atoms placed at the position of either molecular nuclei [1]. Therefore, when the indirect terms are negligible, the interference pattern due to coherent emission from both molecular centers appears explicitly in the 5DCSs expression, and can be closely related with the one appearing in Young's double-slit experiments. This latter can be considered in fact as one of the main virtues of the TEC model.

3. Results

In figure 2, TEC differential cross sections (thick lines) of ionization of H_2 molecules are presented as a function of the emitted angle θ_e for the case of transitions to the *gerade* ground state of the residual H_2^+ molecular ion. These calculations are obtained by using the full transition matrix element given by equation (10). Three particular cases corresponding to molecules aligned at 0° (dotted lines), 45° (dashed lines) and 90° (full lines) relative to the momentum transfer direction are considered in the figure. The impact and emission energies are E_i = 4087 eV and E_e = 100 eV, respectively, whereas the scattering angle is fixed at angle $\theta_s = -1°$. This high energy regime is similar to the one analyzed in [1]. Approximate 5DCS's obtained from equation (13) for *gerade* final H_2^+ state are also included in figure 2 (thin lines). At the energies considered here, differential cross sections are only slightly modified by neglecting the indirect terms contribution in the transition matrix element.

Interference patterns coming from the coherent electron emission from the different scattering centers in the molecule are observed in figure 2. As pointed out previously [1], particular regions in which the electron emission is more likely to happen are revealed in the figure. This behavior comes from the combined contribution of the interference factor $1+\cos(\boldsymbol{\chi}.\boldsymbol{\rho}_0)$ and the differential cross section corresponding to *effective* atoms placed at the position of each nucleus. A strong dependence on the internuclear vector of the 5DCS's is observed also in figure 2. Hence, differential cross sections corresponding to molecules oriented parallel to the momentum transfer are lower than the ones obtained for molecules aligned in the perpendicular direction. This is the opposite behavior to that observed in recent measurements for the case of dissociative ionization [4].

Surprisingly, in the case of molecules oriented in the **K** direction, supression of the binary peak is observed in figure 2. This is a really unexpected behavior because this direction is

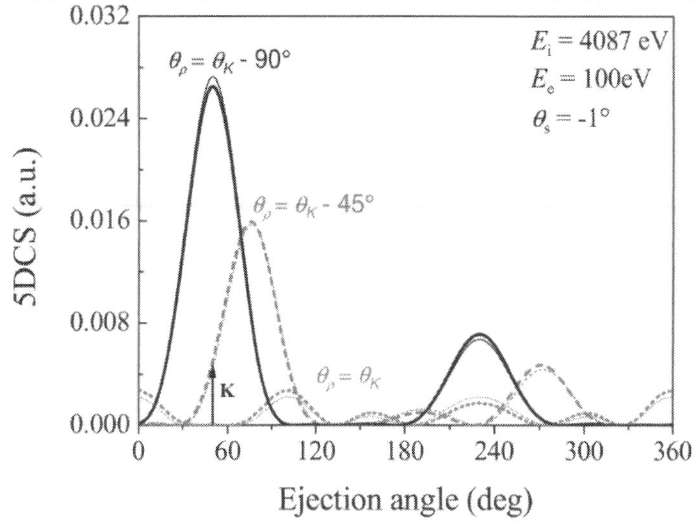

Figure 2. 5DCS's as a function of the ejection angle θ_e for H_2 molecules oriented at angles of 0° (dotted lines), 45° (dashed lines) and 90° (full lines) relative to the momentum transfer direction **K** with polar angle θ_K. The results correspond to final *gerade* state of the residual H_2^+ molecular ion. The impact energy is $E_i = 4087$ eV, the emission energy is $E_e = 100$ eV and the scattering angle is $\theta_s = -1°$. The positive orientation for the polar angles is taken clockwise. Thick lines, full TEC model. Thin lines: approximate TEC model obtained by neglecting indirect contributions (see text).

precisely the more favorable classical one for the electron emission. This effect comes from the interference factor which predicts total destructive interference when the condition $\boldsymbol{\chi}.\boldsymbol{\rho}_0 = l\pi$ is reached for some odd l value. For the kinematical arrangement of figure 2, it can be shown easily that $k_e \sim \pi/\rho_0 + K$ for $\theta_\rho = \theta_K$. This result can be related to the one obtained by Fernández and co-workers [9] for photoionization of H_2 molecules. In this case and when the molecule is aligned parallel to the polarization vector of the incidence radiation, no photoelectrons are emitted under certain similar conditions.

In figure 3, molecular TEC 5DCS's as a function of the emitted electron angle are presented for the particular case in which the final state of the residual H_2^+ molecule corresponds to the $^2\Sigma_u^-$ state. The kinematical conditions here are the same as the ones considered in figure 2. In this case, the indirect terms are more important than in the case of the *gerade* H_2^+ final state. In particular, this contribution seems to be more important for the case of $\theta_\rho = \theta_K$. Also from figure 3, it can be seen that differential cross sections at different molecular orientations are reversed in order compared to the cases studied in the preceding figure. Therefore, cross sections for parallel alignment are in general greater than the ones for the perpendicular direction, in a similar way to that obtained in Ref. [4]. This difference in the relative order of the cross sections with the *gerade* state comes from the electronic distribution of the *ungerade* state. Moreover, this provokes the supression of both the binary and recoil peaks in the case of $\theta_\rho = \theta_K - 90°$ because the *ungerade* state has a nodal plane at those directions.

4. Conclusions

To sum up, 5DCS's of single ionization of H_2 molecules as a function of both electron momenta in the final channel of the reaction, and the internuclear orientation, were computed for both non-dissociative and dissociative final H_2^+ states within the two-effective center approximation. The

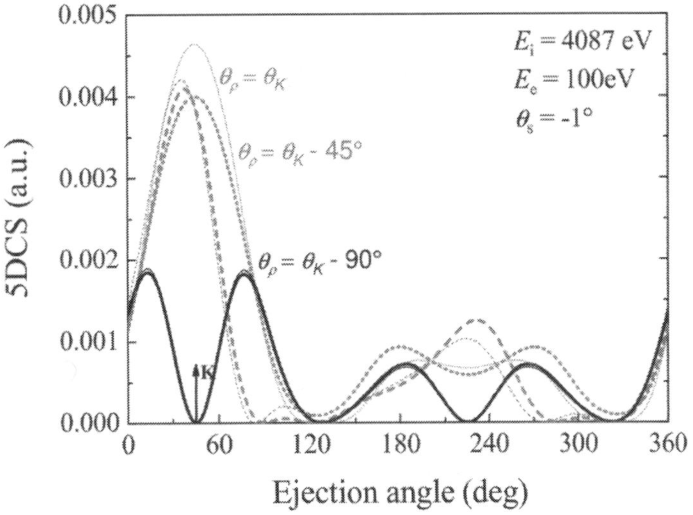

Figure 3. Same as Fig. 2 but for the case of final dissociative $^2\Sigma_u^-$ state of the residual H_2^+ molecular ion.

inclusion of indirect terms barely affects previous calculations [1] for the *gerade* state in which they were neglected. In contrast, some influences may appear for the *ungerade* state studied in this work. Present results at high enough impact energy show that the interference pattern changes strongly with the final state of the residual molecular ion. Moreover, interesting features are observed at the binary and recoil peaks for certain conditions. More elaborated results on this matter will be published soon elsewhere.

5. Acknowledgments

Authors acknowledge financial support from the Agencia Nacional de Promoción Científica y Tecnológica (PICT No. 01912), and the Consejo Nacional de Investigaciones Científicas y Técnicas de la República Argentina (PIP No. 11220090101026).

References

[1] Stia C R, Fojón O A, Weck Ph, Hanssen J and Rivarola R D 2003 *J. Phys. B: At. Mol. Opt. Phys.* **36** L257
[2] Weck P, Fojón O A, Hanssen J, Joulakian B and Rivarola R D 2001 *Phys. Rev.* A **63** 042709
[3] Chatterjee S, Kasthurirangan S, Kelkar A H, Stia C R, Fojón O A, Rivarola R D and Tribedi L 2009 *J. Phys. B: At. Mol. Opt. Phys.* **42** 065201
[4] Senftleben A, Pflüger T, Ren X, Al-Hagan O, Najjari B, Madison D, Dorn A and Ullrich J 2010 *J. Phys. B: At. Mol. Opt. Phys.* **43** 081002
[5] Dey S, McCarthy I E, Teubner P J O and Weigold E 1975 *Phys. Rev. Lett.* **34** 782
[6] Zurales R W and Lucchese RR 1988 *Phys. Rev.* A **37** 1176
[7] Wang S C 1928 *Phys. Rev.* **39** 579
[8] Stia C R, Fojón O A, Weck Ph, Hanssen J, Joulakian B, and Rivarola R D 2002 *Phys. Rev.* A **66** 052709
[9] Fernández J, Fojón O, Palacios A and Martín F 2007 *Phys. Rev. Lett.* **98** 043005

Journal of Physics: Conference Series **288** (2011) 012006 doi:10.1088/1742-6596/288/1/012006

(e,2e) experiments on C_{60}

P Bolognesi[1,*], L Praviça[2,3], J Berakdar[4], Y Pavlyukh[4], R Camilloni[1], D Cvejanovic[2,3] and L Avaldi[1]

1) CNR-IMIP, Area della Ricerca di Roma 1, CP10, 00016 Monterotondo Scalo, Italy
2) Centre for Atomic Molecular and Surface Physics, Physics Department, University of Western Australia, Perth 6009, Australia
3) ARC Centre for Antimatter–Matter Studies, Physics Department, University of Western Australia, Perth 6009, Australia
4) Institute für Physik, Martin-Luther-Universität Halle-Wittenberg, Heinrich-Damerow-Straße 4, 06120 Halle, Germany

E-mail: paola.bolognesi@imip.cnr.it

Abstract. The ionization of C_{60} in gas phase has been studied by (e,2e) experiments at about 1000 eV in asymmetric kinematics and 0.6 a.u. momentum transfer. The binding energy spectrum has been compared with previous photoemission data and theoretical calculations, while the measured coincidence angular distributions corresponding to the ionization of the HOMO and HOMO-1 have been compared with calculations that account for the ionic symmetry and electronic structure of C_{60}.

1. Introduction

Electron impact ionization of atoms and molecules is one of the basic processes in nature. Ionizing collisions induced by electrons are relevant to different scientific and technological areas like gas discharges, cold plasmas used in chemical vapour deposition of materials, lasers and the chemistry of the upper atmosphere. Moreover their theoretical understanding involves the modeling of the few-body Coulomb interaction. As such electron impact ionization has attracted a lot of experimental and theoretical interest. The most complete characterization of the dynamics of an ionization process by electron impact is achieved by detecting the scattered and ejected electrons in coincidence after energy and angular selection [1]. The quantity measured in these experiments, known as (e,2e) experiments, is the triple differential cross section (TDCS) $d^3\sigma / d\Omega_a d\Omega_b dE_a$, i.e. a cross section that is differential in the solid angles of the scattered and ejected electrons, labelled as 'a' and 'b' respectively, and in the energy of one of them. The energy of the second electron is determined by the energy conservation, $E_a+E_b=E_0-IP$, where IP is the ionization potential of the target and E_0 the incoming electron energy. The (e,2e) technique has scored several successes in describing both the ionization mechanisms in one

[*] Author to whom any correspondence should be addressed.

and two electron targets [2] and the structure of molecular orbitals via electron momentum spectroscopy [3].

In this work we present results of the first application of the (e,2e) technique to the study of C_{60} in gas phase. Carbon is the fourth most abundant element in the Universe and forms more compounds than any other element, with almost ten million pure organic compounds described to date. One of carbon allotropic forms, the fullerene discovered in 1985 by Kroto and Smalley [4] is attracting a lot of interest for its amazing properties [5]. Fullerene can be considered as an intermediate form between a molecule and a cluster. This makes it a very special target for the investigation of the electron-electron interactions. Its highly symmetric and stable structure shows, already at the single molecule level, both atomic and collective behaviors. Valence electron properties of C_{60} have been investigated via (e,2e) experiments on thin films [6]. The experimental data have been interpreted by comparison with the calculated orbitals of a C_{60} molecule. To our best knowledge no gas phase (e,2e) data on C_{60} have been reported in the literature, while an extensive series of photoionisation studies have been performed [7-10].

2. Experiments

A crossed-beam apparatus has been used to measure the ionization cross-section of C_{60}. The apparatus has been described in detail elsewhere [11], here only information relevant to the present measurements will be r
hemispherical electrostatic analysers, rotatable independently in the scattering plane and a resistively heated, anti-inductively wounded oven to produce the beam of C_{60} [12]. The compensation of the earth magnetic field is performed by three pairs of orthogonal square coils [11] external to the vacuum chamber and by an internal 0.4 mm thick Skudotech layer [12].

The scattered/ejected electrons are analysed in energy by one of the two electron spectrometers, with an energy resolution of about 1eV. A three element zoom electrostatic lens focuses the electrons from the target region onto the entrance slit of the hemispherical analyser, with an angular acceptance of ±2 and ±4° for the scattered and ejected electrons respectively The electrons, after angle and energy selection, are then detected by a channeltron electron multiplier. The output signals of the detectors are sent to the TAC (Time to Amplitude Converter) through preamplifiers and constant fraction discriminators. Finally, the TAC output is analysed by a triggered ADC PCI card displaying the time spectrum. The typical incident current, monitored by a Faraday Cup, was about 5 μA. A personal computer via a Labview software scans the energy of the incident beam, changes the energy loss of the scattered electron, controls the movement of the turntables, sets the dwell time of the measurements, stores the non-coincidence and coincidence data and monitors the current of the beam during the acquisition.

Even though the role of 'scattered' and 'ejected' electron cannot be uniquely related to the energy of the two electrons, in this paper the convention of labeling the fast electron, 'a', as the scattered electron and the slow electron, 'b', as the ejected one has been adopted. In the measurements the scattered electrons (E_a=1000 eV) were detected at a fixed angle θ_a=4°, while the ejected electrons (E_b=20 eV) were measured in the angular range θ_b varying from 40 to 130°. The θ_a scale was calibrated by determining the symmetry of the scattered electron yield around the incident beam direction. The energy resolution measured as full-width-half-maximum (FWHM) of the coincidence binding energy spectrum of He was about 1.4 eV. A check of any anisotropic response of the set-up was done by measuring the TDCS of He in the same kinematic conditions of the C_{60}.

For the present experiments the crucible was filled by about 300 mg of C_{60} powder from Sigma-Aldrich, with purity >97%, used without further purification. In the typical operating conditions, the oven was kept at about 515 and 475°C on the top and bottom of the furnace respectively, running continuously for at least a month before some recharging of the sample was required. The stability of the oven was checked at the beginning and at the end of each coincidence run by a measurement of the energy loss spectrum in the valence region. A series of these energy loss measurements taken over few weeks is shown in figure 1. The spectrum shows a well resolved excitation peak at 6 eV, and some other structures at about 10, 13 and 17 eV superimposed to a broad feature with its maximum at about 28 eV. All these features are in agreement with the observation by Keller and Coplan [13], who also provided some assignment for the different features and a comparison with a solid state experiment by Lucas et al. [14].

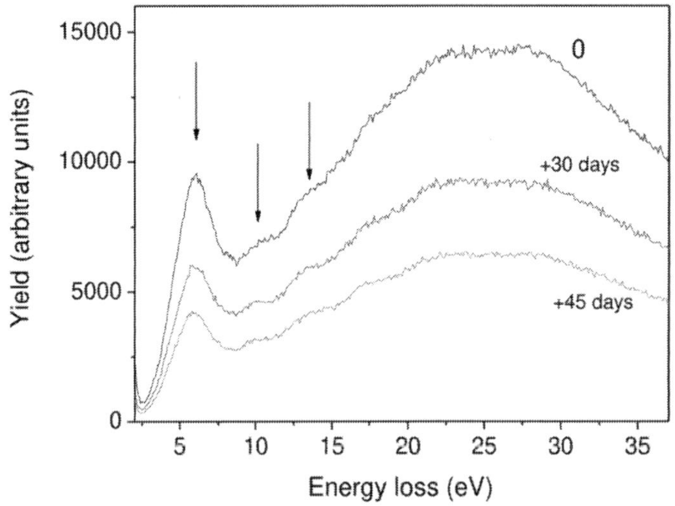

Figure 1. C_{60} energy loss spectra taken over a period of 45 days measured to check the performance of the set-up during the (e,2e) experiments. The arrows indicate the main structures identified in previous works [13, 14].

3. Theory

The (e,2e) process from C_{60} is a challenge for theory and one has to resort to appropriate approximations. For an adequate description of the experiments two aspects should be accounted for: 1) the scattering dynamics in the initial and the final channels; and 2) the electronic and ionic structure of the target. As for the first point the appropriate choice depends on the experimental parameters. For example, for strongly asymmetric energy sharing of the final state electrons (E_a>>E_b), large impact energy, and small momentum transfer, the first Born or even the plane wave approximation may provide a good description. In fact the application of both of these approximations to fullerenes is more justified than for atoms because of the effective screening [15, 19] that makes the interaction of the continuum electrons with the target short ranged. The price to be paid for this is the calculation of the screening which is a challenging task in itself [18]. As shown in [18] however the screening effects are easily described in the high impact energy regime. In the present experiment the arrangements are such that the continuum electrons may be described with decoupled plane waves. We put more effort in the ab-initio description of the target: C_{60} is nearly, but not exactly spherically symmetric. Its lower symmetry (I_h) leads to additional splitting of the energy levels. Also the final results for the TDCS must be averaged over the possible orientations of the molecule. An important ingredient of the theory is the knowledge of the target electronic wave-functions. We computed the single particle wave-functions in the basis of Gaussian functions centered at the ionic positions using the Hartree-Fock method (all technical details can be found in [16]). We note that the density functional theory (DFT) can also be used to obtain the single-particle states. However, we find that both methods provide very consistent results for the space-dependent wave-functions; the difference is mostly in the energy of the states.

As was demonstrated in [16], the angular momentum quantum number ℓ governs the energy spectrum of C_{60}. All valence electrons (240) of the molecule can be grouped into two parabolas (principal quantum numbers n=1,2) with 60 (180) electrons in the upper (lower) branch as a function of ℓ. The highest occupied molecular orbital (HOMO) has h_u symmetry and can be assigned ℓ=5 value. However, this state is only 5-fold degenerate. Six other states with ℓ=5 can be grouped into 2 groups above the Fermi level. The HOMO-1 state with ℓ=4 is split in I_h environment into 5-fold and 4-fold substates with symmetries h_g and g_g, respectively. The energy separation is rather small to be resolved in the present experiments (table 1). On the basis of this electron counting one can already expect that the ratio of signal from HOMO-1 and HOMO states should be 9:5. However, further important differences come from the exact shape of the electron momentum distribution, $|\psi(\mathbf{q})|^2$, of the electronic states. Here q denotes the norm of the bound-state electron momentum in radial direction from the center of molecule.

In the first approximation the electrons in C_{60} can be considered as free particles confined to the surface of the sphere with radius of approximately r=6.75a_B, where a_B is the Bohr radius. For the HOMO state from a simple relation $\ell=|q\times r|$ we obtain an estimate for the maximum of the momentum distribution to be at q~0.74 a.u.. For HOMO-1 we likewise obtain q~0.60 a.u.. These numbers reflect the trend that decreasing the state energy the position of the maximum of the electron momentum distribution is shifted towards lower q values. However, such a simple consideration misses the quantum mechanical nature of the states. Two parabolic branches in the energy dispersion can be distinguished by the principal quantum number, n, which characterizes electronic states in the confining potential [17]. In accordance with the oscillation theorem of quantum mechanics the states in the lowest branch are nodeless, while the states in the upper branch have one node. The nodal surface lies approximately at the potential minimum, i.e. at the radius of the molecule (r). Accordingly, the states from the upper branch have two maxima and one nodal surface in the momentum space.

The maxima in the momentum distribution translate into the positions of the lobes in the angular resolved (e,2e) spectra. Within the first Bohr approximation the spectrum is symmetric with respect to the direction of the momentum transfer $\mathbf{K}=\mathbf{k_0}-\mathbf{k_a}$, where k_0 and k_a are the momenta of the incident and scattered electrons respectively We will see below that for C_{60} this condition does not strictly hold. From general considerations, three sources can be identified for this behavior: i) interactions among the continuum state electrons; ii) effects of dynamical screening, and iii) many-body initial state correlations that go beyond the mean-field correlation incorporated in the calculations below or effective exchange and correlation, as described within the local density approximation to the exchange and correlation functional that enters the DFT calculations. A detailed analysis of these aspects and a comparison with the atomic case is scope of the present work.

4. Results

We have performed two types of (e,2e) experiments on C_{60}. In a first experiment the analyser of the ejected electron was set at fixed θ_b and a binding energy spectrum was measured. In the kinematic conditions of the present experiment (E_a=1000 eV, E_b=20 eV and θ_a=4°) the momentum transfer $\mathbf{K}=\mathbf{k_0}-\mathbf{k_a}$ in the collision is about 0.6 a.u.. In the second type of experiments the TDCS of the two lowest in energy ionic states have been measured. The binding energy spectrum provides both the energetic of the ionic states and allows to establish the relative intensity of their measured TDCS.

In figure 2.a the binding energy spectrum measured at θ_b=80° is shown. The present gas phase (e,2e) data are compared to the results of photoemission experiment [18] of condensed-phase pure C_{60} and to the calculations by Colavita et al. [8

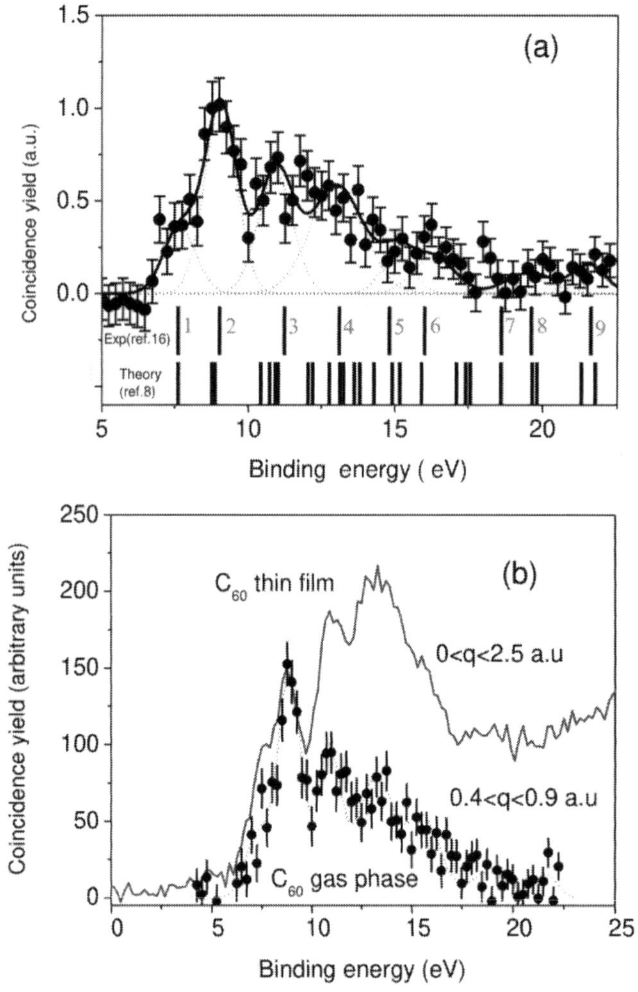

Figure 2. (a) Binding energy spectrum of C_{60} measured at E_a=1000 eV, E_b=20 eV, θ_a=4° and θ_b=80°. The bars below the spectrum are the positions of the electronic bands measured in [18] and calculated in [8]; (b) comparison between the binding energy spectra of C_{60} thin film integrated over the momenta between 0 and 2.5 a.u. [6] and the sum of four binding energy spectra of gas phase C_{60} measured at θ_b=60, 80, 100 and 120°. The dashed line is the best fit to the sum of the spectra with Gaussian functions. The four energy spectra measured in the present work scan a momentum region between 0.4 and 0.9 a.u..

Journal of Physics: Conference Series **288** (2011) 012006 doi:10.1088/1742-6596/288/1/012006

The spectrum has been fitted with a series of Gaussian functions where the intensity, energy position and FWHM were used as fitting parameters. The use of the FWHM as a free parameter is justified by the fact that in certain regions a condensation of several states (see table 1) is expected to produce broader spectral features. Nine Gaussian functions have been used in the fit. The number has been chosen according to the previous experimental observation in photoelectron spectroscopy [18]. Increasing or decreasing the number of Gaussian functions may result in an acceptable "visual" representation of the experimental data, but in unphysical values of the fitting parameters (FWHM smaller than the experimental resolution or uncertainties in the peak positions larger that their relative spacing). An absolute calibration of the binding energy scale has not been done, thus we assign the main feature of the spectrum to the HOMO-1 orbital observed in the photoemission experiment. The comparison of the relative position of the different features in the spectrum is reported in table 1.

Table 1. Experimental and calculated ionization energies of the valence and inner valence states of C_{60}. Peak positions, in eV, are referred to the HOMO-1 state at 9 eV [8, 18]. The labeling of the band is done according to ref [8].

Band	Present work (±0.1 eV)	Experiment (ref.)	Theory (ref. 8)	Symmetry (ref. 8)
HOMO Band 1	-1.2	-1.4	-1.13	π
HOMO-1 Band 2	0	0	0.00 0.09	π π
Band 3	2.0	1.8-2.6	1.67 1.97 2.17 2.26	π σ π σ
Band 4	4.2	3.4-4.8	3.29 3.45 3.76 4.03 4.40 4.52 4.88 5.07	σ π σ σ σ π σ π
Band 5	5.6	5.8	5.55	σ
Band 6	6.6	~6.4-7.6	6.17 6.43 7.17	σ σ σ
Band 7&7′	8.9	~9.6	8.36 8.68 8.83 9.86	σ σ σ σ
Band 8	10.0	10.6	10.9 11.06	σ σ
Band 9	11.8	12.6	12.55 13.03	σ σ
Band 10		14.6	14.29 15.15 16.14 17.10 17.6	σ σ σ σ σ

The relative positions measured in the present work are in reasonable agreement with the ones determined by the photoemission experiment in condensed phase [18]. This result is consistent with the findings of the comparison of the energy loss spectra [13] and is due to the fact that C_{60} molecules weakly interact in the solid. Vos et al. [6] have reported a binding energy spectrum of a C_{60} thin film measured in a high energy transmission (e,2e) experiment integrated over all the momenta **q** ($\mathbf{q}=\mathbf{k}_0-(\mathbf{k}_a+\mathbf{k}_b)$) of the target electron before the collision between 0 and 2.5 a.u.. In the present experiment we have measured the binding energy spectra of C_{60} at four different θ_b's, namely 60, 80, 100 and 120°. The sum of our spectra, which are normalized to each other, is compared with the spectrum measured by Vos et al. in figure 2.b. There is a good agreement between the two spectra as for the position of the bands, with the present data characterized by a slightly better energy resolution, but the relative intensity of the bands is quite different. This can be explained considering that the present measurements map a smaller region of the momentum **q**, extending from about 0.4 to 0.9 a.u., compared to the one on the thin film [6] experiment. According to the calculation of the spherical averaged momentum densities reported in figure 2 of Vos et al. [6] the first two π bands have a quite large intensity in the q region measured in this work, while the bands at IP > 10 eV display a large fraction of their intensity outside this region. This explains the agreement between the two spectra of figure 2.b in the low binding energy region and the quite different intensities in the higher energy region.

The TDCS of the HOMO and HOMO-1 states are shown in figure 3, where they are compared with the calculations. The two TDCS are reported on the same relative scale of counts, via the binding energy spectrum of figure 2. The TDCSs under asymmetric conditions are characterized by the presence of two lobes [1]. The first one, oriented nearly in the direction of $\boldsymbol{K}$, is associated with a binary collision of the incident electron, and it is indeed named as 'binary peak'. The second peak, near the opposite direction, $-\boldsymbol{K}$, is known as the recoil peak and it is commonly associated with a backscattering mechanism. The relative intensity of these lobes, their widths and positions with respect to $\pm \boldsymbol{K}$ are the observables which provide the information on the dynamics of the ionizing process. The measurements have been concentrated on the binary region because the measured recoil/binary ratio in the $\pm\boldsymbol{K}$ direction gave 0.1±0.1, consistent with the expected vanishing recoil peak intensity predicted by the theory. The measured angular distributions of both the HOMO and HOMO-1 bands have a maximum at around θ_b=87°, that is about +10° away from the **K** direction. The ratio between the two TDCS is about 3 and the TDCS of the HOMO is broader than the one of the HOMO-1. The intensity ratio can be traced back to the number of electrons occupying the HOMO-1 (8+10) and HOMO (10) states. Different shapes of angular distributions can be explained by considering their momentum distribution $|\psi_i(q)|^2$. The electronic states also have an angular dependence. However, the final results must be averaged over all possible orientations of the molecule. Therefore in $|\psi_i(q)|^2$ the angular averaging is already performed. As we noted above from very general consideration, the HOMO state is characterized by the presence of two maxima. Our numerical calculations yield their position at q=0.66 a.u. and q=1.28 a.u.. The HOMO-1 state is characterized by the smaller value of the orbital angular momentum (ℓ=4). Correspondingly, the maxima are shifted to lower values (q=0.51 a.u. and q=1.11 a.u.). These numbers are also consistent with momentum distributions obtained for the C_{60} film [6]. In order to qualitatively understand the angular dependence of the (e,2e) signal the kinematics of the experiment should also be taken into account. Only peaks at smaller values of momentum q are relevant in the present measurements, since other peaks lie outside the experimentally observed window of $0.4<q<0.9$ a.u., where the smallest value of momentum is in the direction of $\boldsymbol{K}$. The HOMO-1 state is peaked at the smaller values of q, thus electrons emitted from this state will be predominantly observed in a smaller angular interval around $\boldsymbol{K}$. Reversely, for the HOMO state the peak is shifted towards higher values of q, attained for larger angular deviations (θ_b) from the $\boldsymbol{K}$ direction. Correspondingly, the angular distribution is broader. As for the deviation of the binary peak position from the direction of the momentum transfer and even for the interpretation of the origin of the peak one should note here some differences between the atomic case and the case of a polarizable many-electron system: due to the dynamical screening effects caused by the valence-band electrons of the target, the interaction potential becomes frequency dependent and nonlocal [15]. In fact, as discussed in [15] within the jellium model the potential in an (e,2e) experiment is even geometry dependent.

5. Conclusions

This work reports some pioneering results on the investigation of the ionization of gas phase C_{60} with the (e,2e) technique. It is shown that (e,2e) measurements well reproduce the electronic structure which was established by previous measurements and calculations, mainly in condensed phase, confirming the weak interaction of C_{60} molecules in the solid [13, 6].

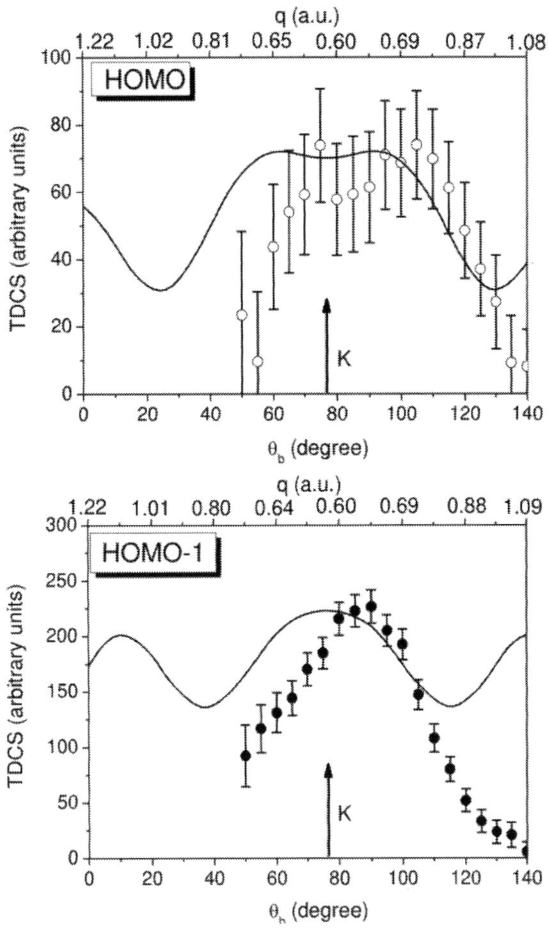

Figure 3. Angular distribution of the HOMO and HOMO-1 states compared with the calculations.

The TDCS of the HOMO and HOMO-1 orbitals have been described by calculations that account for ionic symmetry and electronic structure of C_{60}. The discussion of the differences observed in the comparison between calculations and experiment hints to the role of the screening and its consequences for the interactions between the continuum state electrons. This is the subject of ongoing work.

These first gas phase studies show that a detailed investigation of the dynamics of the electron impact ionization of this prototype molecule of fullerene compounds, which has shown intriguing relationship between dynamic and structure [19] can be undertaken. The comparison of gas phase and solid state studies is expected to help disentangle single molecule and condensed phase properties.

Acknowledgments

LP and DC acknowledge the support of the CNR-Short Term Mobility Program. The authors thank M. Vos for providing the data of the experiment on C_{60} thin film.

References

[1] Avaldi L et al. 1987 *J.Phys. B : At. Mol. Opt. Phys.* **20** 5827
[2] Ehrhardt H, Jung K, Knoth G and Schlemmer P 1986 *Z. Phys. D: At. Mol. Clust.* **1** 3
[3] Weigold E and McCarthy I E 1999 *Electron Momentum Spectroscopy* (new York: Kluwer)
[4] Kroto H W, Heath J R, O'Brien S C, Curl R F and Smalley R F 1985 *Nature* **318** 162
[5] Hebard A F et al 1991 *Nature* **350** 600
[6] Vos M, Canney S A, McCarthy I S, Michalewicz M T and Weigold E 1997 *Phys. Rev. B* **56** 1309
[7] Korica S et al. 2005 *Phys. Rev. A* **71** 013203
[8] Colavita P, De Alti G, Fronzoni G, Stener M and Decleva P 2001 *Phys. Chem. Chem. Phys.* **3** 4481
[9] Berkowitz J 1999 *J. Chem. Phys.* **111** 1446
[10] Aksela S et al 1995 *Phys. Rev. Lett.* **75** 2112
[11] Avaldi L, Camilloni R, Multari R, Stefani G, Zhang X, Walters H R J and Whelan C T 1993 *Phys. Rev.* **A48** 1195
[12] Bolognesi P et al. 2008 *J. Phys. B:At. Mol. Opt. Phys.* **41** 015201
[13] Keller J W and Coplan M A 1992 *Chem. Phys. Lett.* **193** 89
[14] Lucas A et al. 1992 *Phys. Rev B* **45** 13694
[15] Kidun O. and Berakdar J, 2001 *Phys. Rev. Lett.* **87** 263401
[16]Pavlyukh Y. and Berakdar J, *Chem. Phys. Lett.* **468**, 313 (2009)
[17] Pavlyukh Y. and Berakdar J, *Phys. Rev. A* **81**, 042515 (2010)
[18] Weaver J H et al. 1991 *Phys. Rev. Lett.* **66** 1741
[19] Rüdel A et al. 2002 *Phys. Rev. Lett.* **89** 125503

Dynamical (e, 2e) studies using tetrahydrofuran as a DNA analogue

C J Colyer[1], S M Bellm[1], G F Hanne[2], O Al-Hagan[3], D Madison[3], C G Ning[4] and B Lohmann[1,5]

[1] ARC Centre of Excellence for Antimatter-Matter Studies, The University of Adelaide, Adelaide, SA 5005, Australia

[2] Physikalisches Institut, Universität Münster, Wilhelm-Klemm-Straße 10, 48149 Münster, Germany

[3]Department of Physics, Missouri University of Science and Technology, Rolla, MO 65409, USA

[4] Department of Physics and Key Laboratory of Atomic and Molecular NanoSciences of MOE, Tsinghua University, Beijing 100084, People's Republic of China

e-mail: birgit.lohmann@adelaide.edu.au

Abstract. We present dynamical (e, 2e) measurements for the electron impact ionization of formic acid and tetrahydrofuran molecules performed under similar kinematics. The experiments have been performed in the coplanar asymmetric geometry for a range of momentum transfers, at an incident electron energy of 250 eV and with an ejected electron energy of 10eV. The experimental results are compared with theoretical calculations carried out using the molecular 3-body distorted wave (M3DW) model.

1. Introduction

While numerous dynamical (e, 2e) studies exist for atomic targets (see Lahmam-Bennani [1] and Coplan *et al.* [2]) and an impressive number of electron momentum spectroscopy (EMS) measurements exist for molecular systems (see McCarthy and Weigold [3] and Takahashi [4]), (e, 2e) studies dedicated to the dynamics of molecular systems are quite limited. Triply differential cross sections (TDCSs) for molecular targets can be used to elucidate the molecular effects responsible for structures in the cross section. Such structures in the TDCS may be influenced by the non-spherical shape of the molecular target as well as contributions from rotationally and vibrationally excited ground states left after a direct single ionizing collision [5].

Experimental difficulties associated with TDCS measurements for molecular targets arise from the inability of the experiment to resolve different molecular electronic states. Molecular orbitals can, depending on the molecular configuration of the target, be quite closely spaced and contributions from the vibrational and rotational states only add to the challenge. Many of the molecules for which experimental dynamical (e, 2e) studies have been undertaken are summarized in ref. [6]. In addition,

[5] To whom correspondence should be addressed.

(a)

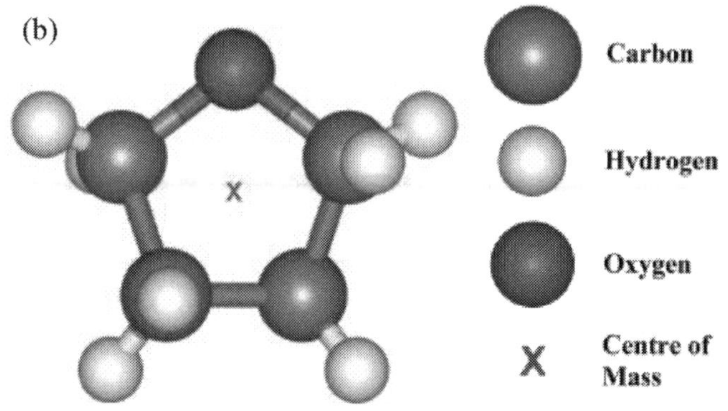

Figure 1. (Colour online) Molecular structures of (a) formic acid and (b) tetrahydrofuran with the centre of mass of each molecule denoted with a (x).

theory has had to deal with modelling polyatomic target species, with many of the approximations that work for the atomic case breaking down due to the polycentric nature of these molecules. Madison and Al-Hagan [7] have reviewed the recent theoretical work that has been performed for some of these molecules.

Ionizing radiation is used widely in medicine as a probe in radio-diagnostic examinations, as a treatment method for many forms of cancer and in industrial production as a means of quantifying internal defects. Traditionally, damage to biological systems by incident high energy radiation was thought to be caused by high-energy ballistic impacts of the incident particle. As the radiation travels through the medium, its energy is redistributed via elastic and inelastic collisions, and produces an abundance of secondary species. In recent years, theoretical and experimental studies have indicated that secondary species generated by the primary ionizing particle also play a significant role in radiation damage to biological media [8]. In the study of Boudaiffa *et al.* [9] it was observed that electrons in the energy range 3-20 eV can cause significant damage to DNA through the process of dissociative attachment, which can occur even at energies well below ionization thresholds. Triple differential cross sections for electron impact ionization of water have been used as an input for a simulation to determine the charged particle track structure of an electron through a biological medium [10]. The challenge is to extend this modelling to other biomolecules, and hence obtain a complete picture for real biological material.

To improve our understanding of the interaction of electrons with biological matter we are investigating this electronic interaction using smaller biomolecules such as formic acid and tetrahydrofuran (THF) as a model for the components of larger biological systems. Formic acid is the simplest organic acid and is thought to play a key role in the formation of larger biologically relevant molecules such as acetic acid and glycine. THF can be regarded as a structural unit in a number of biological molecules including the sugar moiety in the sugar-phosphate backbone of DNA and RNA. This backbone structure can be represented as a series of THF molecules bonded to phosphate units and DNA Bases. Figure 1 shows the structure of (a) formic acid and (b) THF molecules. Also indicated on each figure is the location of the molecules' centre of mass.

Here we describe our recent progress in investigating the electron-impact ionization of formic acid and THF molecules. Our recently published TDCS measurements for the formic acid monomer [6] are directly compared with new measurements for THF under the same kinematics. We also compare the experimental results with distorted wave calculations of the TDCS.

2. Experiment

(e, 2e) events are identified by the relative arrival times of electrons at the two detectors and background events are subtracted using standard statistical methods. The two electron energy analysers are mounted on independent turntables concentric with the interaction region. The higher energy (scattered) electron is detected at a fixed small angle with respect to the incident electron beam direction. Electron angular distributions are measured by scanning the lower energy (ejected) electron energy analyser and detecting electrons at a number of different angles within the scattering plane. The coincidence energy resolution of the system is approximately 1.2 eV full width half maximum (FWHM).

Formic acid and THF are both liquids at room temperature and a vapour target is obtained from the liquid sample held in a glass vial which has been purified prior to use via several freeze-pump-thaw cycles. In the case of formic acid, the vapour is a mixture of monomers and dimers whose ratio is a function of temperature and driving pressure. At higher temperatures, the extra kinetic energy serves to break up most of the dimers into monomers. It was demonstrated in ref. [12] that at temperatures in excess of 120°C, the target is composed of greater than 99% monomers. As a result, the beam-forming needle is held at approximately 135°C, while the associated gas handling system and vacuum chamber are heated to approximately 75°C and 50°C, respectively, to prevent condensation. Similarly for THF the gas handling system and chamber are heated to prevent condensation and to further inhibit THF being deposited on the electron optics and other elements within the vacuum chamber, a liquid nitrogen filled cold finger located directly above the interaction volume was used to trap THF.

3. Results and Discussion

3.1. Triple differential cross sections for formic acid

As discussed previously, closely spaced molecular orbitals are difficult to resolve and with the limited coincident energy resolution of our spectrometer we were unable to fully resolve contributions from the outermost 10a′ and 2a″ valence orbitals of formic acid. Figure 2 shows our recent TDCS data for formic acid which was presented in ref [6]. The experimental data are for electron impact ionization of the summed (10a′ + 2a″) outermost valence orbitals. The energy of the incident electrons (E_0) is 250 eV and the energy of the ejected electrons (E_b) is 10 eV. Results are shown for scattering angles of (a) -5° and (b) -15°. Also shown on the figure are molecular 3-body distorted wave (M3DW) [7] and M3DW-CPE calculations for the 10a′ orbital. While experimental results are shown for the sum of the 10a′ and 2a″ valence orbitals, we are able to calculate results for the 10a′ orbital only. The orientation averaged molecular orbital (OAMO) approximation used in the calculations for formic acid is not valid for the 2a″ orbital since the average is zero for this symmetry, see ref [6] for details. The M3DW-CPE calculations include correlation-polarization-exchange terms in the M3DW model [6].

The angular distributions can be divided into two regions. The first is the binary region which ranges from 0° to 180°. Structure here arises from single binary collisions and depending upon the kinematics it may contain strong signatures of the orbital structure [13]. The second, which extends from 180° to 360°, is referred to as the recoil region. It is aptly named as it contains structure arising from processes in which the ejected electron produced by an initial binary collision undergoes recoil scattering from the target nucleus.

As the scattered electron angle is changed from -5° to -15° the magnitude of the recoil peak relative to the binary peak decreases significantly. The M3DW predicts the shape of the recoil peak quite well,

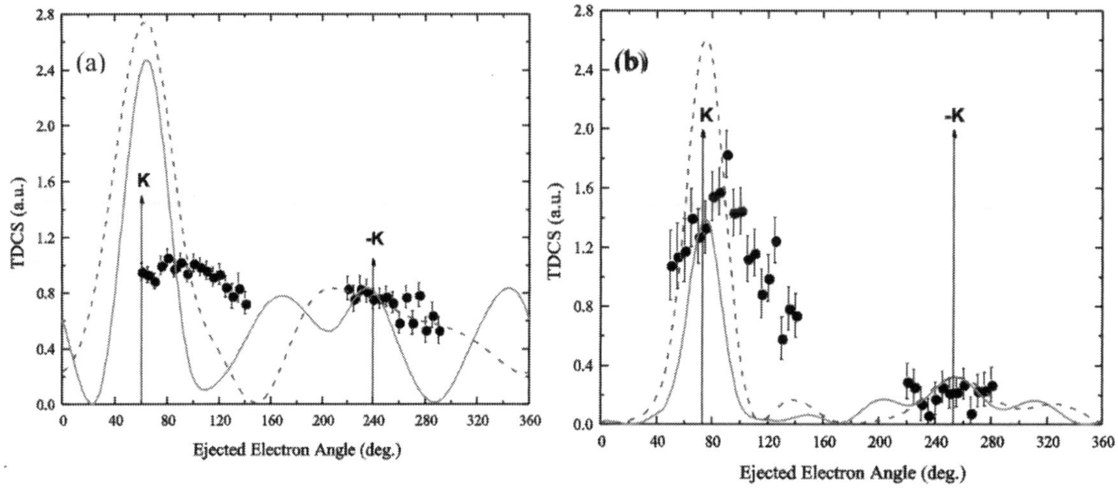

Figure 2. (Colour online) Experimental triple differential cross sections for the summed 10a′ and 2a″ valence orbitals of formic acid (solid circles), with $E_0 = 250$ eV and $E_b = 10$ eV, plotted against the M3DW-CPE (solid line) and M3DW (dashed line) calculations for the 10a′ orbital only. The scattered electron detection angles and corresponding momentum transfers are (a) -5°, |K |= 0.42 au and (b) -15°, |K| = 1.11 au (Adapted from Colyer *et al.* [6].).

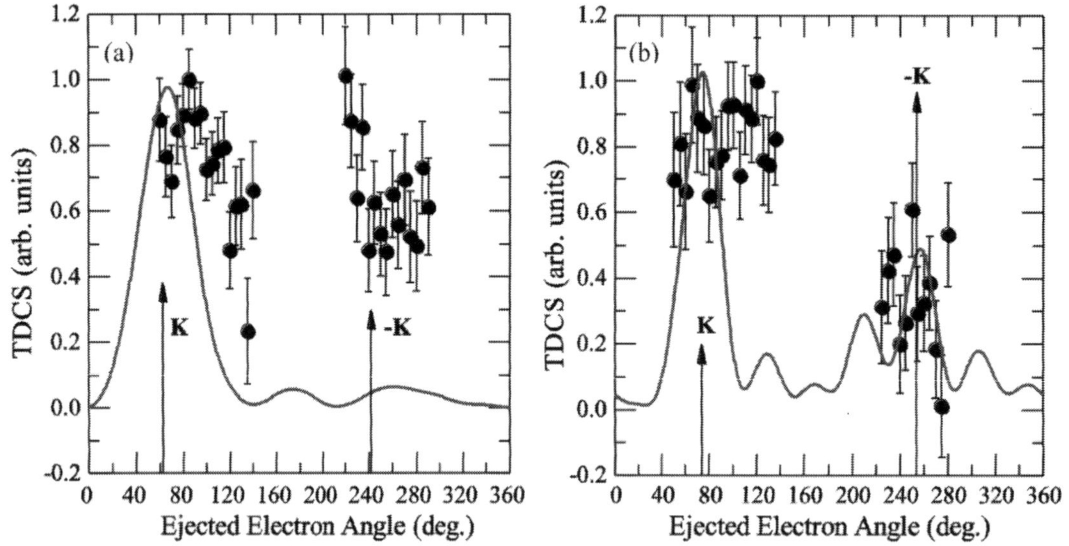

Figure 3. (Colour online) Plot of the triple differential cross sections for ionization of the combined 9b+12a′ valence orbital of tetrahydrofuran, with $E_0 = 250$ eV and $E_b = 10$ eV. The scattered electron detection angles and corresponding momentum transfers are (a) -5°, |K| = 0.40 au and (b) -15°, |K| = 1.11 au. The points are the experimental data and the solid line is the M3DW calculation. The positions of the momentum transfer vector, **K**, and –**K** are indicated by the arrows. in both cases.

however, the relative size of the binary and recoil peaks is better predicted by the M3DW-CPE, which successfully predicts the correct relative magnitudes of the binary peak and the recoil peak at -15°. The experimental data exhibits a broader binary peak when compared to the theory, which is due to contributions from the 2a″ state that are not included in the M3DW calculations.

At room temperature THF exists in two conformers in approximately equal proportions [14] and we are unable to separate the individual contributions from each of these. The two conformers are of C_2 and C_s point group symmetry and their outermost orbitals are assigned as 9b and 12a′, respectively [14]. Figure 3 shows the TDCS for electron impact ionization of THF, again with 250 eV incident electrons and measured for 10 eV ejected electrons. A M3DW calculation for the combined 9b + 12a′ orbital is also shown on the figure. The contribution of the outermost orbital for each conformer in the calculation is 50%. In the binary region, it appears that the experiment has a double peak structure whilst the theory only predicts a single peak which is centred about the momentum transfer direction. It is expected that the second peak is lost in the averaging process used to calculate the molecular wave function.

As is the case for formic acid the magnitude of the recoil peak relative to that of the binary peak is less for -15° (figure 3(b)) than for the -5° scattering angle (figure 3(a)). This is in contrast to our earlier TDCS measurements for H_2O under the same kinematics [15]. Following from arguments presented in ref. [16], it is postulated that a lack of nuclear charge at the molecules' centre of mass results in reduced recoil scattering.

The experimental TDCS's for both formic acid and THF molecules exhibit a very broad binary peak with a minimum centred close to the momentum transfer direction. A similar very broad binary peak has also been observed for ionization of the outer valence orbital of methane [17]. At the -5° scattering angle the intensity of the binary peak is close to that of the recoil peak, as is observed for the formic acid monomer. At -15° the shape of the formic acid binary peak appears sharper than at -5° and closer to the M3DW calculation whereas the THF binary peak is still very broad at this scattering angle. Further THF results and a more detailed analysis will appear in a future publication [18].

4. Conclusions

(e, 2e) TDCSs for the THF molecule, which is a model compound to investigate electron interactions with the deoxyribose molecule, are reported and compared to our recent measurements on the formic acid monomer. Similar to the formic acid monomer, the TDCSs at all scattered electron angles investigated show a broad binary peak. The ratio between the recoil peak magnitude and binary peak magnitude for both formic acid and THF is much smaller than that observed for ionization of water under the same kinematics.

Acknowledgements

This work was supported by the Australian Research Council Centre of Excellence for Antimatter-Matter Studies and the American NSF under grant number 0757749. The author CGN would like to acknowledge the support of the National Natural Science Foundation of China under contract no 10704046. GFH was supported by the Heinrich-Hertz-Stiftung des Landes Nordrhein-Westfalen, Germany, and an Australian International Science Linkages Programme grant. The author OA-H would like to acknowledge the support of the Saudi Ministry of Higher Education's King Abdullah Bin Abdul-Aziz Scholarship and the author CJC would like to acknowledge the support of an Australian Postgraduate Award.

References

[1] Lahmam-Bennani A 1991 *J. Phys. B* **24** 2401
[2] Coplan M A, Moore J H and Doering J P 1994 *Rev. Mod. Phys.* **66** 985
[3] McCarthy I E and Weigold E 1991 *Rep. Prog. Phys.* **54** 789

[4] Takahashi M 2009 *Bull. Chem. Soc. Jpn.* **82** 751-777

[6] Colyer C J, Stevenson M A, Al-Hagan O, Madison D H, Ning C G and Lohmann B 2009 *J.* Phys. B **42** 235207
[7] Madison D H and Al-Hagan O 2010 *J. Atomic Molecular Optical Phys.* **2010** 367180
[8] Cobut V, Frongillo Y, Patau J P, Fraser M J and Jay-Gerin J P 1998 *Radiat. Phys. Chem.* **51** 229
[9] Boudaïffa B, Cloutier P, Hunting D, Huels M A and Sanche L 2000 *Science*, **287** 1658
[10] Champion C 2003 *Phys. Med. Biol.* **48** 2147
[11] Cavanagh S J and Lohmann B 1999 *J. Phys. B* **32** L261
[12] Nixon K L, Lawrance W D, Jones D B, Euripides P, Saha S, Wang F and Brunger M J 2008 *Chem. Phys. Lett.* **451** 18
[13] Ehrhardt H, Jung K, Knoth G and Schlemmer P 1986 *Zeitschrift für Physik D Atoms, Molecules and Clusters* **1** 3
[14] Ning C G, Huang Y R, Zhang S F, Deng J K, Liu K, Luo Z H, and Wang F 2008 *J. Phys. Chem. A* **112** 11078
[15] Milne-Brownlie D S, Cavanagh S J, Lohmann B, Champion C, Hervieux P A and Hanssen J 2004 *Phys. Rev.* A **69** 032701
[16] Al-Hagan O, Kaiser C, Madison D H and Murray A J 2009 *Nat. Phys.* **5** 59
[17] Lahmam-Bennani A, Naja A, Staicu Casagrande E M, Okumus N, Dal Cappello C, Charpentier I and Houamer S 2009 *J. Phys. B: At. Mol. Opt. Phys.* **42** 165201
[18] Colyer C J, Bellm S M, Lohmann B, Hanne G F, Al-Hagan O, Madison D H and Ning C G 2010 *J. Chem. Phys.* (submitted).

Journal of Physics: Conference Series **288** (2011) 012008 doi:10.1088/1742-6596/288/1/012008

Calculation of (e, 2e) triple differential cross sections of Mg in coplanar geometry

G. Purohit*, Vinod Patidar and K. K. Sud

Department of Physics, School of Engineering, Sir Padampat Singhania University, Bhatewar, Udaipur-313 601, India

*g_vpurohit@yahoo.com, ghanshyam.purohit@spsu.ac.in

Abstract. We report the results of triple differential cross section of coplanar (e, 2e) processes on Mg (3s) atom in modified distorted wave Born approximation (DWBA). The standard DWBA formalism has been modified by including the correlation-polarization potential (which is function of electron density) and post collision interaction. We compare our computed results with the available experimental data and observe that the inclusion of polarization potential in the standard DWBA is able to improve the agreement with experimental results.

1. Introduction

Study of electron-impact single ionization (i.e. (e, 2e) processes) of atoms and ions provides important information about target structure, target wavefunction and collision dynamics. Since the experiment of Ehrhardt et al. [1] extensive studies have been performed for the electron impact single ionization of hydrogen [2] and helium [3] atoms in different geometrical arrangements. Apart from hydrogen and helium atoms, (e, 2e) studies have been performed on a number of targets using different theoretical approaches (see review articles [4-8]). There has been recent focus on the study of (e, 2e) process in coplanar symmetric geometry in which the incident and outgoing electrons lie in the same plane. We get a close insight of the (e, 2e) reaction in coplanar symmetric geometry due to clean knockout of the target electron. The symmetric collisions in which both the final state electrons have the same energy and are observed at the same scattering angles on opposite sides of the beam direction in the scattering plane, provide a severe test of the theory. Distorted wave Born approximation (DWBA) [9, 10] and convergent close coupling approximation (CCC) [11-13] have successfully described the salient features of the TDCS of hydrogen and helium atoms. There has been particular interest for the study of (e, 2e) processes in the low incident energy regime in coplanar symmetric geometry [11, 14]. The CCC approximation has been widely successful to explain the features of TDCS, however still there are certain discrepancies [15]. The extended complex scaling method [2] provides an exact framework for solving ionization problems. The time dependent close coupling (TDCC) formalism has been applied to study electron impact ionization of atoms [16-18]. TDCC has been used to investigate the fully differential cross section of He atom at incident electron energy 44.6 eV [17] and good agreement is found between experiment and TDCC results. However, there are certain discrepancies in the coplanar symmetric geometry.

The distorted wave Born approximation (DWBA) is well known and able to produce correct trends of TDCS for a wide variety of atomic targets above incident electron energy 50 eV [8]. For the (e, 2e) processes at low energies different physical effects are very important such as post-collision interaction (PCI), charge cloud polarization, electron exchange etc. and these processes are responsible for the discrepancies between theory and experiment. Biava et al [19] have performed DWBA calculations of TDCS using different exchange effects and concluded that the exchange effects are particularly important for inner-valence ionization. Further improvement of the agreement between experiment and theory of the TDCS at low energy (e, 2e) processes can be achieved by proper treatment of PCI and inclusion of polarization [20, 21]. Various studies [22-26] have shown that inclusion of PCI and polarization potential in DWBA calculations improves the agreement of calculations with the experimental results significantly.

There has been recent focus on the study of (e, 2e) processes on alkali and alkali earth targets initiated by the experimental study of (e, 2e) processes in coplanar geometry by Murray and Cvejanovic [27] and Murray [28] on alkali and alkali earth targets Na, K and Mg, Ca. Study of (e, 2e) processes on these targets is of prime importance since these are having one and two electrons respectively in the valence shell and after ionization they are left as inert core (alkali target) and one electron in the valence shell above inert core (alkali earth target). Following the experimental results of Murray [28] several theoretical results have been reported for the (e, 2e) processes in coplanar geometry on the alkali and alkali earth targets [22, 29-34] and particularly, the study of (e, 2e) processes on Ca atom [22, 29, 32, 33]. Apart from the above mentioned studies Bolognesi et al. [35, 36] have reported a series of (e, 2e) measurements in unequal energy sharing conditions for the ionization of Mg (3s) target. Most of these studies have been performed in DWBA formalism and the first CCC calculation has been reported for the electron-sodium ionization in coplanar symmetric geometry [31]. We report the results of our modified distorted wave Born approximation calculations (DWBACPE) for Mg (3s) atoms. We modify the standard DWBA formalism by inclusion of correlation-polarization potential (as a function of electron density) and post collision interaction. We compare the results of TDCS with the recent available measurements of Murray [28].

2. Theory

The triple differential cross section for (e, 2e) process on a target may be written as

$$\frac{d^3\sigma}{d\Omega_1\, d\Omega_2\, dE_1} = (2\pi)^4 \frac{k_1 k_2}{k_0} \sum_{av} \left|T(k_1, k_2, k_0)\right|^2 \tag{1}$$

with

$$T(k_1,k_2,k_0) = \left\langle k_1 k_2 \left| T \right| \psi_{nl} k_0 \right\rangle.$$

The expression in Eq. (1) includes a sum over final and average over initial magnetic and spin state degeneracy. The T matrix in Eq.(1) is the reaction amplitude, it couples the initial states ψ_i and the final states. T includes interaction between the incident and target electrons and the nucleus. It is the part of the TDCS that is the subject of approximation. The T-matrix element which represent the ionization amplitude is conveniently written in terms of distorted waves as

$$\left\langle k_1 k_2 \left|T\right| \psi_{nl} k_o \right\rangle = \left\langle X_1^{(-)}(k_1) X_2^{(-)}(k_2) \right| v_1 + v_3 - U \left| \psi_i X_0^{(+)}(k_0) \right\rangle \tag{2}$$

The electron-electron potential v_3 is responsible for ionization and U is the distorting potential. Use of orthogonality between orbital ψ_i and distorted wave for ejected electron eliminates v_1 and U from the expression of T matrix element. As mentioned earlier the initial state ψ_i contains an electron bound to the atom core with separation energy E_b. The distorted waves for the incident and scattered electrons, $X_0^{(+)}(k_0)$ and $X_1^{(-)}(k_1)$, respectively, are calculated in a potential $U = \left\langle \psi_i \right| v_1 + v_3 \left| \psi_i \right\rangle$.

Equation (1) in terms of direct and exchange amplitudes may be written as

$$\frac{d^3\sigma}{d\Omega_1\, d\Omega_2\, dE_1} = (2\pi)^4 \frac{k_1 k_2}{k_0} \sum_{av} \left(|f|^2 + |g|^2 - \mathrm{Re}(f^* g)\right) \tag{3}$$

where

$$f = \left\langle X_1^{(-)}(k_1, r_1)\, X_2^{(-)}(k_2, r_2) \middle| v_3 \middle| X_0^{(+)}(k_0, r_1)\, \psi_{nl}(r_2) \right\rangle, \tag{4}$$

$$g = \left\langle X_1^{(-)}(k_1, r_2)\, X_2^{(-)}(k_2, r_1) \middle| v_3 \middle| X_0^{(+)}(k_0, r_1)\, \psi_{nl}(r_2) \right\rangle \tag{5}$$

here $v_3 = \frac{1}{|r_1 - r_2|}$ is the interaction potential between the incident and target electrons responsible for the ionization, $X_0^{(+)}$ is the distorted wavefunction for the incident electron, $X_1^{(-)}$ and $X_2^{(-)}$ represent the distorted wavefunctions for the two outgoing electrons and each is orthogonalized with respect to ψ_{nl}. Equations (4) and (5) are direct and exchange amplitudes for ionization from the (n,l) shell of the target atom where ψ_{nl} is the corresponding target orbital from which ionization is taking place and n and l are the principal and orbital quantum number respectively. We have used Hartree-Fock orbitals of Clementi and Roetti [37] for ψ_{nl}. The distorted wavefunction for the incident electron is generated in the equivalent local ground state potential of the atom whereas the distorted wavefunctions for the outgoing electrons are generated in the equivalent local ground state potential of the ion. For the work reported here we have made a careful check to ensure that the cross sections converge satisfactorily. The spin-averaged static-exchange potential [38] as modified by Riley and Truhlar [39] with the exchange potential taken in the equivalent local approximation is given as

$$V_E(r) = 0.5[E_0 + V_D(r) - \{[E_0 + V_D(r)]^2 - 2\pi\rho(r)^2\}^{1/2}] \tag{6}$$

where $\rho(r)$ is the electron density. The direct distorting potential $V_D(r)$ for the incident electron is obtained from the target radial orbital $u_{nl}(r)$ [40] as

$$V_D(r) = -\frac{Z}{r} + \sum_{nl} N_{nl} \int dr' [u_{nl}(r')]^2 / r_>, \tag{7}$$

where $r_>$ is the greater of r and r'. The details about the integration of the radial matrix element are described by McCarthy [40] and we have not given here for brevity. The equivalent local ground state potential V_{00}, which is the distorting potential, is the sum of exchange and direct potentials and is expressed as follows;

$$U = V_D(r) + V_E(r), \tag{8}$$

$$U = V_D(r) + 0.5[E_0 + V_D(r) - \{[E_0 + V_D(r)]^2 - 2\pi\rho(r)^2\}^{1/2}]. \tag{9}$$

We have modified the distorting potential (eq. (9)) used to calculate distorted waves $X_0^{(+)}$, $X_1^{(-)}$ and $X_2^{(-)}$ by addition of correlation-polarization potential V_{CP} to see the effect of electron correlation and target polarization on the trend of TDCS (DWBACPE calculations). The fundamental form of the short range correlation plus long range polarization potential may be approximated by means of local density functional theory (Padial and Norcross [41], Perdew and Zunger [42] and Yuan and Zhang [43]) as follows

$$V_{CP}(r) = V_{SR}^{Corr}(r), \quad r \le r_0$$
$$= -\frac{\alpha_d}{2r^4}, \quad r > r_0 \tag{10}$$

where α_d is dipole polarizability of the target and we have used short range correlation potential similar to [41].

We have also included PCI in our DWBA calculations using the Ward-Macek factor (M_{ee}) [44]. The M_{ee} is defined as

$$M_{ee} = N_{ee} \left| {}_1F_1(-i v_3, 1, -2 k_3 r_{3ave}) \right|^2$$

Where $v_3 = -\frac{1}{|k_1 - k_2|}$, $r_{3ave} = \frac{\pi^2}{16\varepsilon}\left(1 + \frac{0.627}{\varepsilon}\sqrt{\varepsilon}\ln\varepsilon\right)^2$, $N_{ee} = \frac{\gamma}{e^{\gamma} - 1}$ here $\gamma = \frac{2\pi}{|k_1 - k_2|}$

ε is the total energy of the two exiting electrons.
The TDCS incorporating PCI through M_{ee} is now written as

$$\frac{d^3\sigma}{d\Omega_1\, d\Omega_2\, dE_1} = M_{ee} (2\pi)^4 \frac{k_1 k_2}{k_0} \sum_{av} \left(|f|^2 + |g|^2 - \mathrm{Re}(f^* g) \right) \tag{11}$$

3. Results and discussion

The results of triple differential cross sections (TDCS) for the electron impact ionization of Mg (3s) atoms are presented in Figures 1 and 2. The theoretical and experimental results have been normalized to unity at symmetric scattering angle $\theta_1 = 45^0$. The solid curve represents our modified DWBA calculations which include correlation-polarization potential and post collision interaction (DWBACPE calculations), the dashed curve represents our standard DWBA calculations. The solid circles are experimental data of Murray [28]. Till date there are few attempts, as mentioned in the introduction section of this paper, to explain the features of TDCS of doubly symmetric (e, 2e) processes [22, 34] and unequal energy sharing condition [35, 36] on Mg (3s) target. All of these calculations have been done in DWBA formalism and have mix degree of agreement with the experimental results. Hitawala et al. [22] have calculated TDCS using spin averaged static exchange potential [36] and included target polarization in the standard DWBA formalism. Khajuria et al. [34] have reported the results of TDCS using modified semi classical exchange potential [45] and also included PCI using Gamow factor in the standard DWBA formalism.

We present the TDCS results of our DWBA and DWBACPE calculations for the ionization of Mg atom at incident electron energies 17.65, 22.65, 27.65, 37.65, 47.65, 57.65 and 67.65 eV which are 5.0 eV to 30.0 eV above ionization threshold. A two peak structure is observed in the TDCS profile for Mg atom in our calculations as well as in the experimental measurement. The two peak structure is observed up to incident electron energy 37.65 eV and at the other incident electron energies a shallow peak is observed near scattering angle $\theta_1 = 80^0$. This feature of TDCS is visible in our DWBA calculations as well in the experimental results. We observe that the DWBA results including correlation-polarization potential and PCI are able to produce correct position of the forward peak at incident electron energies 17.65, 22.65 and 27.65 eV, which is similar to the direction of forward peak in the experimental results (see solid curve in Fig. 1). We also observe that the DWBACPE calculations are also able to produce the correct position of the dip at the above mentioned incident electron energies, however still there are some discrepancies in the magnitude and direction of backward scattering peak at the incident electron energies 17.65, 22.65 and 27.65 eV. Thus we can state that the polarization and post collisional interaction effects are important at the low energies in the coplanar symmetric geometry and inclusion of these effects in standard DWBA formalism improves the agreement of theoretical results with the experimental data. The DWBA as well as the DWBACPE calculations are in very good agreement with the experimental results in the forward peak region at the other incident electron energies (see Fig. 2). However there are some discrepancies in our theoretical results and experimental data in the backward peak direction. The overall agreement of our present DWBA calculations including correlation-polarization potential and PCI and experimental data is very good.

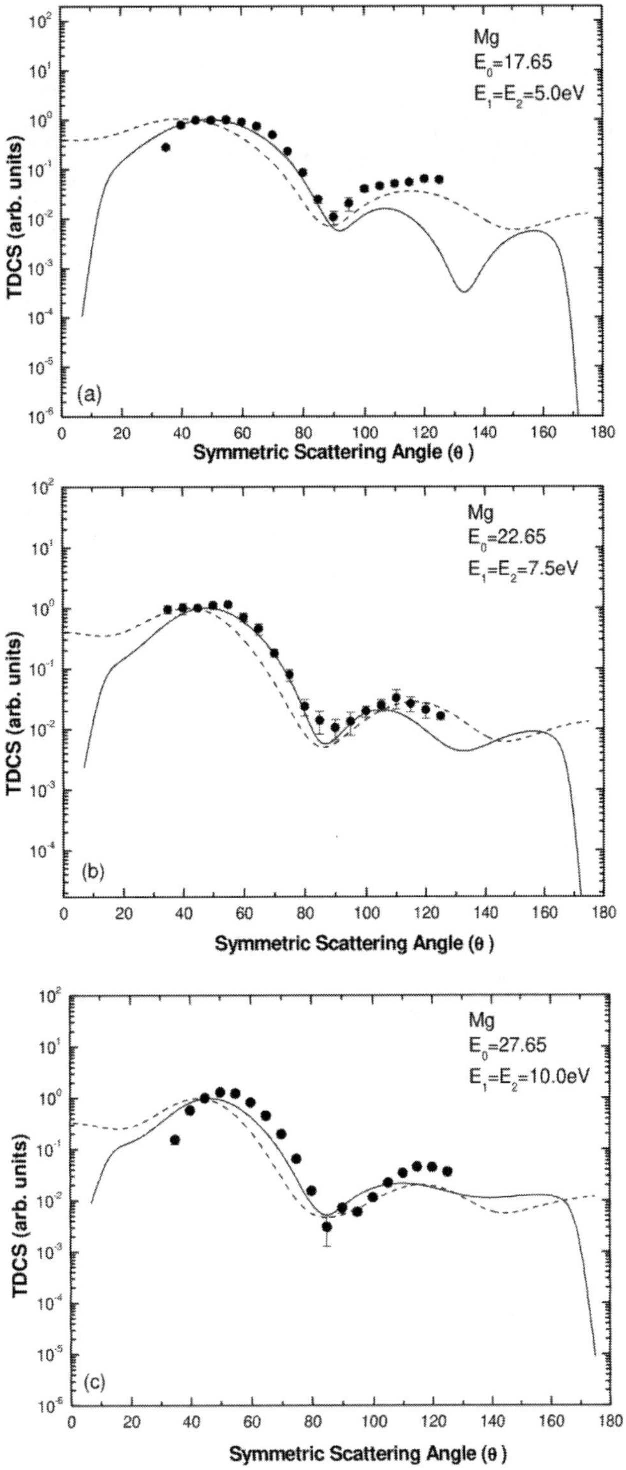

Figure 1: Triple differential cross section of Mg atom as a function of symmetric angle $\theta_1 = -\theta_2 = \theta$ in the coplanar symmetric geometry; symbol (•): experimental data of Murray [28]; Dashed curve: present standard DWBA calculation; Solid curve: present DWBA calculation including correlation-polarization potential and PCI. The experimental and theoretical results have been normalized to unity at symmetric scattering angle $\theta_1 = 45^0$.

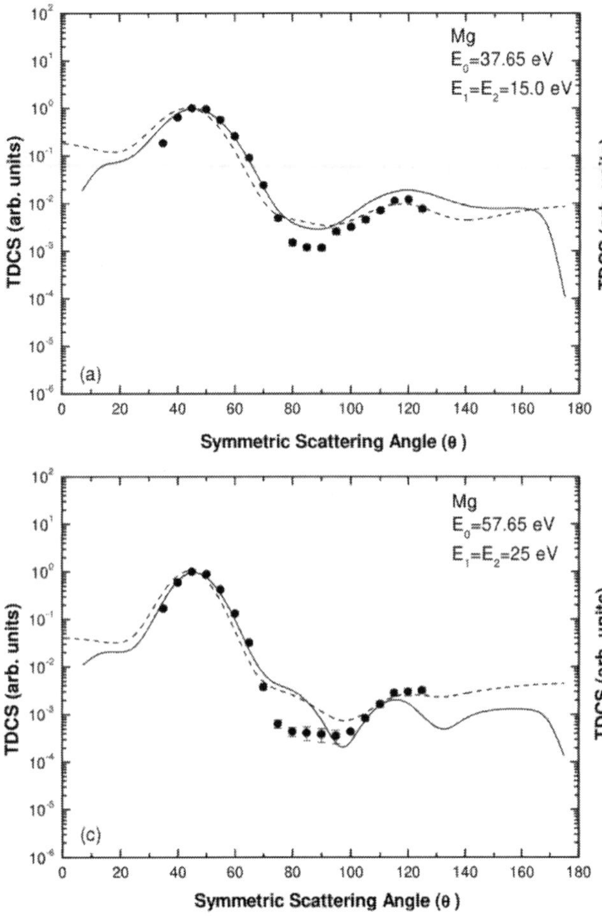

Figure 2: Triple differential cross section of Mg atom as a function of symmetric angle $\theta_1 = -\theta_2 = \theta$ in the coplanar symmetric geometry; symbol (•): experimental data of Murray [28]; Dashed curve: present standard DWBA calculation; Solid curve: present DWBA calculation including correlation-polarization potential and PCI. The experimental and theoretical results have been normalized to unity at symmetric scattering angle $\theta_1 = 45^0$.

We have presented our results for seven incident electron energies (from excess energy 5 eV to 30 eV) on which the measurements are available. We observe that our modified DWBA results (which include correlation-polarization potential and PCI) are able to produce most of the trends of TDCS as observed in the experimental data but still there exist few discrepancies at very low incident electron energies. The polarization effect plays an important role at the incident electron energies used in the present investigation in the coplanar symmetric geometry; however the effect of PCI is only significant for few lower incident electron energies. All the previously available calculations [22, 34] have reasonable amount of agreement with the experimental data but the present attempt further improves the agreement between theory and measurements. Finally, we conclude that further studies with the proper treatment of polarization, exchange, electron correlation and higher order effects are required to improve the understanding of collision dynamics of alkali atom magnesium in the coplanar symmetric geometry.

Acknowledgements

One of us (VP) acknowledges to the Science and Engineering Research Council (SERC), Department of Science Technology (DST), Govt. of India for the Fast Track Young Scientist Research Grant (SR/FTP/PS-17/2009). We are thankful to Professor Andrew James Murray for providing the experimental data.

References

[1] Ehrhardt H, Schulz M, Tekaat T and Willmann K 1969 Phys. Rev. Lett. **22**, 89.
[2] Rescigno T N, Baertschy M, Isaacs W A and McCurdy C W 1999 Science **286,** 2474.
[3] Bray I, Fursa D V, Kheifets A S and Stelbovics A T 2002 J. Phys. B **35**, R117.
[4] Whelan C T, Allan R J, Walters H R J and Zhang X 1993 (e, 2e) and related processes ed. C. T. Whelan et al. (Dordrechat: Kluwer) p1.
[5] McCarthy I E and Weigold E 1976 Phys. Rep. **C27**, 275.
[6] Byron Jr F W and Joachain C J 1989 Phys. Rep. **179**, 211.
[7] Nakel W and Whelan C T 1999 Phys. Rep. **315**, 409.
[8] Lahmam-Bennani A 1991 J Phys B **24,** 3645.
[9] Whelan C T, Allan R J, Walters H R J and Zhang X 1993 (e, 2e) and related processes ed. C. T. Whelan et al. (Dordrechat: Kluwer).
[10] Whelan C T, Allan R J, Rasch J, Walters H R J, Zhang X, Roder J, Jung K and Ehrhardt H 1994 Phys. Rev. A **50**, 4394.
[11] Bray I, Fursa D V, Roder J and Ehrhardt H 1998 Phys. Rev. A **57**, R3161.
[12] Stelbovics A T, Bray I, Fursa D V and Bartschat K 2005 Phys. Rev. A **71**, 052716.
[13] Bray I 2002 Phys. Rev. Lett. **89**, 273201.
[14] Rioual S, Roder J, Rouvellou B, Ehrhardt H, Pochat A, Bray I and Fursa D V 1998 J. Phys. B **31**, 3117.
[15] Bray I 2000 J. Phys. B **33**, 581.
[16] Pindzola M S et al. 2007 J. Phys. B **40**, R39.
[17] Colgan J, Pindzola M S, Childers G and Khakoo M 2006 Phys. Rev. A **73**, 042710.
[18] Colgan J, Foster M, Pindzola M S, Bray I, Stelbovics A T and Fursa D V 2009 J. Phys. B **42**, 145002.
[19] Biava D A, Saha H P, Engel E, Dreizler R M, McEachran R P, Haynes M A, Lohmann B, Whelan C T and Madison D H 2002 J. Phys. B **35**, 293.
[20] Prideaux A and Madison D H 2003 Phys. Rev. A **67**, 052710.
[21] Prideaux A, Madison D H and Bartschat K 2005 Phys. Rev. A **72**, 032702.
[22] Hitawala U, Purohit G and Sud K K 2008 J. Phys. B **41**, 035205.
[23] Purohit G, Bhullar A S and Sud K K 2003 Indian J. Phys. B **77(2),** 177.
[24] Allan R J, Whelan C T and Walters H R J 1993 J. Phys. IV **3**, 39.
[25] Rioual S, Pochat A, Gelebart F, Allan R J, Whelan C T and Walters H R J 1995 J. Phys. B **28**, 5317.
[26] Whelan C T, Allan R J, Rasch J, Walters H R J, Zhang X, Roder J, Jung K and Ehrhardt H 1994 Phys. Rev. A **50**, 4394.
[27] Murray A J and Cvejanovic D 2003 J. Phys. B **36**, 4875.
[28] Murray A J 2005 Phys. Rev. A **72**, 062711.
[29] Purohit G, Patidar Vinod and Sud K K 2009 Physica Scripta **80,** 065301.
[30] Purohit G, Patidar Vinod and Sud K K 2010 Physics Letters A **374,** 2654.
[31] Bray I, Fursa D V and Stelbovics A T 2008 J. Phys. B **41**, 215203.
[32] Khajuria Y and Deshmukh P C 2008 Phys. Rev. A **78**, 024702.

Journal of Physics: Conference Series **288** (2011) 012008 doi:10.1088/1742-6596/288/1/012008

[33] Chauhan R K, Srivastava M K and Srivastava R 2005 Phys. Rev. A **71**, 032708.
[34] Khajuria Y, Sunil Kumar S and Deshmukh P C 2007 Phys. Rev. A **75**, 022708.
[35] Bolognesi P et al. 2008 J. Phys. B 41, 015201.
[36] Bolognesi P et al. 2008 J. Phys. B 41, 065203.
[37] Clementi E and Roetti C 1974 At. Dat. Nucl. Dat. Tab. **14**, 187.
[38] Furness J B and McCarthy I E 1973 J. Phys. B **6**, 2280.
[39] Riley M E and Truhlar D G 1975 J. Chem. Phys. **63**, 2182.
[40] McCarthy I E 1995 Aust. J. Phys. **48** 1.
[41] Padial N T and Norcross D W 1984 Phys. Rev. A **29**, 1742.
[42] Perdew J P and Zunger A 1981 Phys. Rev. B **23**, 5048.
[43] Yuan J and Zhang Z 1989 J. Phys. B **22**, 2751.
[44] Ward S J and Macek J H 1994 Phys. Rev. A **49**, 1049.
[45] Gianturco F A and Scialla S 1987 J. Phys. B **20**, 3171.

Journal of Physics: Conference Series **288** (2011) 012009 doi:10.1088/1742-6596/288/1/012009

Theory of laser-assisted electron momentum spectroscopy: Beyond the Volkov wave Born approximation

Konstantin A. Kouzakov
Department of Nuclear Physics and Quantum Theory of Collisions, Faculty of Physics, Moscow State University, Moscow 119991, Russia
E-mail: kouzakov@srd.sinp.msu.ru

Yuri V. Popov
Institute of Nuclear Physics, Moscow State University, Moscow 119991, Russia
E-mail: popov@srd.sinp.msu.ru

Masahiko Takahashi
Institute of Multidisciplinary Research for Advanced Materials, Tohoku University, Sendai 980-8577, Japan
E-mail: masahiko@tagen.tohoku.ac.jp

Abstract. The $(e, 2e)$ spectroscopy of an atomic system at high impact energy and large momentum transfer is considered theoretically in the case of the presence of a laser field. The $(e, 2e)$ transition is supposed to take place due to electron-electron interaction. The incident electron is described by a Volkov wave. Several approximations for the three-body final state are formulated, which include the effects unaccounted by the Volkov wave Born approximation.

1. Introduction

The term "electron momentum spectroscopy" (EMS) [1, 2, 3, 4] usually refers to the $(e, 2e)$ method at high impact energy and large momentum transfer, when the latter is almost entirely absorbed by an ejected electron. EMS provides direct information on momentum distribution of electrons in various systems, ranging from atoms [5, 6] and molecules [7, 8, 9] to clusters [10] and solids [11, 12, 13]. This feature is due to the validity of the plane wave Born approximation (PWBA) for theoretical treatment of the EMS coincident differential cross sections (so-called momentum profiles). And when there are more or less appreciable effects beyond the scope of PWBA, one usually can take them into account by means of the distorted wave Born approximation (DWBA) (see, for instance, Weigold and McCarthy [3] and references therein) or, for instance, the 3C approach [14, 15].

Recently we have carried out the first theoretical consideration of the EMS process on an atomic system in the presence of laser radiation [16]. We found that the laser-assisted EMS method has a rich potential for investigating the laser-field effect on the target states even in such situations where the laser electric-field amplitude is much weaker than a typical intra-atomic field. In particular, when the laser frequency is resonant to the frequency of the transition from the ground to the first excited target state, one observes characteristic EMS momentum profiles from the ground and excited states, populations of which depend on laser parameters, and the momentum profile from the excited state directly reflects the laser polarization. Our theoretical analysis has been carried out within the framework of the Volkov wave Born approximation (VWBA) which naturally generalizes the PWBA model to the case of the presence of a laser field. At the same time, in future laser-assisted EMS studies one might have a necessity for generalization of the models beyond PWBA. It is the aim of the present work to develop several such generalizations. Namely, we formulate analogs of the plane wave impulse approximation (PWIA), DWBA, and 3C approach in the case when the laser field is present.

The paper is organized as follows. In section 2, we deliver a general formulation of the problem and specify the S matrix for the laser-assisted EMS process on an atomic target. Different models for the final state of the colliding system, which respectively determine different approximations to the S matrix, are presented in section 3. Section 4 summarizes this work. Atomic units (a.u., $e = \hbar = m_e = 1$) are used throughout unless otherwise stated.

2. General formulation

We focus on the laser-assisted $(e, 2e)$ reaction in an atomic system having only one active electron. High impact energy and large momentum transfer are assumed. The laser field switches on and off adiabatically at $t \to \mp\infty$, respectively, and is determined by the vector potential $\mathbf{A}(t)$, which is treated in the dipole approximation. The laser intensity is supposed to be not so high as to induce any appreciable ionizing effect. The incident, scattered, and ejected electron energies and momenta are specified below by respectively $(E_0, \mathbf{p}_0)$, $(E_s, \mathbf{p}_s)$, and $(E_e, \mathbf{p}_e)$.

The S matrix for the process under consideration can be presented as

$$S = -i \int_{-\infty}^{\infty} dt \Big\langle \Phi_{\mathbf{p}_s \mathbf{p}_e}(\mathbf{r}_0, \mathbf{r}_1, t) \Big| \frac{1}{r_{01}} \Big| \chi_{\mathbf{p}_0}(\mathbf{r}_0, t) \psi_T(\mathbf{r}_1, t) \Big\rangle, \tag{1}$$

where $r_{01} = \mathbf{r}_0 - \mathbf{r}_1$. $\chi_{\mathbf{p}_0}(\mathbf{r}_0, t)$ is a nonrelativistic Volkov wave [17, 18] describing the motion of the incident electron under the action of the laser field. $\psi_T(\mathbf{r}_1, t)$ is a wave function of the field-dressed target electron state. $\Phi_{\mathbf{p}_s \mathbf{p}_e}(\mathbf{r}_0, \mathbf{r}_1, t)$ is the final scattering state of the colliding system embedded in a background laser field.

The incident Volkov wave solves the following Schrödinger equation (hereafter we use the velocity gauge):

$$i \frac{\partial}{\partial t} \chi_{\mathbf{p}_0}(\mathbf{r}_0, t) = \frac{1}{2} \left[\hat{\mathbf{p}}_0 + \frac{1}{c} \mathbf{A}(t) \right]^2 \chi_{\mathbf{p}_0}(\mathbf{r}_0, t). \tag{2}$$

Thus, we have (see, for instance, Ref. [19])

$$\chi_{\mathbf{p}_0}(\mathbf{r}_0, t) = \exp\left\{ i \left[\mathbf{p}_0 \cdot \mathbf{r}_0 - \alpha(\mathbf{p}_0, t) - Et - \zeta(t) \right] \right\}, \tag{3}$$

where $E = p^2/2$ and

$$\alpha(\mathbf{p}_0, t) = \frac{1}{c} \int_{-\infty}^{t} \mathbf{p}_0 \cdot \mathbf{A}(t') dt', \qquad \zeta(t) = \frac{1}{2c^2} \int_{-\infty}^{t} A^2(t') dt'.$$

The target wave function is a solution to the time-dependent Schrödinger equation

$$i \frac{\partial}{\partial t} \psi_T(\mathbf{r}_1, t) = \left\{ \frac{1}{2} \left[\hat{\mathbf{p}}_1 + \frac{1}{c} \mathbf{A}(t) \right]^2 + V(\mathbf{r}_1) \right\} \psi_T(\mathbf{r}_1, t), \tag{4}$$

with the boundary condition

$$\psi_T(\mathbf{r}_1, t \to -\infty) \to \exp(-iE_g t)\psi_g(\mathbf{r}_1),$$

where E_g and $\psi_g(\mathbf{r}_1)$ are the energy and wave function of the undressed ground state of the target.

The wave function $\Phi_{\mathbf{p}_s\mathbf{p}_e}(\mathbf{r}_0, \mathbf{r}_1, t)$ describes the motion of two outgoing electrons (scattered and ejected) in the field of the ionized target. It is given by the solution of the Schrödinger equation

$$i\frac{\partial}{\partial t}\Phi_{\mathbf{p}_s\mathbf{p}_e}(\mathbf{r}_0, \mathbf{r}_1, t) = H\Phi_{\mathbf{p}_s\mathbf{p}_e}(\mathbf{r}_0, \mathbf{r}_1, t), \tag{5}$$

with the Hamiltonian

$$H = \frac{1}{2}\left[\hat{\mathbf{p}}_0 + \frac{1}{c}\mathbf{A}(t)\right]^2 + \frac{1}{2}\left[\hat{\mathbf{p}}_1 + \frac{1}{c}\mathbf{A}(t)\right]^2 + V(\mathbf{r}_0) + V(\mathbf{r}_1) + \frac{1}{\mathbf{r}_{01}}, \tag{6}$$

where $V(\mathbf{r}) \to -1/r$ at $r \to \infty$, and the boundary condition

$$\Phi_{\mathbf{p}_s\mathbf{p}_e}(\mathbf{r}_0, \mathbf{r}_1, t \to +\infty) \to \exp[-i(E_s + E_e)t]\Phi^{(-)}_{\mathbf{p}_s\mathbf{p}_e}(\mathbf{r}_0, \mathbf{r}_1), \tag{7}$$

where $\Phi^{(-)}_{\mathbf{p}_s\mathbf{p}_e}(\mathbf{r}_0, \mathbf{r}_1)$ is the final stationary state of the colliding system in the field-free case.

3. Final-state models

In this section, several approximations to the final state of the colliding system are formulated. Note in this connection that the VWBA model of [16] amounts to neglecting in (6) interactions of outgoing electrons with the ionized target and between each other, which yields

$$\Phi_{\mathbf{p}_s\mathbf{p}_e}(\mathbf{r}_0, \mathbf{r}_1, t) = \chi_{\mathbf{p}_s}(\mathbf{r}_0, t)\chi_{\mathbf{p}_e}(\mathbf{r}_1, t). \tag{8}$$

In the field-free case, the Volkov wave reduces to the plane wave,

$$\chi_{\mathbf{p}}(\mathbf{r}, t) \to \exp[i(\mathbf{p}\cdot\mathbf{r} - Et)],$$

and hence VWBA reduces to PWBA.

3.1. Volkov wave impulse approximation

If in (6) we neglect the interactions between outgoing electrons and the ionized target, then we obtain

$$\Phi_{\mathbf{p}_s\mathbf{p}_e}(\mathbf{r}_0, \mathbf{r}_1, t) = \chi_{\mathbf{P}_{se}}(\mathbf{R}_{01}, t)\varphi^{(-)}_{\mathbf{p}_{se}}(\mathbf{r}_{01}), \quad \mathbf{R}_{01} = \frac{1}{2}(\mathbf{r}_0 + \mathbf{r}_1), \quad \mathbf{P}_{se} = \mathbf{p}_s + \mathbf{p}_e, \quad \mathbf{p}_{se} = \frac{1}{2}(\mathbf{p}_s - \mathbf{p}_e), \tag{9}$$

where

$$\chi_{\mathbf{P}_{se}}(\mathbf{r}_0, t) = \exp\{i[\mathbf{P}_{se}\cdot\mathbf{R}_{01} - \alpha(\mathbf{P}_{se}, t) - (E_s + E_e)t - 2\zeta(t)]\} \tag{10}$$

can be regarded as a Volkov wave describing the motion of the electron pair as a whole in the laser field, and $\varphi^{(-)}_{\mathbf{p}_{se}}(\mathbf{r}_{01})$ is an outgoing Coulomb wave which describes the relative motion of the scattered and ejected electrons. It is readily seen that in the field-free case the function (9) is given by

$$\Phi_{\mathbf{p}_s\mathbf{p}_e}(\mathbf{r}_0, \mathbf{r}_1, t) = \exp\{i[\mathbf{P}_{se}\cdot\mathbf{R}_{01} - (E_s + E_e)t]\}\varphi^{(-)}_{\mathbf{p}_{se}}(\mathbf{r}_{01}),$$

which amounts to the PWIA model traditionally employed in the EMS studies (see, for instance, Weigold and McCarthy [3]).

3.2. Distorted-Volkov wave Born approximation
Neglecting in (6) the electrn-electron interaction, we get

$$\Phi_{\mathbf{p}_s\mathbf{p}_e}(\mathbf{r}_0,\mathbf{r}_1,t)=\psi^{(-)}_{\mathbf{p}_s}(\mathbf{r}_0,t)\psi^{(-)}_{\mathbf{p}_e}(\mathbf{r}_1,t). \tag{11}$$

The function $\psi^{(-)}_{\mathbf{p}}(\mathbf{r},t)$ describes the state of the outgoing electron under the combined action of the potential $V(\mathbf{r})$ and the laser field. Finding its exact form is, in general, an intractable task. The following approximation can be useful:

$$\psi^{(-)}_{\mathbf{p}}(\mathbf{r},t)=\exp\left\{-i\left[\alpha(\mathbf{p},t)+Et+\zeta(t)\right]\right\}\psi^{(-)}_{\mathbf{p}}(\mathbf{r}), \tag{12}$$

where $\psi^{(-)}_{\mathbf{p}}(\mathbf{r})$ is a distorted wave in the potential $V(\mathbf{r})$. When switching off the potential $V(\mathbf{r})$, the function (12) gives a Volkov wave. And in the field-free case, it gives the distorted wave, thus yielding, in accordance with (11), the DWBA model.

3.3. 3C-Volkov function
In the field-free case, the 3C function [14] is an approximation to the exact three-body stationary wave function $\Phi^{(-)}_{\mathbf{p}_s\mathbf{p}_e}(\mathbf{r}_0,\mathbf{r}_1)$. It is a product of three Coulomb waves, or, more specifically,

$$\Phi_{3\mathrm{C}}(\mathbf{r}_0,\mathbf{r}_1)=e^{-i\mathbf{p}_{se}\cdot\mathbf{r}_{01}}\varphi^{(-)}_{\mathbf{p}_s}(\mathbf{r}_0)\varphi^{(-)}_{\mathbf{p}_e}(\mathbf{r}_1)\varphi^{(-)}_{\mathbf{p}_{se}}(\mathbf{r}_{01}), \tag{13}$$

where $\varphi^{(-)}_{\mathbf{p}_{s(e)}}(\mathbf{r}_{0(1)})$ is an outgoing Coulomb wave in the potential $-1/r_{0(1)}$. It obeys correct asymptotic conditions at r_0, r_1, and $r_{01}\to\infty$. One can incorporate the laser-field effect into the 3C model by replacing $\exp(-iE_st)\varphi^{(-)}_{\mathbf{p}_s}(\mathbf{r}_0)$ and $\exp(-iE_et)\varphi^{(-)}_{\mathbf{p}_e}(\mathbf{r}_1)$ with the so-called Coulomb-Volkov waves [20]. Such a replacement yields

$$\Phi_{\mathbf{p}_s\mathbf{p}_e}(\mathbf{r}_0,\mathbf{r}_1,t)=\exp\left\{-i\left[\alpha(\mathbf{P}_{se},t)+(E_s+E_e)t+2\zeta(t)\right]\right\}\Phi_{3\mathrm{C}}(\mathbf{r}_0,\mathbf{r}_1). \tag{14}$$

It can be noted that models (9) and (11) (the latter if the distorting potential is $V(\mathbf{r})=-1/r$) are particular cases of (14).

4. Summary
In this work, we have considered the laser-assisted EMS of an atomic system from a theoretical viewpoint. A general definition of the corresponding S matrix has been given that allows to take into account the dynamical effects beyond the VWBA approach. Several approximations to the S matrix, based on different models of the final three-charged-particle scattering state, have been formulated. These approximations are generalizations of the usual "field-free" ones to the situation where the laser field is present. They can be useful in theoretical treatments of laser-assisted EMS measurements, which are feasible in near future.

References
[1] Coplan M A, Moore J H and Doering J P 1994 *Rev. Mod. Phys.* **66** 985
[2] Neudatchin V G, Popov Yu V and Smirnov Yu F 1999 *Phys. Usp.* **42** 1017
[3] Weigold E and McCarthy I E 1999 *Electron Momentum Spectroscopy* (New York: Kluwer Academic/Plenum)
[4] Takahashi M 2009 *Bull. Chem. Soc. Jpn.* **82** 751
[5] Watanabe N, Khajuria Y, Takahashi M, Udagawa Y, Vinitsky P S, Popov Yu V, Chuluunbaatar O and Kouzakov K A 2005 *Phys. Rev. A* **72** 032705
[6] Watanabe N, Takahashi M, Udagawa Y, Kouzakov K A and Popov Yu V 2007 *Phys. Rev. A* **75** 052701
[7] Luo Z H, Ning C G, Liu K, Huang Y R and Deng J K 2009 *J. Phys. B: At. Mol. Opt. Phys.* **42** 165205
[8] Li Z J, Chen X J, Shan X, Liu T and Xu K Z 2009 *J. Chem. Phys.* **130** 054302
[9] Hajgató B, Deleuze M S and Morini F 2009 *J. Phys. Chem. A* **113** 7138

[10] Nixon K L, Lawrance W D, Jones D B, Euripidies P, Saha S, Wang F and Brunger M J 2008 *Chem. Phys. Lett.* **451** 18
[11] Chen C, Gale M N, Kheifets A S, Vos M and Went M R 2005 *J. Phys.-Condens. Mat.* **17** 7689
[12] Vos M, Bowles C, Kheifets A S and Went M R 2006 *Phys. Rev. B* **73** 085207
[13] Nixon K L, Vos M, Bowles C and Ford M J 2006 *Surf. Interface Anal.* **38** 1236
[14] Brauner M, Briggs J and Klar H 1989 *J. Phys. B: At. Mol. Opt. Phys.* **22** 2265
[15] Kouzakov K A, Vinitsky P S, Popov Yu V and Dal Cappello C 2007 *J. Electron Spectrosc.* **161** 35
[16] Kouzakov K A, Popov Yu V and Takahashi M 2010 *Phys. Rev. A* **82** 023410
[17] Volkov D M 1935 *Z. Phys.* **94** 250
[18] Gordon W 1926 *Z. Phys.* **40** 117
[19] Fedorov M V 1997 *Atomic and free electrons in a strong light field* (Singapore: World Scientific)
[20] Cavaliere P, Ferrante G and Leone C 1980 *J. Phys. B: At. Mol. Phys.* **13** 4495

Journal of Physics: Conference Series **288** (2011) 012010 doi:10.1088/1742-6596/288/1/012010

Ionization of liquid water by fast electron impact: multiple differential cross sections for the $1B_1$ orbital

O. A. Fojón[1], M. L. de Sanctis[1], R. Vuilleumier[2], C. R. Stia[1], and M.-F. Politis[3]

[1] Instituto de Física Rosario (CONICET-UNR), Av. Pellegrini 250, 2000 Rosario, Argentina.
[2] LPTMC, UMR-CNRS 7600, Université P M Curie, 4 place Jussieu, 75005 Paris, France.
[3] IMPMC, Université P M Curie, Campus Boucicaut, 140 rue de Lourmel, 75015 Paris, France.

E-mail: `fojon@ifir-conicet.gov.ar`

Abstract. We present a theoretical study of single ionization of water molecules in liquid phase by impact of fast electrons in a coplanar geometry. Multiple differential cross sections are obtained through a first order model obtained within the framework of an independent electron approximation in which relaxation of the target is not taken into account. The wavefunctions for a single water molecule in the liquid phase are obtained through a Wannier orbital formalism and the ejected electron is described by means of Coulomb functions. We also present averaged calculations over all molecular orientations. A comparison with previous theoretical and experimental results, the latter corresponding to water in gaz phase, shows a good agreement. The main physical features of the reaction (such as binary and recoil peaks) present in measurements for vapor are also observed in the present theoretical predictions.

1. Introduction

Ionization of water molecules in liquid phase is of relevance in several domains such as radiobiology and medical physics. As the biological matter is composed mainly of water, the analysis of this reaction is crucial to understand the damage provoked to the living tissue by the ionizing radiations. In particular, the production of low energy secondary electrons resulting from a primary ionization reaction of water is of importance to elucidate the mechanisms that lead to cell alteration. In track structure analysis, the liquid water molecules in living cells are treated often as gaz water molecules to simplify the study. So, a more realistic description is required. Moreover, recent studies on (e, 2e) processes with liquid phase [1, 2] show little differences with the gaz one in spite of the fact that they have very diferent physical properties. This needs an explanation. Therefore, we initiate the study of the single ionization of water molecules in liquid phase by fast electron impact, i.e., impact energies of some hundreds of eV.

The ionization by electron impact of water in gaz phase has been studied since long time ago ([3], [4], see [5] for a review). On the contrary, only a few works deal with the liquid phase [1, 2]. In order to treat properly this phase, one has to describe in a realistic way the electronic states of the molecules. This is a difficult task that taken in conjunction with those of the collision itself (among them the correct description of the Coulomb interactions between charged fragments in the final channel of the reaction) turns the problem practically irresoluble. Therefore, several approximations are required in order to obtain the observables of the reaction (multiple differential cross sections, MDCS) to have a first approach to the dynamics of the (e,2e) process for water molecules in liquid phase.

Our aim is to study in a simple manner the dynamics involved in the (e,2e) reaction with water in the liquid phase. We present a simple first-order model developed within the framework of an independent electron approximation that constitutes the departure point for more elaborated ones. Thus, we employ an independent electron approximation in which the passive electrons of the molecule (those not ionized during the collision) remain as frozen in their initial orbitals. In this way, the multielectronic problem is reduced to a monoelectronic one. This approximation was used in many works dealing with multielectronic targets ([6], [7], see also [8] for a review) and even for liquid water [2]. In these works, it is assumed that the effective collision time at high impact energies is much smaller than the relaxation time. In analogous way and for the sake of simplicity, we assume that this condition is valid.

We compute MDCS for the $1B_1$ orbital of the water molecule as a function of the ejection angle at definite scattering angle, incident and ejected energies, and for fixed orientations of the molecule. In particular, we consider coplanar geometries in which the incident, scattered and ejected electrons lie in the same plane. Moreover, we analyze asymmetric kinematic conditions for the scattered and ejected electrons (ejected energies of some eV). To circumvent the challenges presented by the description of the liquid phase, we employ an interesting Wannier orbital formalism that transforms the wavefunction of the whole liquid system into electronic orbitals localized over each water molecule in the liquid phase [9]. As experimental results of MDCS at fixed molecular orientations for liquid water are not available, we compare our theoretical predictions with other theoretical ones obtained for the gaz phase [10]. In addition, as molecules are randomly oriented in experiments with water vapor [11], we also present molecular cross sections averaged over all the molecular orientations.

As is well known, electronic emission is most probable in the direction given by the momentum transfer that may be obtained by conservation laws from the projectile electron-active electron subsystem. Consequently, this fact may be ascribed to a binary collision between the two mentioned particles. This classical ionization mechanism may be traced in the quantum MDCS by the presence of a prominent peak, the so called *binary peak* being one of the most important physical features of the MDCS. It is worthy to mention that the presence of the peak is verified in the case of active electrons coming from initial s orbitals whereas for those coming from p ones, the binary peak may be split up into two lobes symmetric with respect to the direction of the momentum transfer. Another important physical characteristic of the MDCS is the so called *recoil peak* observed in measurements at almost the opposite direction at which appears the binary peak. This may be related to a two-step mechanism in which the active electron suffers a binary collision with the projectile and then the electron is back scattered elastically by the target nucleus. This gives rise to another peak in the MDCS, in general less prominent than the binary peak that can be thus associated to this mechanism in two steps. The presence of these characteristic peaks will be studied in the liquid phase.

Atomic units are used except otherwise explicitly stated.

2. Theory

In the framework of an independent electron model, we develop a simple monoelectronic approximation considering that the passive electrons of the reaction (electrons of the target that are not ionized) remain as frozen during the reaction in their initial orbitals. In this way, MDCS for the active electron (the one to be ionized) for a coplanar geometry at a fixed molecule orientation defined by de Euler angles are obtained as,

$$\sigma^{(8)}(\alpha,\beta,\gamma) = \frac{\mathrm{d}\sigma}{\mathrm{d}\Omega_{\mathrm{mol}}\mathrm{d}\Omega_{\mathrm{e}}\mathrm{d}\Omega_{\mathrm{s}}\mathrm{dE_e}} = \mathrm{N}(2\pi)^4\frac{\mathrm{k_e k_s}}{\mathrm{k_i}}\left|\left\langle \Psi_{\mathrm{f}}^{-} \mid \mathrm{V_i} \mid \psi_{\mathrm{i}}\right\rangle\right|^2 \qquad (1)$$

Journal of Physics: Conference Series **288** (2011) 012010 doi:10.1088/1742-6596/288/1/012010

where ψ_i and Ψ_f^- are the initial and the final wavefunctions (the latter with correct boundary conditions), respectively, and V_i is the perturbation in the initial channel. $\mathbf{k}_i$, $\mathbf{k}_s$ and $\mathbf{k}_e$ are the incident, scattered and ejected electron momenta, respectively, and E_e is the ejected electron energy. $d\Omega_e$ and $d\Omega_s$ denote the solid angles corresponding to $\mathbf{k}_e$ and $\mathbf{k}_s$, respectively, whereas $d\Omega_{mol} = \sin\beta \, d\alpha \, d\beta \, d\gamma$ where α, β, γ are the Euler angles of the water molecule. As exchange is not taken into account, N = 2 gives the number of electrons in the considered molecular orbital.

The initial wavefunction is chosen as,

$$\psi_i = \frac{e^{i\mathbf{k}_i.\mathbf{R}}}{(2\pi)^{3/2}} \Phi_i(\mathbf{r}) \tag{2}$$

where $\mathbf{R}$ and $\mathbf{r}$ are the position vectors of the incident and active electron, respectively, with respect to the center of mass of the molecule. A plane wave describes the incident electron and Φ_i represents the initial molecular orbital of the water molecule in the liquid phase. Briefly, this phase is simulated with a box of approximately 30 a.u. at 300K with periodic boundary conditions containing 128 molecules of water. The water orbitals are constructed from the occupied extended (Kohn-Sham) ones using the maximally localized Wannier functions. In this way, effective molecular orbitals localized on single molecules are obtained [9]. Band positions are related to average orbital energies. This procedure gives for the $1B_1$ orbital for the liquid phase an average ionization energy equal to 8.2 eV.

The final wavefunction is chosen as,

$$\Psi_f^- \cong \frac{e^{i\mathbf{k}_s.\mathbf{R}}}{(2\pi)^{3/2}} C(\mathbf{k}_e, \mathbf{r}, \nu) \tag{3}$$

where the continuum Coulomb wave function

$$C(\mathbf{k}, \mathbf{r}, \nu) = \Gamma(1 - i\nu) \, e^{-\pi\nu/2} \, \frac{e^{i\mathbf{k}_e.\mathbf{r}}}{(2\pi)^{3/2}} \, {}_1F_1[i\nu; 1; -i(kr + \mathbf{k}.\mathbf{r})] \tag{4}$$

describes the ionized electron in the final channel at asymptotically large distances, being ${}_1F_1$ the confluent hypergeometric function and $\nu = -Z^*/k_e = -1/k_e$ the corresponding Sommerfeld parameter. The charge $Z^* = 1$ takes into account the fact that the passive electrons produce a complete screening of the nuclear charges of the molecule at asymptotically large distances giving as a result a residual target with a net charge equal to unity.

The perturbation in the entrance channel V_i is given by

$$V_i = \frac{1}{|\mathbf{r} - \mathbf{R}|} - \frac{1}{R} \tag{5}$$

in accordance with the choice of the initial wavefunction and with the following approximation. We neglect the spatial distribution of the nuclei and passive electrons in the water molecule, obtaining an average for the interaction of the projectile with these particles. Thus, we consider the interaction of the projectile with a net charge equal to unity that is equivalent to consider a total screening of the charge of the nuclei by the passive electrons at all distances. So, the short range part of the real interaction is neglected. This approximation was used by several authors, see for instance [6] for the gaz and [2] for the liquid phase.

Integrating the MDCS given by Eq. (1) over the Euler angles, we obtain five-fold differential cross sections (5DCS) averaged over all the possible molecular orientations,

$$\sigma^{(5)} = \frac{1}{8\pi^2} \int \sigma^{(8)}(\alpha, \beta, \gamma) \sin\beta \, d\alpha \, d\beta \, d\gamma \tag{6}$$

To compute the cross sections, numerical quadratures are performed within the simulation box.

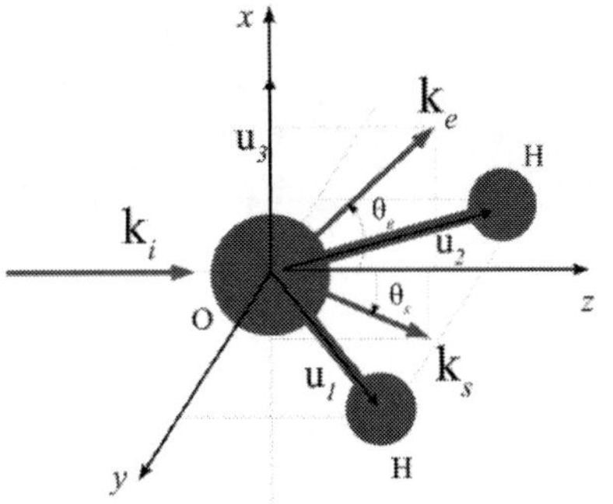

Figure 1. Perpendicular orientation of the water molecules considered in Fig. 2

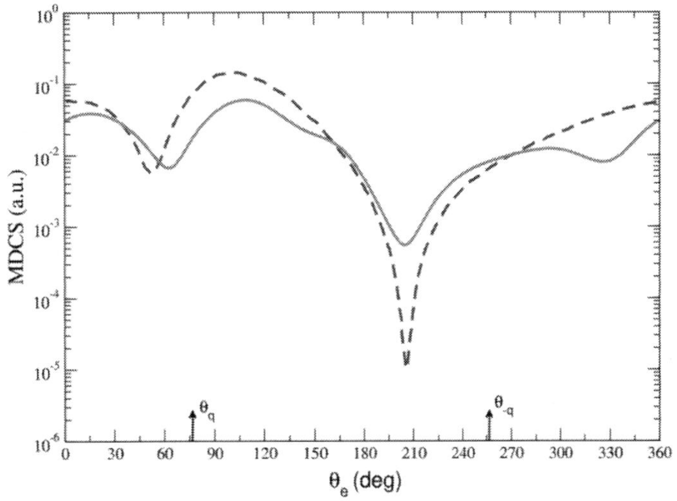

Figure 2. MDCS *per electron* for perpendicular orientation of the water molecules as in Fig. 1 for an incident energy $E_i = 250$ eV, ejected energy $E_e = 5$ eV and scattering angle $\theta_s = 15°$. Full line, present results. Dashed line, gaz phase results [10]. Binary and recoil peak angles are indicated by arrows (see text)

3. Results

We present coplanar geometry results of MDCS for the ionization of the $1B_1$ orbital of liquid water. We consider first an incident energy $E_i = 250$ eV, ejected energy $E_e = 5$ eV and scattering angle $\theta_s = 15°$. As experimental results of MDCS at fixed molecular orientations for liquid phase are not available, we compare our theoretical predictions with other theoretical ones obtained for water molecules in gaz phase [10]. These previous calculations correspond also to a first order model in which the water orbitals in gaz phase are obtained through a development centered on the heavy oxygen atom as given by Moccia [12]. We consider the case of molecules oriented with its molecular plane perpendicular to the collision one being the latter defined by the incident and scattering directions (see figure 1). In order to make the comparison possible, we present results in which the part $1/R$ of the perturbation is not taken into account in our calculations as the previous theoretical calculations [10] do not include this term. However, as will be seen below,

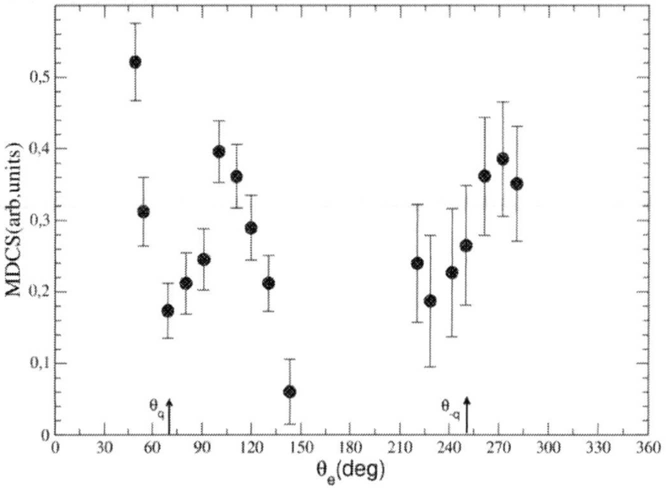

Figure 3. Summed experimental MDCS on a arbitrary scale for the $1b_1$ and $3a_1$ orbitals of randomly oriented water molecules in gaz phase for an incident energy $E_i = 250$ eV, ejected energy $E_e = 10$ eV and scattering angle $\theta_s = 15°$ extracted from Ref. [11]. Binary and recoil peak angles for the $1b_1$ orbital are indicated by arrows.

this term is of relevance to properly represent the physical features of the reaction observed in measurements. In Fig. 2, it can be seen that our predictions agree well in a qualitative way with the ones corresponding to the gaz phase. In particular, both predictions are of the same order. Results for the gaz phase exhibit a deep minimum around $\theta_e = 210°$ that also appears for the liquid phase although in a less pronounced way. The theoretical results predict a binary peak two-lobe structure at the expected classical angular value of about $\theta_q = 76.5°$ and $\theta_q = 75.7°$ for the liquid and the gaz phases, respectively. This values are indicated in the figure but they are almost undistinguishable. The θ_{-q} recoil angular position is also shown. The $1B_1$ orbital has a strong p character and accordingly to this, a two-lobe structure is observed in the binary peak region.

In Fig. 3, we show the experimental results on a relative scale for the gaz phase corresponding to the sum of the independent $1b_1$ and $3a_1$ orbital contributions as the experiments are not able to resolve their difference in energy especially in the recoil region. It can be seen that experiments show the main physical features associated to the ionization reaction, i.e., the binary and the recoil peaks.

In Fig. 4 we present our results of MDCS averaged over all molecular orientations for the $1B_1$ orbital for an incident energy $E_i = 250$ eV, ejected energy $E_e = 10$ eV and scattering angle $\theta_s = 15°$. Normalized experiments for vapor from [11] corresponding only to the $1b_1$ orbital are shown. As mentioned above, the experimental resolution of the measurements is not able to resolve the contributions from the $1b_1$ and $3a_1$ orbitals in the recoil peak region, so experimental results belonging only to the binary region are presented in Fig. 4. It can be seen that our results present a good qualitative agreement with the experiments in the binary region describing the characteristic two-lobe structure of a p orbital. Moreover, our results are also in qualitative good agreement with a previous first-order calculations for water in gaz phase [13]. In this calculation the initial state is represented as in Ref. [10] by using a plane wave for the incident electron and the Moccia's molecular wavefunction for the bound state of the target in the gaz phase. The final state is described as the product of a plane wave for the scattered electron and a Coulomb wave function with effective ionic charge $Z^* = 1$ (1CW model) for the

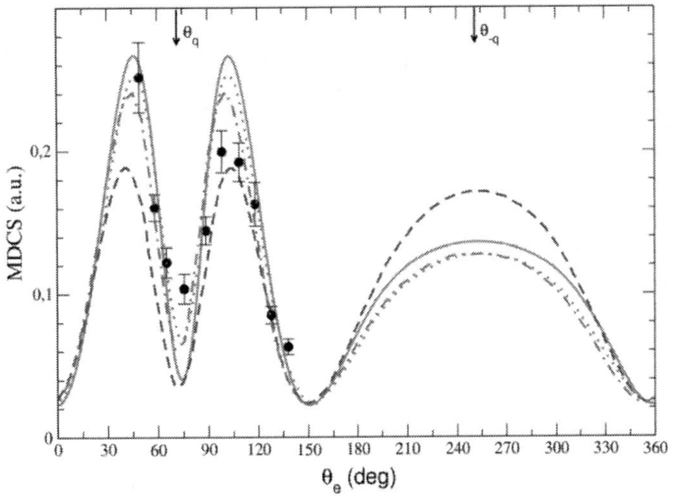

Figure 4. MDCS averaged over all molecular orientations for an incident energy $E_i = 250$ eV, ejected energy $E_e = 10$ eV and scattering angle $\theta_s = 15°$. Experiments for the gaz phase are extracted from Ref. [11] and normalized conveniently. Full line, present results for the liquid phase. Dashed line, 1CW model for the gaz phase [13]. Dotted line, FBA-CW for liquid phase [2]. Dash-dotted line, FBA-CW for the gaz phase [2]. Binary and recoil peak angles for the $1B_1$ orbital are indicated by arrows.

ionized electron. In addition, the $1/R$ term is included in the perturbation. This model use the independent electron approach to reduce the multielectronic problem to a monoelectronic one by neglecting the relaxation of the target. We also show in the figure the results for the gaz and liquid phase obtained with a single-center partial-wave approach [2]. In this model (FBA-CW), water molecules in both thermodinamical phases are described by means of single-center molecular wavefunctions developed in a Gaussian basis. It is worthy to mention that this model and the 1CW one are similar to ours except in the description of the bound state. Although molecules in the liquid phase do not exhibit the same symmetries as in the gaz phase, the $1B_1$ orbital used in this work is almost symmetric. Therefore, the MDCS to first order present also an almost symmetric curve with respect to the momentum transfer direction as in the case of molecules in the gaz phase. All the theoretical predictions present binary and recoil structures in qualitative good agreement. However, it is surprising that the results for the gaz and liquid phase given by the FBA-CW model are practically the same. On the contrary, quantitative differences are observed between our results and the ones corresponding to the gaz phase given by the 1CW and FBA-CW models. For instance, the height of the lobes in the binary region given by the 1CW model is almost 50% lower than the one of our results for liquid whereas the situation is reserved for the recoil peak. At last, there is a very good agreement between our predictions and the ones of the FBA-CW approximation. In fact, only slight differences are seen in the magnitude of the two-lobe structure and the recoil region. In general, results for the liquid phase in the binary region are higher than the ones for the gaz indicating that the $1B_1$ orbital is more diphuse in the liquid phase leading to greater cross sections. This fact is attributed in [2] to polarization effects provoked by the surrounding molecules.

4. Conclusions

We have studied ionization of water molecules in the liquid phase by means of simple first order model by using an independent electron approximation. The wave function of a single molecule in the liquid phase has been obtained through the use of Wannier orbitals techniques [9]. The results obtained so far show that the main physical features observed in experiments for vapor (such as binary and recoil peaks) are similar to the ones observed in our results. Moreover, a very good qualitative agreement with previous result for the liquid is found. These findings are encouraging to extend our work to the other orbitals of liquid water. This task is in progress and once finished may be useful to understand the basics of the dynamics involved in the reaction of interest. If the trend for MDCS for the other orbitals is similar to the one for the $1B_1$ orbital, then we will be in position to improve the approximations used in our simple model. For instance, we assume that the relaxation of the target is negligible during the collision time as was assumed previously for reactions with molecules in both phases. However, we cannot establish at this very first stage of our studies if this condition really holds for the liquid as experiments for this phase are not available. A more detailed study on this point is out of the scope of the present work. Previous results for the gaz and liquid phase obtained with a single-center partial-wave approach [2] are practically the same. This is surprising taking into account the important differences in the physical properties between both thermodinamical phases. Although our results for the liquid are in qualitative agreement with previous predictions for the gaz, some quantitative differences are observed. However, it is not possible to determine to what extent these differences will remain for the integrated MDCS. All the subjects pointed out here will be matter of analysis in future works.

5. Acknowledgments

Authors acknowledge financial support from the Agencia Nacional de Promoción Científica y Tecnológica (PICT No. 01912), and the Consejo Nacional de Investigaciones Científicas y Técnicas de la República Argentina (PIP No. 11220090101026).

References

[1] Hafied H, Eschenbrenner A, Champion C, Ruiz-López R F, Dal Cappello C, Charpentier I, Hervieux P -A 2007 *Chem. Phys. Lett.* **439** 55.
[2] Champion C 2010 *Phys. Med. Biol.* **55** 11.
[3] Opal C B, Beaty E C, Peterson W K 1972 *At. Data* **4** 209.
[4] Straub H C et al 1996 *Phys. Rev. A* **54** 2146.
[5] Itikawa Y, Mason N J 2005 *J. Phys. Chem. Ref. Data* **34** 1.
[6] Dal Cappello C, Hervieux P A, Charpentier I, Ruiz-Lopez F 2008 *Phys. Rev.* A **78** 042702.
[7] Champion C, Lekadir H, Galassi M E, Fojón O, Rivarola R D and Hanssen J, 2010 *Phys. Med. Biol.* **55** 6053.
[8] Stolterforht N, DuBois R D and Rivarola R D 1997 *Electron Emission in Heavy Ion-Atom Collisions* ed G Ecker, P Lambropoulos, I I Sobelman and H Walther (Berlin: Springer)
[9] Hunt P, M. Sprik and R. Vuilleumier 2003 *Chem. Phys. Lett.* **376** 68.
[10] Champion C et al 2001 *Phys. Rev.* A **63** 052720; Champion C. et al 2005 *Phys. Rev.* A **72** 059906.
[11] Milne-Brownlie D. S. et al 2004 *Phys. Rev.* A **69** 032701.
[12] Moccia R 1964 *J. Chem. Phys.* **40** 2186.
[13] Champion C, Dal Cappello C, Houamer S and Mansouri A 2006 *Phys Rev.* A **73** 012717.

Int. Conf. on Many Particle Spectroscopy of Atoms, Molecules, Clusters and Surfaces IOP Publishing
Journal of Physics: Conference Series **288** (2011) 012011 doi:10.1088/1742-6596/288/1/012011

Differential Positron and Electron Ionization Studies: How a Plus or Minus Sign Makes a Difference

R.D. DuBois

Department of Physics, Missouri University of Science and Technology, Rolla, MO 65409 USA

dubois@mst.edu

Abstract. Doubly and triply differential data for single ionization of argon, neon and molecular nitrogen have been measured for positron and electron impact energies of 200, 250, and 500 eV. Ratios of triply to doubly differential ionization yields as functions of the projectile scattering angle and energy loss are compared in order to investigate differences associated with the sign of the projectile charge. For all three systems, systematic differences are observed in the relative contributions of binary interactions. These differences are shown to increase with increasing scattering angle. Nearly identical recoil intensities are found for positron and electron impact ionization of the atomic targets but not for the molecular target.

1. Introduction

Since the advent of atomic physics, a multitude of experimental and theoretical investigations have studied inelastic interactions between atomic particles. The vast majority of these studies employed electrons and ions as one of the collision partners and many studies measured ionization probabilities or studied the inelastic interaction kinematics. With the discovery of the positron, and later of the antiproton, similar studies have been performed using antiparticles. These studies are performed in part to provide information about similarities and differences between matter-matter and antimatter-matter interactions and in part to isolate and study specific interaction channels and processes in order to test theoretical models in more detail.

Several commonly used theoretical models treat the interaction using first-order perturbation theory. For these models, since the interaction probabilities are proportional to the square of the projectile charge, identical total–and differential cross sections are predicted at higher energies for matter and antimatter impact. In contrast, more sophisticated treatments predict distinct differences in the differential electron emission measured as a function of the momentum transfer for positron and electron impact. Examples of this are the triply differential ionization cross section calculations for ionization of helium and atomic hydrogen [1-3]. With respect to a first Born model, these models predict that the binary intensities in fully differential positron impact studies are enhanced while the recoil intensities are reduced. For electron impact, just the opposite effects are predicted, e.g., a reduction in the binary intensity and an increase in the recoil intensity. These models also predict that the directions of the binary and recoil lobes will be shifted with respect to the momentum transfer direction, with the direction of the shift depending on the sign of the charge. More recent calculations predict similar features for ionization of H_2 [4].

Unfortunately, testing these predictions was not possible until recently. This is because limited positron beam intensities restricted experimental measurements of individual atomic interactions to integral type measurements, for example, to total cross section measurements of single and multiple target ionization or to total probabilities for positronium formation. Detailed information such as that obtained from differential type studies and as required to stringently test theory was available in only a small handful of cases. For example, Falke et al. [5] measured probabilities for single and double ionization, positronium formation, and transfer ionization, all as functions of observation angle for 120 eV positron impact on argon. At MS&T (Missouri University of Science and Technology), DuBois and coworkers measured energy loss spectra for single and double ionization of argon and krypton at 750 eV [6,7]. Moxom et al. [8] measured the energy spectrum for electron emission from argon resulting from 50-150 eV positron impact. All of these were singly differential measurements, capable of testing theory in some detail but unable to test the predictions discussed above. Doubly differential studies provide more stringent tests, but here again data for positron impact are very limited. Schmitt et al. [9] measured 15 eV electron emission between 20 and 90 degrees at 100 eV and Kövér, Larricchia and coworkers at University College-London (ULC) measured electron emission from argon at 30° and 45° for 60 and 100 eV impact energies [10,11].

The first triply (fully) differential studies for positron impact were performed by the ULC group 15 years ago [12-14]. The reason that triply differential studies are particularly useful in testing our theoretical understanding of atomic interactions is that the interaction kinematics are fully defined and because there is no need to integrate theory over energies or angles of one of the particles before comparing with experiment. This results in a much more stringent test of theory. But, as the primary interest of the ULC group was in studying electron capture to the continuum processes, their studies were all performed for 0° electron emission and projectile scattering. As a result, these triply differential data can only test limited aspects of theory. With regard to testing theory, it should be mentioned that very sophisticated triply differential data can be obtained for electron impact. See for example, ref. 15. But, uncertainties in both the experimental magnitudes and the theoretical models precludes using only electron impact data to test the theoretical predictions of the changes in magnitudes of the binary and recoil lobes as compared to those predicted using perturbation theory. For this purpose, data for both positron and electron impact are required and, ideally, such data should be obtained under identical conditions in order to minimize or eliminate any experimental uncertainties.

At MS&T we have developed an apparatus and techniques precisely for this purpose. In the next section, a brief description of the apparatus and techniques being used will be given. Following this, examples of triply differential data which we have obtained for both positron and electron impact are presented and compared. The purpose of this manuscript is to illustrate differences associated with the sign of the projectile charge rather than to test specific theoretical models such as was done in a recent Letter [16].

2. Experimental Method

The method being used in these studies is to ionize a gas jet of target atoms (or molecules) by a beam of leptons and to measure coincidences between the various ionization products. The beams crossed at the center of two biased plates used to produce an electric field perpendicular to the beam direction. By using a single apparatus for these studies and as identical conditions as possible, experimental uncertainties between the positron and electron impact data are minimized or totally eliminated.

Our positron beam was produced using a ^{22}Na source, a tungsten moderator, and an electrostatic transport system. To switch to electron impact, an electron gun plus a 90° deflector were inserted into the positron beamline such that the deflected electron beam entered the scattering chamber via the same input aperture and trajectory used for positron impact. Target ions were extracted by a weak electric field and detected by a channeltron. Time of flight techniques established their charge states and masses. Scattered projectiles were measured as a function of their scattering angle and energy loss using an electrostatic energy analyzer and a position sensitive channelplate. Electrons ejected from the target were measured as a function of their detection angle using a second position sensitive channelplate positioned close to the interaction region and at 90° with respect to the beam direction. List mode data collection allowed correlations between the various particles, angles, and energies to be established. For example, coincidences between singly ionized target atoms (or molecules) and scattered projectiles provided doubly differential cross section (DDCS) information as a function of scattering angle and energy loss. Coincidences between target ions, scattered projectiles, and ejected electrons provided triply differential cross section (TDCS) information which could be generated either for the scattered projectile or for the ejected electron. It is important to note that the TDCS data presented here are for all electrons of a particular energy that are detected between roughly 30 and 150 degrees with respect to the beam direction. Thus, they represent the integral intensity of the portions of the binary and recoil electron emission lobes that we can observe.

Examples of DDCS and TDCS data for 250 eV positron impact on molecular nitrogen are shown in Fig. 1. Here, the vertical axis is the projectile scattering angles, in degrees, which the horizontal axis is the energy loss in eV. Negative scattering angles imply that the projectile is scattered vertically downwards while positive angles imply upward scattering. Because our electron detector is located above the interaction region, "upward" emitted electrons are detected. Thus, the TDCS intensities for negative scattering angles, e.g., correlated downward scattered projectiles and upward emitted target electrons, are a direct indication of binary events since the scattered projectile and ejected electron are detected in opposite hemispheres. Likewise the intensities for positive scattering angles indicate recoil events since here both particles are detected in the same hemisphere.

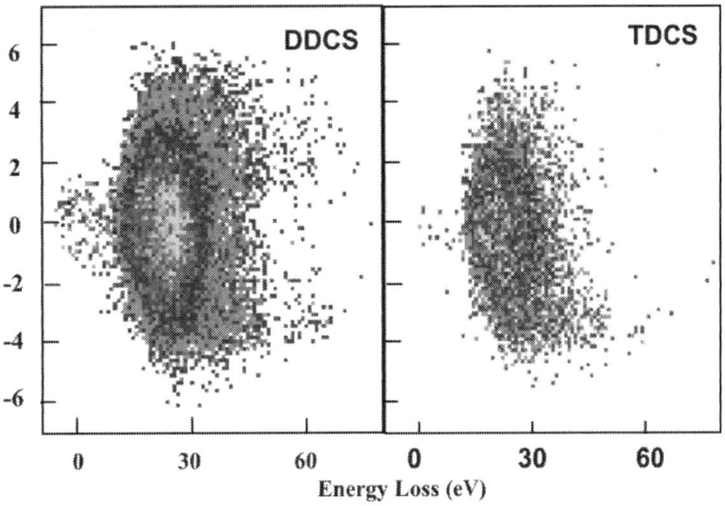

Figure 1. Doubly and triply differential data for single ionization of molecular nitrogen by 250 eV positron impact. The vertical and horizontal axes are the projectile scattering angles (in degrees) and energy loss (in eV). For the DDCS data, the z (color) axis is linear between 5 and 50; for the TDCS data, the axis is linear between 2 and 15.

Forward scattered projectiles were limited to a horizontal scattering range of 0° ± 2.4° by a slit at the entrance to the energy analyzer and to vertical scattering angles less than ± 7° because of the projectile channelplate size and distance from the interaction region. Various energy loss ranges could be set by adjusting the spectrometer voltages. But typical energy losses were less than 25% of the impact energy because of insufficient statistics at higher energy loss. With the present setup, impact energies between 200 and 1000 eV have been investigated. The ejected electron channelplate was sensitive to geometric emission angles between 30° and 150° along and perpendicular to the beam direction. No energy analysis of the ejected electrons was used. Rather, their energies were determined using coincidences with projectiles that suffered a particular energy loss which for single ionization unequivocally defines the ejected electron energy.

However, the energies, angles, and probabilities for the ejected electrons are subject to various experimental conditions. For example, the electric field used to extract the target ions modifies both the directions and acceptance ranges of the ejected electrons that are detected. This is particularly true for very low ejected electron energies which, depending upon their initial energy and direction, can even be turned around by the field. To minimize this problem, a small extraction field of 1.2 V/cm was used. SIMION simulations and a detailed model of the interaction volume and fields showed that such problems were minimal for emission energies above a few eV. But, to achieve sufficient statistics for positron impact, a relative large beam diameter (approximately 5-6 mm) and projectile scattering angle and energy loss bins of 1° and several eV were used. This results in a range of emission energies (energy losses), rather than a single energy, which contribute for any particular energy loss data point. Because a range of electron energies is involved, a range of electric field effects and uncertainties in correlating the observation and emission angles of the detected electrons results. This inhibits a direct solid angle adjustment of the raw data in order to compare with theory. Because of this, at present theory must be convoluted over our experimental parameters in order to compare with our data. A program for doing this was written and briefly described in a recent publication.[16] Additional experimental details can be found in references 16-18.

3. Results

Using background subtracted 2D spectra such as shown in Fig. 1, doubly and triply differential intensities were determined for scattering angles between -6° and +6° and for a range of energy losses. To reduce the statistical uncertainties, data within angular and energy loss bins of 1° and 7-17 eV were summed. Identical conditions were used for positron and electron impact. Ratios of triply to doubly differential cross sections were then calculated from the integrated intensities. Using ratios removes any problems associated with electric field effects on the ejected electron trajectories as well as any solid angle effects since these influence both the doubly and triply differential data in the same fashion. It also removes any experimental asymmetries between positive and negative scattering angles since by definition the DDCS must be symmetric with respect to zero degrees.

In Figure 2, we present three examples of these ratios and their full statistical uncertainties, all for an energy loss of approximately 37 eV. (To relate these data in terms of the ejected electron energy, the appropriate single ionization potentials must be subtracted.) Positron impact data are shown by the filled red circles; electron impact data by the open blue triangles. In order to provide a picture as to how impact energy and target species influence these data, the left figure is for 200 eV impact single ionization of argon, the middle figure is for 250 eV ionization of molecular nitrogen, and the right figure is for 500 eV ionization of neon. The curves in the argon figure are polynomial fits to the data, serving only to guide the eye.

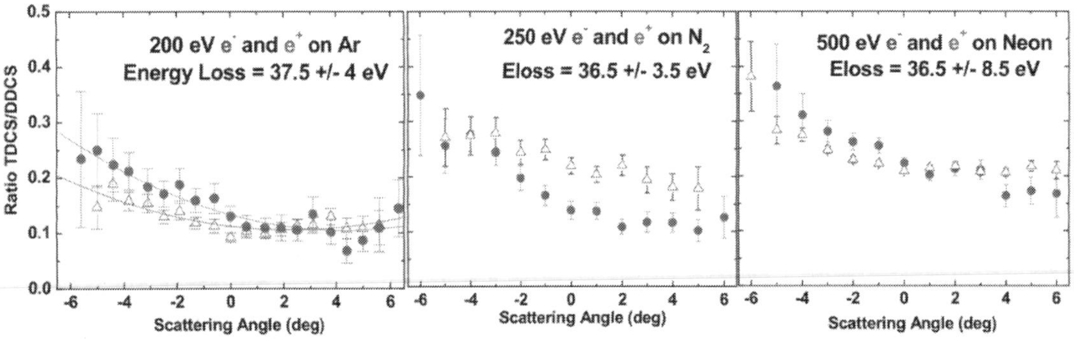

Figure 2. Ratios of triply to doubly differential cross sections for positron (red filled circles) and electron (blue open triangles) impact single ionization of argon, neon, and molecular nitrogen. The

solid curves in the left figure are polynomial fits to the data. Negative (positive) scattering angles correspond to binary (recoil) interactions.

As seen, the general trend in all cases is relatively isotropic ratios for positive scattering angles and a monotonic increase for increasing negative scattering angles. The reader is reminded that negative scattering angles indicate binary interactions while positive angles indicate recoil interactions. Thus, these data show that the relative number of binary to recoil interactions increases as a function of scattering angle, i.e., as a function of increasing momentum transfer. This appears to be independent of target species and also of impact energy. This increase may be in part due to a rotation of the binary lobe to larger angles with increasing momentum transfer, see for example figure 3 in reference 15. However, the present data also show that the increase is larger for positron impact. Larger binary intensities for positron impact are predicted by theory but theory also predicts that the binary lobe for positron impact should be shifted toward smaller angles whereas for electron impact it should be shifted toward larger angles. Thus, if the observed increase is simply due to a rotation of the binary lobe one would expect a larger increase for electron impact, just the opposite as what the data show.

Theory also predicts smaller recoil lobe intensities for positron impact as opposed to electron impact. But, our argon and neon data show nearly identical recoil lobe contributions for positron and electron impact. A possible interpretation of this is that recoil interactions involve an additional interaction between the ejected electron and the nuclear core and this interaction may smear out any initial differences associated with the projectile charge.

For ionization of nitrogen the relative recoil lobe contributions are different for positron and electron impact. Whether this is associated with differences between a single or multiple nuclear core or not will require further studies.

4. Comments and summary

Doubly and triply differential data for single ionization of argon and neon atoms plus nitrogen molecules have been measured for both positron and electron impact. Comparisons of the positron and electron impact data show systematic differences associated with the sign of the projectile charge. Specifically, binary interactions for positron impact have larger intensities than for electron impact. This is consistent with long standing theoretical predictions. Theory also predicts that just the opposite effect will be observed for the recoil lobe intensities. However, for argon and neon, any such effects are observed to be quite small. Comparisons made for ionization of molecular nitrogen yield similar qualitative results but show quantitative differences.

Although not shown or discussed in the present manuscript, it should be noted that our experimental method has also been used to investigate projectile charge effects on double ionization. In this case, by combining positron and electron impact data, information about first- and second-order doubly ionization plus about how they interfere, can be obtained. [19]

Acknowledgements: This work supported by the National Science Foundation.

References

[1] Sharma, S. and Srivastiva, M. K., Phys. Rev. A **38**, 1083 (1988).
[2] Brauner, M. and Briggs, J. S., J. Phys. B: At. Mol. Opt. Phys. **26**, 2451 (1993).
[3] Berakdar, J.; Briggs, J. S. and Klar, H., J. Phys. B **26**, 285 (1993).
[4] Benedek, Á. and Campeanu, R.I., Nucl. Inst. and Meth. B **266,** 458 (2008).
[5] Falke, T.; Brandt, T.; Kühl, O.; Raith, W.; Weber, M., J. Phys. B: At. Mol. Opt. Phys. **30**, 3247 (1997).
[6] DuBois, R. D.; Doudna, C.; Lloyd, C.; Kahveci, M.; Khayyat, K.; Zhou, Y.; Madison, D. H., J. Phys. B At. Mol. Opt. Phys. **34**, L783-L789 (2001).
[7] Santos, A. C. F.; Hasan, A.; DuBois, R. D., Phys. Rev. A **69**, 032706 (2004).
[8] Moxom, J.; Laricchia, G.; Charlton, M.; Jones, G. O.; Kövér, Á., J. Phys. B: At. Mol. Opt. Phys.

25, L613 (1992).
[9] Schmitt, A.; Cerny, U.; Möller, H.; Raith, R.; Weber, M., Phys. Rev. A **49**, R5 (1994).
[10] Kövér, Á.; Laricchia, G.; Charlton, M., J. Phys. B: At. Mol. Opt. Phys. **27**, 2409 (1994).
[11] Kövér, Á.; Finch, R. M.; Charlton, M.; Laricchia, G., J. Phys. B: At. Mol. Opt. Phys. **30**, L507-L512 (1997).
[12] Kövér, Á.; Laricchia, G.; Charlton, M., J. Phys. B: At. Mol. Opt. Phys. **26**, L575 (1993).
[13] Kövér, Á.; Laricchia, G., Phys. Rev. Lett. **80**, 5309 (1998).
[14] Kövér, Á.; Paludon, K.; Laricchia, G., J. Phys. B: At. Mol. Opt. Phys. **34**, L219-L222 (2001).
[15] Ren, X.; Senftleben, A.; Pflüger, T.; Dorn, A.; Bartschat, K. and Ullrich, J., J. Phys. B: At. Mol. Opt. Phys. **43**, 035202 (2010).
[16] de Lucio, O.G.; Otranto, S.; Olson, R.E. and DuBois, R.D., Phys. Rev. Lett. **104**, 163201 (2010).
[17] de Lucio, O. G.; Gavin, J. and DuBois, R. D., Phys. Rev. Lett. **97**, 243201 (2006).
[18] DuBois, R. D.; de Lucio, O. G. and Santos, A. C. F., *Radiation Physics Research Progress*, edited by A. N. Camilleri (Nova, New York, 2008).
[19] DuBois, R.D.; de Lucio, O.G. and Gavin, J., Europhys. Lett. **89** 23001 (2010).

Journal of Physics: Conference Series **288** (2011) 012012 doi:10.1088/1742-6596/288/1/012012

Surface-sensitive measurements of excitonic processes in rare gas clusters by electron energy loss spectroscopy

Hayato Kubotera[1], **Tatsuya Sekitsuka**[1], **Satoshi Jinno**[2] **Takayuki Tachibana**[2,a] **and Takato Hirayama**[1,2]

[1]Department of Physics, Rikkyo University, Toshima-ku, Tokyo 171-8501, JAPAN
[2]Research Center for Measurement in Advanced Studies (RCMAS), Rikkyo University, Toshima-ku, Tokyo 171-8501, JAPAN
[a]Present address: Department of Physics, Tokyo University of Science, Shinjuku-ku, Tokyo 162-8601, JAPAN

E-mail: hirayama@rikkyo.ac.jp

Abstract. We present the electron energy loss spectra of Ar clusters as a function of cluster size ranging from 120 to 3500 atoms/cluster. The intensity of the bulk excitation peak decreases for cluster sizes larger than 500 atoms/cluster. This characteristic feature can be explained based on a simple calculation that takes into account the mean free path of the incident electrons. The present study demonstrates the promise of surface-sensitive measurements of the excitation processes of clusters.

1. Introduction

The physics of clusters has attracted significant interest, because they not only bridge the gap between atoms and solids, but also have specific physical properties [1]. One unique aspect of clusters is their large surface-to-volume ratio, which enables sensitive measurements of surface processes. In addition, clusters traveling through a vacuum have very clean surfaces, as opposed to solid-substrate surfaces; this is especially true for condensed solids such as rare-gas solids [2]. This property is important because small amounts of impurities may affect dynamic processes on the surface [3].

In a previous paper [4], we published electron energy loss spectra of Ar clusters as a function of both incident electron energy and cluster size (average cluster size, $\langle N \rangle = 120 \sim 270$ atoms/cluster). These results clearly showed the surface and bulk excitation peaks, and the appearance cluster size of the bulk excitation peak was found to be dependent on the incident electron energy. This characteristic feature was qualitatively explained by a simple model that takes into account the mean free path, λ, of the incident electrons in Ar clusters.

Here, we report experimental results for larger clusters up to $\langle N \rangle = 3500$ atoms/cluster. We show that the bulk excitation peak decreases for larger clusters and discuss the promise of surface-sensitive measurements.

2. Experiments

Details of the experimental apparatus and procedure have been described elsewhere [4]. Briefly, a pulsed cluster beam is generated by adiabatic expansion through a conical nozzle attached to a pulse valve (General Valve Co.). The average cluster size is controlled by changing the stagnation pressure and nozzle temperature. The size distribution is determined experimentally by measuring the kinetic energy of the ionized cluster beam [5].

A mono-energetic electron beam is generated using a simulated hemispherical energy selector [6]. The electron beam is deflected by 90° by a set of parallel plates and subsequently merges with the cluster beam. The length of the merging region is approximately 200 mm. After passing through the merging region, the electrons are analyzed in terms of their energy and detected by a channel electron multiplier. The overall energy resolution of our setup is 150–180 meV.

Under the present experimental conditions (typical duration and frequency of the cluster beam pulse of 10 ms and 2–4 Hz, respectively), the acquisition time of one spectrum is about 50 hours. The merging-beam geometry is essential to obtain energy loss spectra with adequate statistics, because of the low target (cluster) density.

3. Experimental results

Figure 1 shows electron energy loss spectra of Ar clusters for average cluster sizes, $\langle N \rangle$, ranging from 120 to 3500 atoms/cluster at an incident electron energy $E_0 = 100$ eV. The observation angle is 0° with respect to the direction of the incident electron beam. Vertical lines indicate the energy positions of surface and bulk excitons and the band gap energy, E_g, in solid Ar [7].

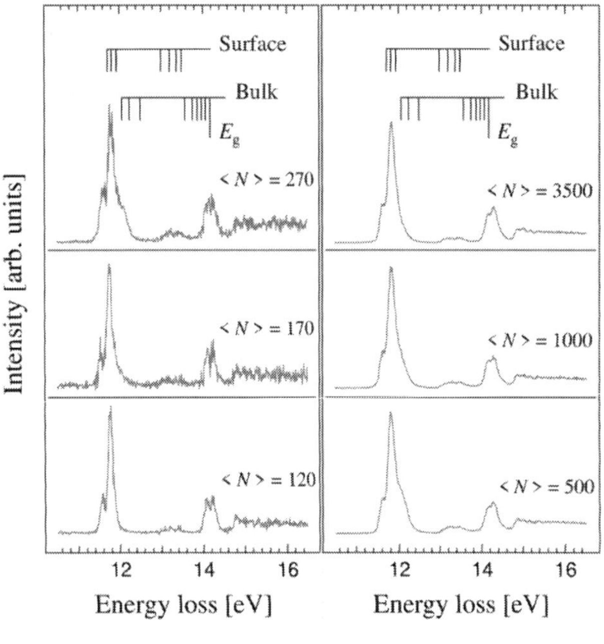

Figure 1. Electron energy loss spectra of Ar clusters with $\langle N \rangle$ ranging from 120 to 3500 atoms/cluster. The incident electron energy is 100 eV and the observation angle is 0°. Vertical lines show the energy positions of surface and bulk excitons and the band gap energy, E_g, in solid Ar [7].

The peak (shoulder) of the $3p^5 4s$ bulk excitation at 12.1 eV is found in the spectrum for a cluster size of $\langle N \rangle = 170$ atoms/cluster, indicating that bulk excitation can occur in clusters with diameters corresponding to approximately five atoms [4]. We also find that the intensity of the bulk peak relative to the surface peak increases with cluster size up to $\langle N \rangle = 500$ atoms/cluster, but decreases for sizes larger than 500 atoms/cluster, as shown in Fig. 2. This characteristic feature is discussed in the following section.

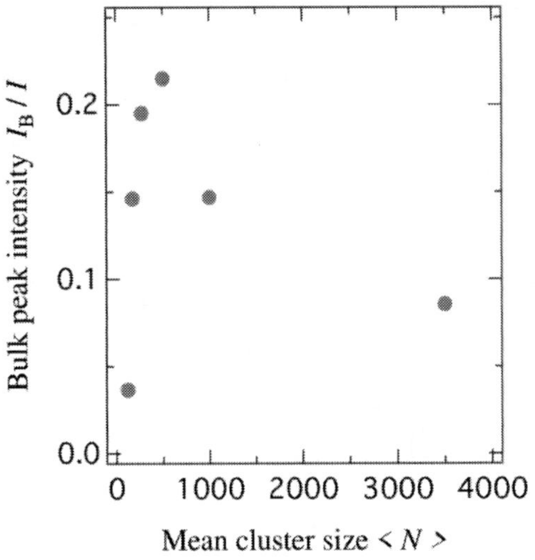

Figure 2. Dependence of the $3p^54s$ bulk peak intensity, I_B, at 12.1 eV on mean cluster size, $\langle N\rangle$. The bulk peak intensity is normalized to $I = I_B + I_S$, where I_S is $3p^54s$ surface peak intensity.

4. Discussion

We now discuss the dependence of the surface and bulk excitation peak intensities on cluster size (cf. our previous work [4]). The ratio of the number of surface atoms, N_S, to the total number of atoms, N, in a cluster is given by [8]

$$\frac{N_S}{N} = \frac{4\pi R^2/\pi r^2}{(4\pi R^3/3)\,/\,(4\pi r^3/3)} = 4\frac{r}{R} = 4N^{-1/3} \tag{1}$$

and

$$\frac{N_S}{N}N^{1/3} = 4, \tag{2}$$

where r and R are the radii of a single atom and a cluster, respectively. Assuming that the intensity of surface excitation, I_S, is proportional to the number of surface atoms, N_S, and that the total intensity, $I = I_S + I_B$, is proportional to the experimentally determined cluster size $\langle N\rangle$, we can write Eq. (2) as

$$\frac{I_S}{I_S + I_B}\langle N\rangle^{1/3} = 4. \tag{3}$$

The peak intensities of the surface excitation I_S and those of the bulk excitation I_B are obtained by fitting the measured spectra to three Gaussian peaks, as shown in Fig. 3.

Figure 4 shows a plot of $(I_S/I)\langle N\rangle^{1/3}$ versus $\langle N\rangle^{1/3}$. The experimental results for $E_0 =$ 100 and 250 eV (spectra not shown) show a value around 4 for smaller cluster sizes, but this increases with increasing cluster size. We attribute this characteristic feature to the fact that the mean free path of the incident electrons in Ar clusters is smaller than the cluster diameter and that λ varies with the incident electron energy. In other words, the numbers of surface atoms, N'_S, and bulk atoms, N'_B, that can be excited by an incident electron and detected by our 0° spectrometer are smaller than the numbers of surface and bulk atoms in a cluster.

We calculate N'_S and N'_B using the following simple model. We assume that the electron cannot be detected by our spectrometer if the length of its trajectory within a cluster is longer than 2λ; i.e., an electron excites an atom at a depth of λ in the cluster and it can then pass through the cluster without exciting another atom in the cluster within a distance of λ. Figure 5 shows the model schematically. An electron coming from the left passes through the cluster

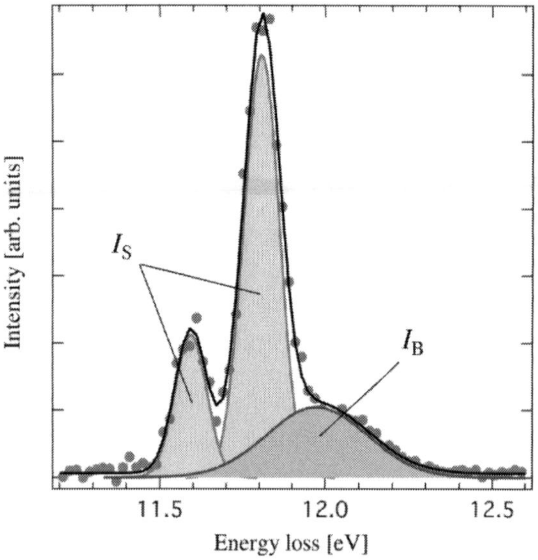

Figure 3. Electron energy loss spectrum of Ar clusters from 11 eV to 13 eV. I_S and I_B are the intensities of surface and bulk excitations, respectively. See text for details.

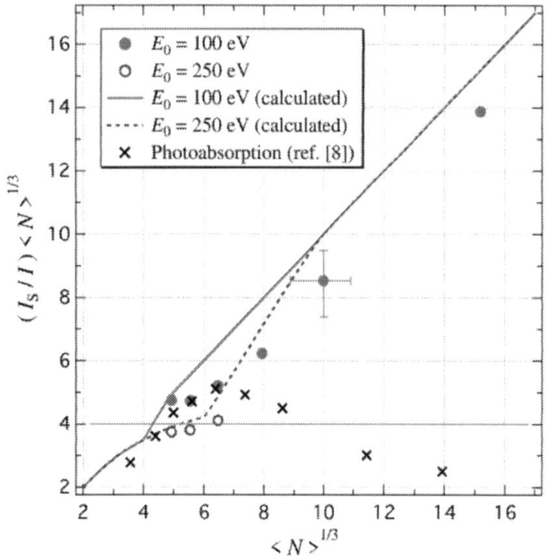

Figure 4. $(I_S/I)\langle N\rangle^{1/3}$ for incident electron energy $E_0 = 100$ eV (●) and 250 eV (○). Also shown are $[N'_S/(N'_S + N'_B)]\langle N\rangle^{1/3}$, calculated using the model taking into account the mean free path of incident electrons for electron energies of 100 eV (——) and 250 eV (- - - -). Photoabosorption results (×) [8] are also plotted for comparison. See text for details.

and is detected by the 0° detector on the right-hand side of the cluster. If the mean free path is much longer than the cluster radius, as shown in Fig. 5(a), all surface and bulk atoms can contribute to the electron energy loss spectra; i.e., electrons scattered by any atom within the cluster can be detected by our spectrometer ($N'_S = N_S$ and $N'_B = N_B$). In the case shown in Fig. 5(b), where λ is slightly smaller than R, some of the surface and bulk atoms can be invisible to our spectrometer; i.e., $N'_S < N_S$ and $N'_B < N_B$, because multiple collision may occur in the cluster. Finally, for $\lambda \ll R$, as in Fig. 5(c), all of the bulk atoms are invisible and only some of the surface atoms are visible; i.e., $N'_S < N_S$ and $N'_B = 0$. In the latter case, it is expected that the spectrum is composed of the surface excitation peaks only.

The solid and dashed lines in Fig. 4 show the calculated results for incident electron energies, E_0, of 100 and 250 eV, respectively. The calculation has been done in a similar manner described

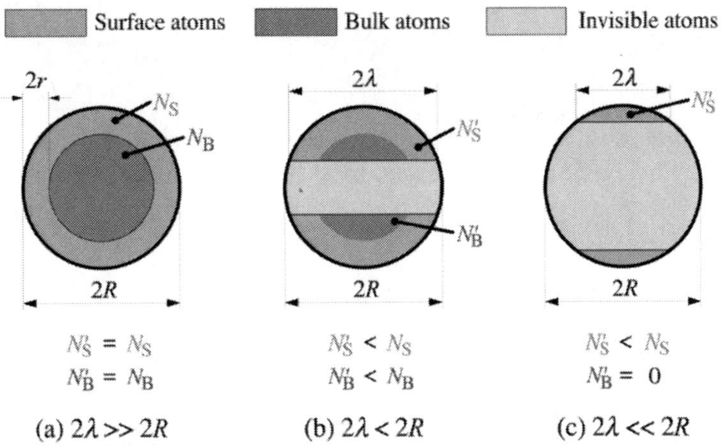

Figure 5. Schematic of the model used in our calculation. R, r, and λ are the radii of a cluster and a single atom, and the mean free path of an electron in a cluster, respectively. N_S and N_B are the numbers of surface and bulk atoms in a cluster, respectively, and N'_S and N'_B are the numbers of surface and bulk atoms visible to the spectrometer, respectively. See text for details.

in [4]. It should be noted that in the present calculation, we have estimated both N'_S and N'_B, while only N'_B has been estimated in our previous work. We use the fact that the nearest-neighbor distance in solid Ar is 0.376 nm [9] and that the mean free paths of incident electrons with $E_0 = 100$ and 250 eV in solid Ar are approximately 0.7 and 1.2 nm, respectively [7]. As shown in the figure, the model calculation qualitatively reproduces the trend in the experimental results.

For cluster sizes larger than 1000 atoms/cluster, the slope of the experimental results approaches unity, which is well reproduced by the calculation. This finding indicates that most of the bulk atoms are invisible to our spectrometer, as shown schematically in Fig. 5(c). In turn, this suggests that surface excitation can be observed effectively by selecting the cluster size and the incident electron energy even for very large clusters. Note that this is not the case in the photo-absorption study, because the penetration depth of a VUV photon into the cluster is generally longer than the electrons' mean free path, resulting in a surface-to-bulk intensity ratio similar to the geometrically calculated ratio. Therefore, the photo-absorption results [8] in Fig. 2 show a roughly constant value of 4 for cluster sizes between 50 and 3000 atoms/cluster. It can be expected that the surface sensitivity will be higher for heavier rare gases (such as Kr and Xe) because of their longer nearest-neighbor distances and shorter mean free paths.

Surface-sensitive measurements are currently underway. Electron energy loss spectra of Ar clusters are measured with a small amount of gaseous Ne simultaneously introduced into the main chamber. The peak at 16.848 eV of the Ne ($2p^5 3s\ ^1P_1$) atomic line is used for energy calibration. The blue shift of approximately 30 meV was observed in our preliminary results of surface excitation for Ar clusters with sizes from 120 to 3500 atoms/cluster. This shift is consistent with previously reported values [10, 11]. Systematic measurements are in progress.

5. Summary

We have studied electron energy loss spectra of Ar clusters as a function of cluster size. The intensity of the bulk excitation peak decreases for larger clusters. This characteristic feature has been qualitatively reproduced based on a simple calculation, taking into account the mean free path of the incident electrons in Ar clusters. We have shown that the surface-excitation process can be effectively observed with an appropriate choice of cluster size and incident electron energy.

Acknowledgments
The authors are grateful to Mr. Satoshi Sakai for his significant contribution in the initial stages of this work. This work was partly supported by the Rikkyo University Special Fund for Research.

References

[1] for example, 1990 *Atomic and Molecular Clusters* edited by E. R. Bernstein (Amsterdam: Elsevier Science)
[2] Hirayama T and Arakawa I 2006 *J. Phys. Cond. Matt.,* **18** S1563
[3] Kuninobu T, Hayama A, Hirayama T and Arakawa I 1997 *Surf. Sci.,* **390** 272
[4] Kubotera H, Sakai S, Sekitsuka T, Tachibana T and Hirayama T 2009 *Appl. Surf. Sci.,* **256** 1046
[5] Hirayama T, Kanehira A and Arakawa I 1993 *Rev. Sci. Instrum.* **64** 962
[6] Jost K 1979 *J. Phys. E: Sci. Instrum.* **12** 1006
[7] Schwentner N, Koch E.-E. and Jortner J 1985 *Electronic excitations in condensed rare gases* (Berlin: Springer-Verlag)
[8] Stapelfeldt J, Wörmer J and Möller T 1989 *Phys. Rev. Lett.,* **62** 98
[9] Sonntag B 1977 *Rare Gas Solids* Vol. 2, p1022 (New York: Academic Press)
[10] Wörmer J, Karnbach R, Joppien M and Möller T 1996 *J. Chem. Phys.* **104** 8369
[11] Burose A, Becker C and Ding A 1991 *Z. Phys.* **D20** 35

Journal of Physics: Conference Series **288** (2011) 012013 doi:10.1088/1742-6596/288/1/012013

Double Differential Cross Sections and Generalized Oscillator Strength Distributions of Ammonia

Karin Yamamoto, Keisuke Nogami, Yuta Hino, and Yasuhiro Sakai

Department of Physics, Toho University, Miyama 2-2-1, Funabashi, Chiba, 2748510 Japan

E-mail: karin@ph.sci.toho-u.ac.jp

Abstract. The absolute double differential cross section (DDCS), the generalized oscillator strength distribution (GOSD), and the ionization efficiency of ammonia (NH_3) were investigated from the threshold to 40 eV under the condition of 200 and 400 eV incident electron energies and 6 and 8 degree scattering angles using electron energy-loss spectroscopy and electron- ion coincidence techniques. To determine the absolute values, we used a mixture of helium (He) and NH_3 and normalized the measured relative DDCS spectrum by the differential cross section for 2^1P excitation of He. Our results are in close agreement with previous dipole (e, e) spectroscopy, although the incident electron energy is lower. The ionization efficiency curve obtained from coincidence measurements indicated the existence of doubly excited states that cause neutral dissociation.

1. Introduction

Di erential cross sections (DCSs) and generalized oscillator strengths (GOSs) are fundamental quantities obtained from electron energy-loss spectra. They provide information on the electronic structure of matter and its interaction with electromagnetic radiation. The concept of GOS was introduced by Bethe [1, 2] in 1930 based on the rst Born approximation. Essentially the GOS is a function of the magnitude of momentum transfer only. When the rst Born approximation is not valid, a similar calculated quantity depends on both the magnitude of momentum transfer and the incident electron energy. This is sometimes known as the apparent GOS. Here, we use the term GOS regardless of whether the rst Born approximation was valid. Much theoretical and experimental work on optically forbidden transitions and excited states in continuum regions remains to be done, but a wealth of DCS and GOS data exist for various excited states lying in discrete region in atoms and molecules. Determination of the absolute DCS is difficult in the region where the rst Born approximation is not valid. In general, when the incident electron energy is high, we can use the TRK sum rule [3] to normalize the GOS from the electron energy-loss spectrum. If the incident electron energy is not high enough, the di erence between the summation of total scattered signal and the TRK sum rule [3] cannot be ignored. Therefore, the energy regions and potentials of these states are not well understood.

We have developed a new normalization method for the determination of absolute DCS under the condition that the incident energy is not high enough, and we have studied the excitation and ionization processes of ammonia (NH_3). NH_3 plays an important role in the physics and chemistry of the upper atmosphere of the Earth and other planets. The study of doubly excited states in NH_3 has mainly entailed photo-ionization experiments [4]. In these studies, the ionization efficiency (ratio of photo-ionization to photo-absorption cross section) was obtained as a function of impact photon energy. The doubly excited states and their decay processes were discussed using ionization efficiency curves. In electron-impact experiments, oscillator strengths for the photo-absorption, photo-ionization and photo-dissociation of NH_3 were determined by Wight *et al.* using much higher electron energies [5]. However, in these studies, information on optically forbidden doubly excited states was not obtained because of the high-energy electron impact. Additionally, the cross sections for the emission of uorescence by neutral fragments in the photo-excitation of NH_3 were measured as a function of impact photon energy by Kato *et al.* [6] Most recently, the electron energy-loss spectrum of NH_3 tagged by Lyman-α photons was demonstrated for 100 eV incident electron energy and at 8 deg scattering angle by Ishikawa *et al.* [7] These measurements are also tools for obtaining information on doubly excited states in molecules. Two optically forbidden doubly excited states of NH_3 were predicted by Ishikawa *et al.* [7].

In electron-impact experiments, smaller incident energies (larger momentum transfer K^2) reveal information on optically forbidden states. We have demonstrated the measurement of the DCSs and GOSs of NH_3 for 200 and 400 eV incident electron energies and 6 degree scattering angle. Furthermore, to obtain information on the optically forbidden doubly excited states of NH_3, we obtained the ionization efficiency curve by scattered electron- ion coincidence measurements for 200 eV incident electron energy and 8 degree scattering angle.

2. Experimental setup

The experimental apparatus used was an (e, e+ ion) spectrometer, which could measure the intensity of the formed ions at the same time as the scattered electrons, as a function of excitation energy.

The apparatus consisted of a pulsed electron gun, an electron-energy analyzer, and a time-of- ight (TOF) mass analyzer, as shown in Fig.1. The components were enclosed in a vacuum chamber, which was evacuated by a 520 L/s turbo molecular pump (TMP). We furnished the chamber with a 230 L/s TMP for the TOF mass analyzer as a di erential pumping system. The ultimate vacuum pressure was lower than several 10 ^{7} Pa, and the pressure was kept to 3 10 ^{4} Pa or less during the measurements.

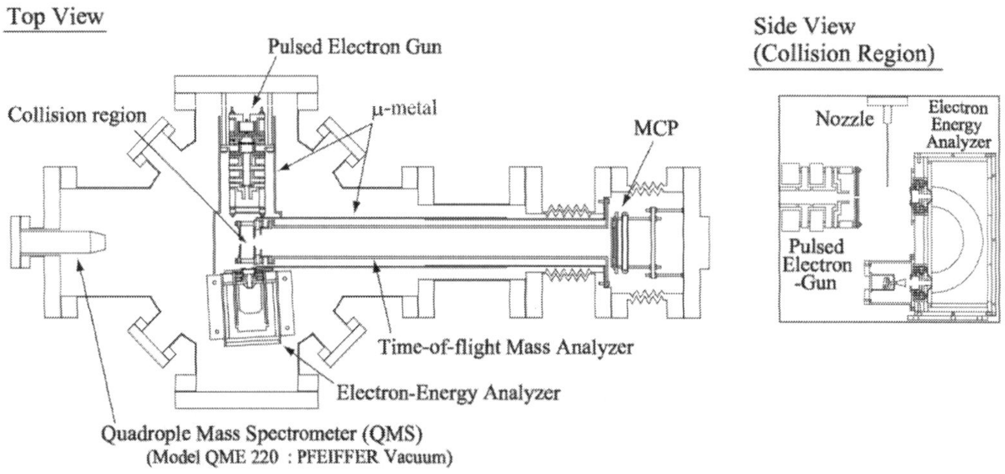

Figure 1. Schematic of the experimental apparatus.

An electron beam emitted from a hairpin-type tungsten lament was accelerated and collimated by an electrostatic lens system, and was pulsed before emerging into the collision region. A target molecular beam was continuously generated from an e usive nozzle. The pulsed electron beam and the molecular beam crossed each other at the center of the 20 mm wide collision region. Hence, the pulsed electrons collided at right angles with the target molecules (crossed-beam geometry). The collision energies of the incident electrons were from 100 to 400 eV. The primary (continuous) electron current was about 1 nA, on average. After colliding with the target molecules, scattered electrons were analyzed with the electron-energy analyzer, and the formed ions were extracted from the collision region by application of a pulsed electric eld. They were then analyzed using a Willy-McLaren type [8] TOF mass analyzer. The electron-energy analyzer was a simulated hemispheric electrostatic type [9] with a mean trajectory radius of 52 mm. The typical energy resolution was about 0.5 eV (full width at half-maximum: FWHM) for the present experiments. After acceleration along a 10 mm long second acceleration region, the extracted ions passed through a 480 mm eld-free region. They were detected using a microchannel plate (MCP). We used a mixture of helium (He) and ammonia (NH_3) as the sample, to normalize the di erential cross sections (DCSs) of NH_3 in the electron energy-loss spectra. A quadrupole mass spectrometer (QMS) installed in our apparatus next to the original TOF mass analyzer was used to con rm the mixing ratio M of NH_3 to He. The mixing ratio M is

$$M = \frac{I_{\mathrm{NH}_3^+}}{I_{\mathrm{He}^+}} \frac{\sigma_{\mathrm{He}^+}}{\sigma_{\mathrm{NH}_3^+}}, \tag{1}$$

where I is the observed signal intensity in the mass spectrum, σ is the partial ionization cross section, and the subscripts denote the species.

Using the mixing ratio M, we estimated the DCSs of the molecules of interest. Some peaks and structures corresponding to the mixing ratio appeared in the electron energy-loss spectrum, which originated from NH_3 and He. Because the DCS for excitation of the 2^1P state in He is well known, the DCSs for the NH_3 excitations of interest could be estimated from

$$\left(\frac{d\sigma}{d\Omega}\right) = \frac{I_{\mathrm{NH}_3}}{I_{\mathrm{He}}} \left(\frac{d\sigma}{d\Omega}\right)_{2^1\mathrm{P}} / M, \tag{2}$$

where I is the relative intensity of the electron energy-loss spectrum, and $d\sigma/d\Omega$ indicates the DCS. Therefore, the generalized oscillator strength (GOS) F was determined using this absolute DCS value: [10]

$$F\left(K^2, E_i\right) = \frac{W}{2}\frac{k_i}{k_s} K^2 \left(\frac{d\sigma}{d\Omega}\right), \tag{3}$$

where W is the excitation energy and k_i and k_s are the momentum of incident and scattered electrons, respectively. K^2 is the magnitude of momentum transfer

$$K^2 = k_i^2 + k_s^2 - 2k_i k_s \cos\theta, \tag{4}$$

where θ is the scattering angle. All quantities are in atomic units. For excitation energies greater than the rst ionizing energy, the relationchip between the DCS and GOS is

$$\frac{dF\left(K^2, E_i\right)}{dE} = \frac{W}{2}\frac{k_i}{k_s} K^2 \left.\frac{d^2\sigma}{dEd\Omega}\right), \tag{5}$$

where dF/dE is the generalized oscillator strength distribution (GOSD) and $d^2\sigma/(dEd\Omega)$ is the double di erential cross section (DDCS).

3. Results and discussion

3.1. Absolute double differential cross sections and generalized oscillator strength distributions of ammonia

The electric con guration of the ground state in C_{3v} symmetry of NH_3 is $(1a_1)^2$ $(2a_1)^2$ $(1e)^4$ $(3a_1)^2$. The ionization threshold for the $(3a_1)$, $(1e)$, and $(2a_1)$ states are 10.85, 16.4, and 27.3 eV, respectively [11, 12].

Figure 2 shows typical electron energy-loss spectra of the He-NH_3 mixture gas and the pure NH_3 gas under the condition of 200 eV incident electron energy and 6 degree scattering angle. We found two peaks around 20 eV that are not seen in pure NH_3 gas in the spectrum of the mixture gas. These were the 2^1P and 3^1P excitation peaks of He. The mixing ratio M was obtained from (1) as 0.75. The intensities $I_{NH_3^+}$ and I_{He^+} were measured with QMS for 70 eV impact energy, and each of the partial cross sections was substituted as $\sigma_{NH_3^+}$ and σ_{He^+}. As $\sigma_{NH_3^+}$ and σ_{He^+}, we used an experimental value measured by Rejoub *et al.* [13] and a theoretical one calculated by Kim *et al.* [14] respectively. Using the mixing ratio M, the intensity, and the DCS for the 2^1P excitation peak of He, we determined the absolute DCSs of NH_3 from energy-loss spectra under the condition of 200 and 400 eV incident electron energies and 6 degree scattering angle.

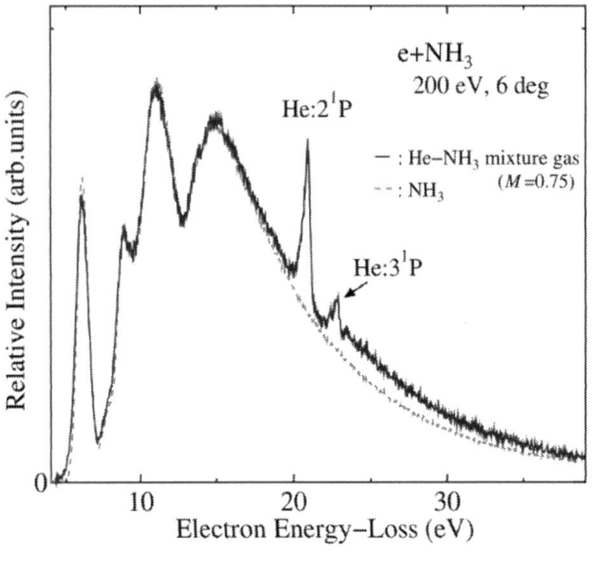

Figure 2. Typical electron energy-loss spectra of the He-NH_3 mixture gas and the pure NH_3 gas: for 200 eV incident electron energy at 6 degree scattering angle. In the spectrum of the He-NH_3 mixture, the mixing ratio M was 0.75. The peaks at 21 and 23 eV are ^1P peaks of He, which were the main optically allowed transitions. The optically forbidden transitions contributed little to peak formation under the present experimental conditions.

The absolute GOSDs of NH_3 are shown in Figs. 3 (a) and (b), which were determined at 6 degrees for 200 eV and 400 eV incident electron energies, respectively. Figure 3 (c) shows the absolute optical oscillator strength determined by dipole (e, e) spectroscopy [15]. The magnitude of momentum transfer K^2 calculated by the energy loss and the scattering angle is given on the u-axis (upper horizontal axis). In these experiments, the values of K^2 ranged from about 0.15 to 0.3 a.u. for 200 eV, and from about 0.3 to 0.4 a.u. for 400 eV. The shape structures observed around 6.5 and 9.5 eV corresponded to excitations of valence $(3a_1)$ and $(1e)$ electrons to a

Rydberg state, respectively [15]. Although both peaks around 12 and 16 eV corresponded to the excitation of (1e) electrons, the ionization of ($3a_1$) electrons contribute to form these structures, because they appeared over the rst ionization threshold. These features are common to these two results and the optical oscillator strengths from dipole (e, e) spectroscopy [15]. We were not able to detect any non-dipole transitions.

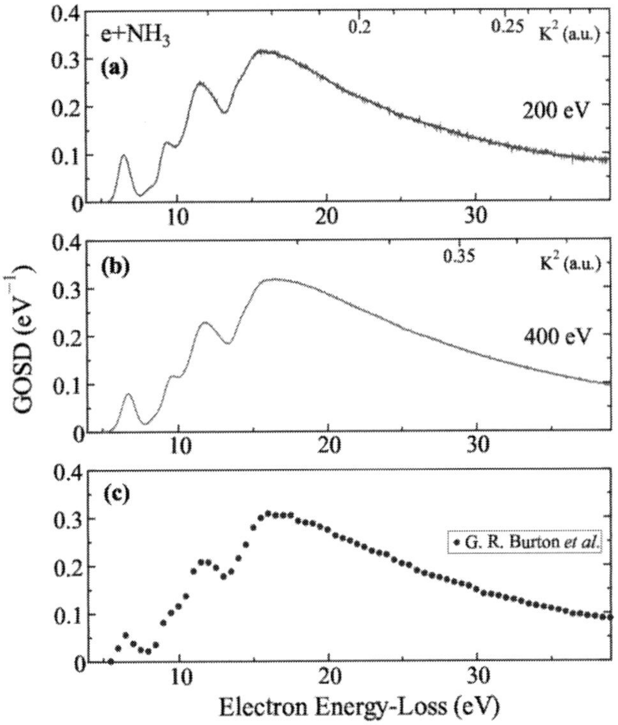

Figure 3. Absolute generalized oscillator strength distributions of NH_3: (a) our results for 200 eV incident electron energy at 6 degree scattering angle, (b) our results for 400 eV incident electron energy at 6 degree scattering angle, and (c) the dipole (e, e) results of Burton *et al.* [15]

3.2. Ionization efficiency curve of ammonia

To obtain the relative ionization efficiency, we divided the total ionic GOSD determined by the electron- ion coincidence by the total GOSD determined from the electron energy-loss spectrum. The ionization efficiency should be 1 when all processes are associated with ionization at a given excitation energy. As it is known that there are no doubly excited states in the 19-21 eV electron energy-loss region [7], we normalized the relative ionization efficiency to 1 in that region.

The ionization efficiency curve of NH_3 for 200 eV incident electron energy at 8 degree scattering angle is shown in Fig. 4. Both the photo-ionization results by Samson *et al.* [4] and the dipole (e, e+ ion) experiment results by Wight *et al.* [5] are shown in Fig. 4 for comparison. Clearly, the present efficiencies are in close agreement with the previous ones over the 10-20 eV energy range. However, our efficiency curve is slightly lower from 22 to 28 eV and from 30 to 36 eV than previous results [4, 5]. This means that optically forbidden doubly excited states exist in the range of 22-28 eV and 30-36 eV that decay by neutral dissociation.

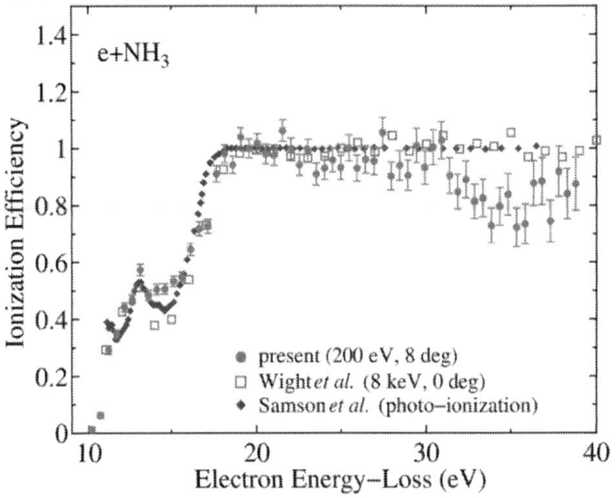

Figure 4. Ionization efficiency curves of NH_3: our results for 200 eV incident energy at 8 degree scattering angle (•); the dipole (e, e+ion) results of Wight *et al.* (□) [5]; the photo-ionization results of Samson *et al.* (♦) [4].

The existence of doubly excited states that cause neutral dissociation in this region has already been proposed by Ishikawa *et al.* [7], and is con rmed by our experiments.

4. Summary

We have investigated the DDCSs, GOSDs and ionization efficiency of NH_3. We used a mixture of He and NH_3 as the sample to determine the absolute DDCSs and GOSDs of NH_3. With this method, we could determine the absolute values without using the TRK sum rule. This means that we could discuss the GOS and GOSD even though the rst Born approximation was not valid. Both GOSDs determined with 200 and 400 eV incident electron energies and 6 degree scattering angle have shown similar behavior to the optical oscillator strength obtained in previous dipole (e, e) spectroscopy. The ionization efficiency curve obtained by the coincidence measurements was slightly lower from 22 to 28 eV and 30 to 36 eV than the photo-ionization and dipole (e, e+ ion) results. Therefore, the existence of doubly excited states that cause neutral dissociation was con rmed by our experiments in these ranges. We will study a wider range of incident electron energies, scattering angles, and molecules in near future.

Acknowledgments

This work was partially supported by Grant-in-Aid for Japan Society for the Promotion of Science (JSPS) Fellows (22·5682).

References

[1] Bethe H 1930 *Ann. Phys.* **5** 325
[2] Bethe H 1932 *Z. Physik* **76** 293
[3] Fano U and Cooper J W 1968 *Rev. Mod. Phys.* **40** 441
[4] Samson J A R, Haddad G N and Kilcoyne L D 1987 *J. Chem. Phys.* **87** 6416
[5] Wight G R, Van der Wiel M J and Brion C E 1977 *J. Phys. B: At. Mol. Opt. Phys.* **10** 1863
[6] Kato M, Odagiri T, Kameta K, Kouchi N and Hatano Y 2003 *J. Phys. B: At. Mol. Opt. Phys.* **36** 3541
[7] Ishikawa L, Odagiri T, Yachi K, Nakazato T, Kurokawa M, Kitajima M and Kouchi N 2008 *J. Phys. B: At. Mol. Opt. Phys.* **41** 195204

[8] Wiley W Cand McLaren I H 1955 *Rev. of Sci. Instrum.* **26** 1150
[9] Jost K 1979 *J. Phys. E. Sci. Instrum.* **12** 1006
[10] Inokuti M 1971 *Rev. Mod. Phys.* **43** 297
[11] Piancastelli M N, Cauletti C and Adam M-Y 1987 *J. Chem. Phys.* **87** 1982
[12] Banna M S and Shirley D A 1975 *J. Chem. Phys.* **63** 4759
[13] Rejoub R, Lindsay B G and Stebbings R F 2001 *J. Chem. Phys.* **115** 5053
[14] Kim Y -K and Rudd M E 1994 *Phys. Rev. A* **50** 3954
[15] Burton G R, Chan W F, Cooper G and Brion C E 1993 *Chem. Phys.* **177** 217

Elastic electron scattering from the DNA bases: cytosine and thymine

C J Colyer[1], S M Bellm[1], F Blanco[2], G Garcia[3] and B Lohmann[1,4]

[1]ARC Centre of Excellence for Antimatter-Matter Studies, The University of Adelaide, Adelaide, SA 5005, Australia

[2]Departamento de Física Atómica Molecular y Nuclear, Facultad de Ciencias Físicas. Universidad Complutense, Avda. Complutense s/n, E-28040 Madrid, Spain

[3]Instituto de Matemáticas y Física Fundamental, Consejo Superior de Investigaciones Científicas, Serrano 113-bis, 28006 Madrid, Spain

e-mail: birgit.lohmann@adelaide.edu.au

Abstract. Relative elastic differential cross sections for elastic scattering from cytosine and thymine have been measured using the crossed-beam method. The measurements have been performed at two electron energies of 500 and 100 eV, and cover the angular range of 10°-130°. Calculations of elastic differential cross sections have been performed via the screen corrected additivity rule method, and agreement is quite good with the present experimental results. The results obtained are important for the modelling of energy deposition in living tissue.

1. Introduction

Ionizing radiation is widely used in medicine as a diagnostic probe in radiology and as a palliative treatment in radiotherapy. It is now known that it is not just the primary ionizing particle that is responsible for cell and tissue damage, a significant amount has been shown to be due to secondary species that arise from the primary particle [1]. Upon entering the body, the primary particle thermalizes quickly through various scattering processes, liberating large numbers of low energy (0-20 eV) secondary electrons. These electrons can then interact with different biological molecules, and have been shown to cause significant damage to DNA via dissociative electron attachment [2, 3]. This leads to single or double DNA strand breakage, either directly or via the formation of free radicals.

Due to the difficult nature of reliable measurement or calculation of cross sectional data for electron scattering from large molecules, there is limited information on species of biological relevance. However, there is great interest in electron scattering data from large biological molecules such as the pyrimidine bases as an input for studying the effects of radiation damage via Monte Carlo based charged particle track structure simulations. Experimentally, elastic differential cross sections (DCS) for electron scattering from water have been around for some time [4], while data has only

[4] To whom correspondence should be addressed.

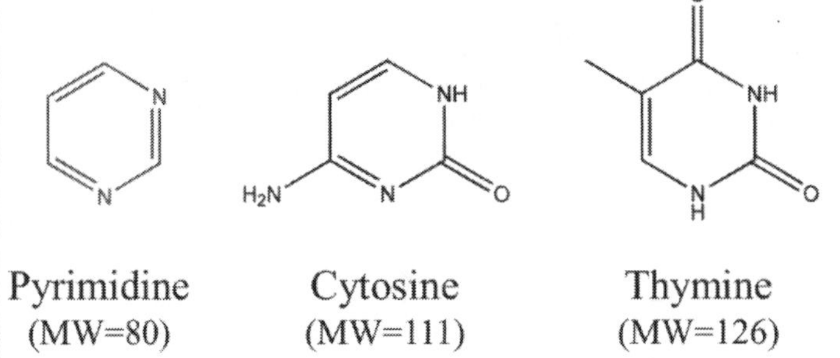

Figure 1. (Colour online) Chemical structure of the pyrimidine molecule and the pyrimidine nucleobases; thymine and cytosine. The molecular weight of each compound is also noted.

recently appeared concerning formic acid [5], variations of the tetrahydrofuran ring [6, 7], alanine [8] and pyrimidine [9]. Due to the difficulties in handling these sorts of targets, theoretical calculation for the large biomolecules is actually preceding experimental investigation. Theoretical studies concerning electron scattering from the DNA bases have been performed by Możejko and Sanche [10] and Blanco and García [11, 12] who have calculated elastic cross sections for all of the DNA and RNA bases, differential in energy as well as in angle. An interesting outcome of both of these studies was the realization that the angular distribution at a given energy for any one of the DNA bases can be related quite accurately to the others by the ratio of their molecular weights or the number of electrons in the target.

The present study reports the first experimental cross sections for the elastic scattering of electrons from two of the DNA bases; cytosine and thymine. Angular distributions are presented for 500 and 100 eV incident electrons, and for angles between 10° and 130°. Theoretical cross sections calculated by the screen-correction additivity rule (SCAR) method are also presented for cytosine and thymine under the same conditions (see [11, 12] for details of the calculation). The measured elastic scattering cross sections are relative, and are only attributed an absolute scale by normalization to theory. Comparison is made with the existing theoretical results by Możejko and Sanche [10]. Due to the structural similarities between the two DNA bases and pyrimidine (see figure 1), cross sections are also compared to the previous experimental study of Maljković *et al.* [9]

2. Experiment

This study has been conducted in a conventional (e, 2e) spectrometer, which has been described previously in the literature [13]. As such only the elements pertinent to the present study shall be documented here.

The elastic differential cross sections are measured via the crossed beam method. Electrons from a conventional electron gun are crossed perpendicularly with the gas target before being energy selected via a hemispherical analyser and detected by a channel electron multiplier. Angular distributions are measured by varying the in-plane detection angle of the elastically scattered electrons whilst maintaining a fixed incident electron energy. As the normal mode of operation for the apparatus is to measure electron impact ionization via coincidence techniques, there is no provision in the experiment for the use of the relative flow technique to establish absolute values for the cross sections.

Both cytosine and thymine are powders at room temperature, and gaseous cytosine and thymine are produced in a molecular beam oven, centrally located in the spectrometer. Through the use of two independent Thermocoax heating elements temperatures in excess of 220°C are achievable with good stability. To avoid deposition of the target materials on elements inside the spectrometer, target gases are trapped via a cold finger that has been installed in the chamber, concentrically mounted above the interaction region. Manufactured from 310-stainless steel with an OFHC copper collection disk at the finger's end, the cold finger is able to be filled externally with liquid nitrogen, and has been observed to be quite effective in the trapping of the DNA bases.

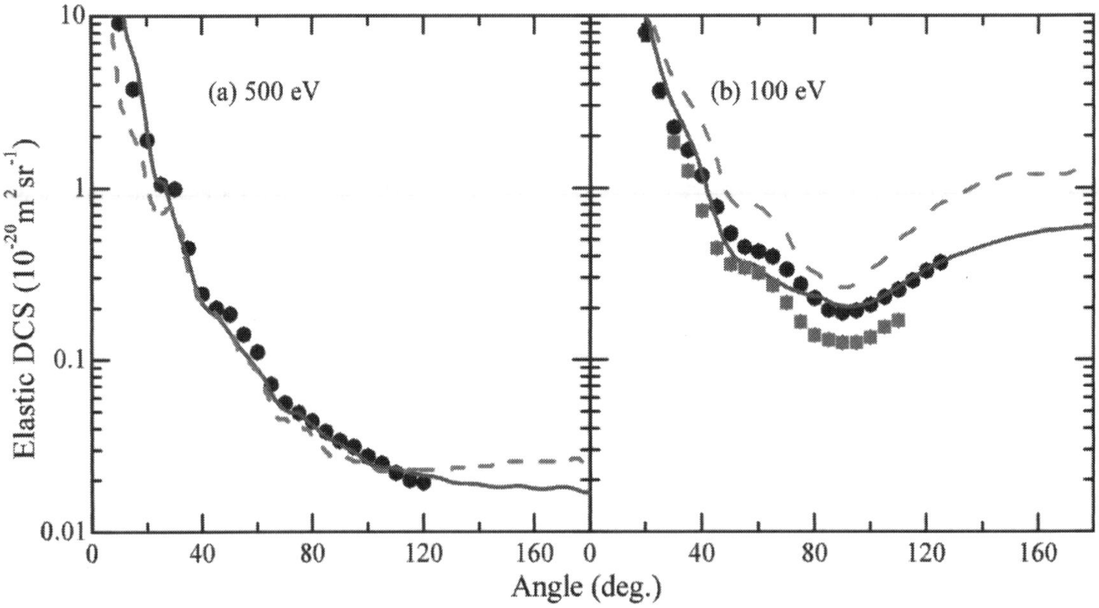

Figure 2. (Colour online) Relative DCS for electron scattering from cytosine (circles) at energies of (a) 500 eV and (b) 100 eV. The present data is attributed absolute value via normalisation to the SCAR calculations (solid line). Also shown are the IAM calculations [10] (dashed line) and the data for pyrimidine [9] (squares). The error associated with the experimental data is less than 10%.

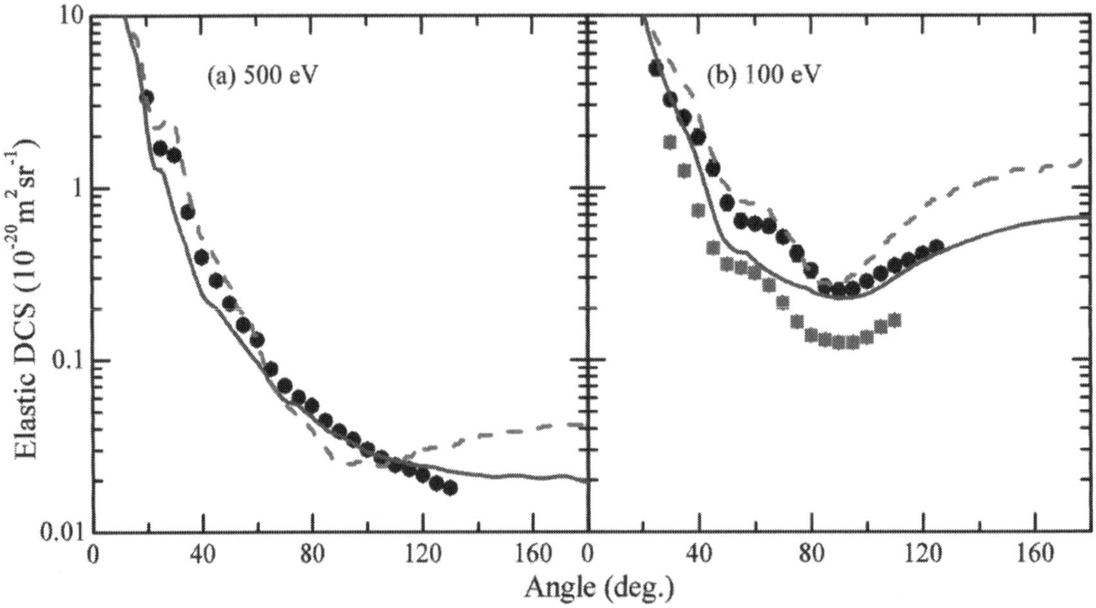

Figure 3. (Colour online) DCS for electron scattering from thymine (circles) at energies of (a) 500 eV and (b) 100 eV. The present data is attributed absolute value via normalisation to the SCAR calculations (solid line). Also shown are the IAM calculations [10] (dashed line) and the data for pyrimidine [9] (squares). The error associated with the experimental data is less than 10%.

3. Results

Relative DCS's for elastic scattering from cytosine and thymine are presented in figures 2 and 3 respectively, for 500 eV (panel (a)) and 100 eV (panel (b)) incident electrons. The data is presented alongside other available experimental and theoretical cross sections.

In figure 2, comparison is made between the present cytosine experimental data and the SCAR calculation. Absolute values are assigned to the measured relative cross sections via normalization to the SCAR calculation. Excellent agreement is seen between the calculated and measured data sets, particularly with regard to the depth and location of the minimum in the 100eV angular distribution (panel (b)). Perhaps the one failing of the SCAR calculation is the underestimation of the shoulder feature in the 100 eV DCS around 60°. The present results are also compared to the published independent atom method (IAM) calculation of Możejko and Sanche [10]. Good agreement is also seen with the IAM calculation; although it tends to overestimate the size of the cross section at backward angles. In contrast to the SCAR method, the IAM results overestimate the size of the shoulder around 60°. The measured cytosine distributions are finally compared to previous absolute experimental data for pyrimidine of Maljković *et al.* [9] at 100 eV (panel (b)). Clearly, the shape of the pyrimidine cross section mimics the recently measured cytosine distribution quite well. The absolute magnitude of the cytosine cross section, assigned by the SCAR calculation, is slightly greater than the pyrimidine cross section, particularly around the DCS's minimum, which can likely be attributed to the previously mentioned difference in molecular weight [11].

Figure 3 compares the present thymine experimental data with the results of the SCAR calculation, with absolute values attributed to the experimental data via normalization to the theory. The SCAR calculation is again quite successful at reproducing the measured cross sections; however the issue with regards to the underestimation of a shoulder in the cross sections at 60° is again present. The thymine results are also compared to the IAM calculation [10], and again good agreement is observed, however the predicted cross sections are still overestimated at backward angles. The IAM results reproduce the shoulder in the cross sections near 60° slightly better in the thymine case. We again compare to the previously published experimental cross sections for pyrimidine [9], for 100 eV incident energy (panel (b)). The shape of the presently measured thymine cross section again correlates quite well with the pyrimidine data; however it is worth noting that the strength of the shoulder near 60° is more pronounced than for both cytosine and pyrimidine. The absolute magnitude of the thymine cross section, as assigned by the SCAR calculation, is greater than both the cytosine and pyrimidine cross sections, which is expected due to its greater molecular weight.

There is a greater difference between the SCAR and IAM calculations for both cytosine and thymine at 100 eV than at 500 eV. As is discussed in detail in ref [11] the IAM does not take into account interactions of the incident electron with more than one atom at a time and ignores geometrical screening corrections for each atom from the rest of the molecule. These are incorporated in an approximate way in the SCAR calculations through the screen corrected additivity rule. At larger electron incident energies atomic cross sections become smaller and the overlapping corrections become smaller. Hence, the inclusion of screening effects becomes less important at higher incident electron energies.

4. Conclusions

The present study reports experimental relative DCS's for elastic electron scattering from cytosine and thymine, at incident electron energies of 500 and 100 eV. Theoretical cross sections produced via the SCAR method are also presented with the experimental cross sections, which are in good agreement with each other, and are also used to attribute absolute scale. Reasonably good agreement is also seen with the theoretical cross sections of Możejko and Sanche [10]. As has been found previously in theoretical studies [10, 11], the relative angular distributions of thymine and cytosine are very similar in shape and they also appear similar to the recent experimental cross sections of Maljković *et al.* [9] on pyrimidine. A shoulder in the angular distributions is shown to be present at approximately 60°, and observed to be stronger for thymine than it is for cytosine. This likeness in the angular dependence of DCS values may be expected due to the similarities in the molecular structures of pyrimidine, thymine and cytosine.

Cross section values for electron impact ionization of DNA are needed for the modelling of radiation damage in biological systems. The present data can be used for energy deposition modelling in biological tissue and studies of radiation damage to biological systems. Further investigation into other incident energies is currently under way.

Acknowledgements
This work was supported by the Australian Research Council Centre of Excellence for Antimatter-Matter Studies. The author CJC would like to acknowledge the support of an Australian Postgraduate Award.

References

[1] Cobut V, Frongillo Y, Patau J P, Goulet T, Fraser M J and Jay-Gerin J P 1998 *Radiat. Phys. Chem.* **51** 229-43
[2] Boudaïffa B, Cloutier P, Hunting D, Huels M A and Sanche L 2000 *Science* **287** 1658-60
[3] Martin F, Burrow P D, Cai Z, Cloutier P, Hunting D and Sanche L 2004 *Phys. Rev. Lett.* **93** 068101-4
[4] Varella M T D N, Bettega M H F, Lima M A P and Ferreira L G 1999 *J. Chem. Phys.* **111** 6396
[5] Vizcaino V, Jelisavcic M, Sullivan J P and Buckman S J 2006 *New J. Phys.* **8** 85
[6] Colyer C J, Vizcaino V, Sullivan J P, Brunger M J and Buckman S J 2007 *New J. Phys* **9** 41
[7] Vizcaino V, Roberts J, Sullivan J P, Brunger M J, Buckman S J, Winstead C and McKoy V 2008 *New J. Phys.* **10** 053002
[8] Marinković B P, Blanco F, Šević D, Pejčev V, García G, Filipović D M, Pavlović D and Mason N J 2008 *International Journal of Mass Spectrometry* **277** 300-4
[9] Maljković J B, Milosavljević A R, Blanco F, Šević D, García G and Marinković B P 2009 *Physical Review A* **79** 052706
[10] Możejko P and Sanche L 2003 *Rad. Env. Biophys.* **42** 201-11
[11] Blanco F and García G 2007 *Physics Letters A* **360** 707-12
[12] Blanco F and García G 2009 *Journal of Physics B: Atomic, Molecular and Optical Physics* **42** 145203
[13] Milne-Brownlie D S, Cavanagh S J, Lohmann B, Champion C, Hervieux P A and Hanssen J 2004 *Phys. Rev. A* **69** 032701-4

Spin-orbit effects in the (e,2e) scattering from a W(110) surface and thin gold layer

SN Samarin[1], OM Artamonov[2], AD Sergeant[1], L Pravica[1], D Cvejanovic[1], P Wilkie[1], P Guagliardo[1], AA Suvorova[3] and JF Williams[1].

[1]ARC Centre of Excellence for Antimatter-Matter Studies, The University of Western Australia, Perth, WA 6009, Australia,
[2]Research Institute of Physics, St. Petersburg University, St.Petersburg, Russia.
[3]CMCA, The University of Western Australia, Perth, WA 6009, Australia

samar@physics.uwa.edu.au

Abstract. Spin-effects were studied in the (e,2e) scattering from W(110) single crystal and thin gold layer on W(110). Intensity asymmetry observed in the spectra using spin-polarised incident beam were identified as spin-orbit interaction in the valence electron states involved in the scattering with the contribution of orbital momentum of incident electrons.

1. Introduction

Incident kinetic electrons impinging on a solid surface can be scattered elastically and inelastically from the surface. The energy distribution of scattered and ejected electrons forms a secondary emission spectrum, which consists of: i) a sharp maximum of elastically scattered electrons, ii) a part of the spectrum that contains electrons scattered with some energy loss, and finally iii) a broad and almost featureless maximum of "true secondary" or "ejected" electrons. In conventional single-electron spectroscopy all these parts of the spectrum are analyzed separately and independently. However there might be an internal relationship between parts ii) and iii). This relationship can be revealed if two electrons resulting from the scattering of an incident electron from a surface are detected and analyzed. This type of experiment is referred to as (e,2e) scattering experiment [1,2] and the electron spectroscopy involving detection and analysis of two electrons represents a simplest example of a multi-electron spectroscopy – two-electron spectroscopy of surfaces. When the incident electron beam is spin-polarized then two spectra can be recorded: i) (I^+) - with the polarization vector of the primary beam parallel to the quantization axis and ii) (I^-) – with polarization vector anti-parallel to the quantization axis. We chose the quantization axis to be perpendicular to the scattering plane which contains the normal to the surface and two detectors (figure 1).

In the case of a heavy metal (for example, W or Au) the spin-orbit interaction in the incident electron state inside the sample as well as the spin-orbit interaction in the valence electron state of the sample can be visualized by the spin-polarized two-electron spectroscopy. It allows to study the spin-dependent electron scattering dynamics as well as spin-dependent electronic structure of surfaces.

In this paper we present a comparative analysis of spin-dependent features in the (e,2e) spectra of the W(110) crystal and Au(111) film on the W(110) substrate.

2. Experimental details

Experimental set-up consists of a scattering chamber and a spin-polarized electron source chamber connected through a valve. In both chambers the vacuum is maintained in the 10^{-11} Torr range. The scattering chamber is equipped with instruments for preparation and characterization of surfaces and thin films such as: Low Energy Electron Diffraction (LEED), Auger Electron spectrometer, ion sputtering gun, electron beam evaporators, mass-spectrometer, and quartz microbalance.

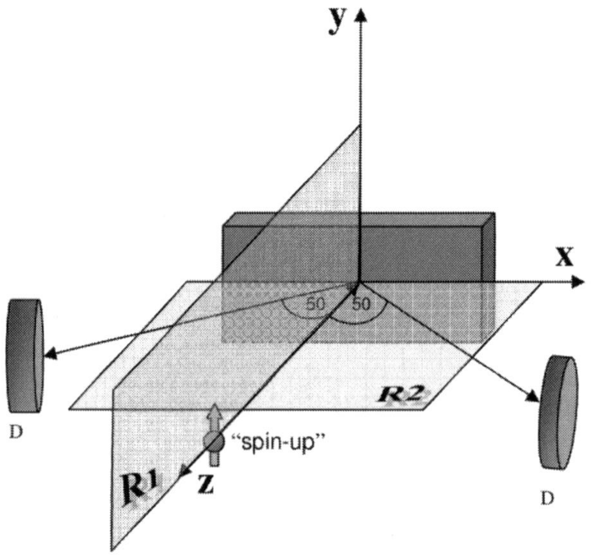

Figure 1. Geometry of the experiment. **R1** and **R2** are the symmetry planes. The $[1\bar{1}0]$ direction of the crystal runs along **y** axis.

The geometrical arrangement of the experiment is shown in figure 1. The (e,2e) spectrometer is described elsewhere [3]. In brief, the incident electron impinges onto the sample surface and two outgoing electrons are detected in coincidence and their momenta are measured using time-of-flight technique [4].This technique requires a reference point on the time scale. Therefore the incident beam is pulsed with the pulse width of less than 1 ns and repetition rate of 4 MHz. The coincidence electronics allows the detection of two time-correlated electrons within the time window of 200 ns. To ensure that the two electrons are resulting from an interaction of a single incident electron with the sample a very low incident current (~ 10^{-14} A) was used providing less than one electron per pulse in average.

For the electron detection we used two position-sensitive detectors based on 75 mm in diameter Micro Channel Plates (MCP). Position sensitivity allows the measurement of the electron angular distribution and the correction of the electron flight distance for the time-of-flight energy measurement [4].

The spin-polarized electron source is based on the photo-emission from a strained GaAs photo-cathode with negative affinity. Photoelectrons excited by circularly polarized light with the wavelength 836 nm are initially longitudinally polarized. They pass through a 90° deflector such that the emergent beam is now transversely polarized. The polarization P of the electron beam is measured in a separate experiment and is estimated to be (66 ± 2) %. It can be reversed from P to –P by changing the sense of circular polarization of the laser light incident on the GaAs photo-cathode. The polarization of the incident beam is chosen to be perpendicular to the scattering plane.

When the coincidence conditions between the two arms of the (e,2e) spectrometer are switched off each of its arms can be used as a Spin-Polarized Electron Energy Loss Spectrometer for recording electron energy loss spectra at two mirror symmetric angles.

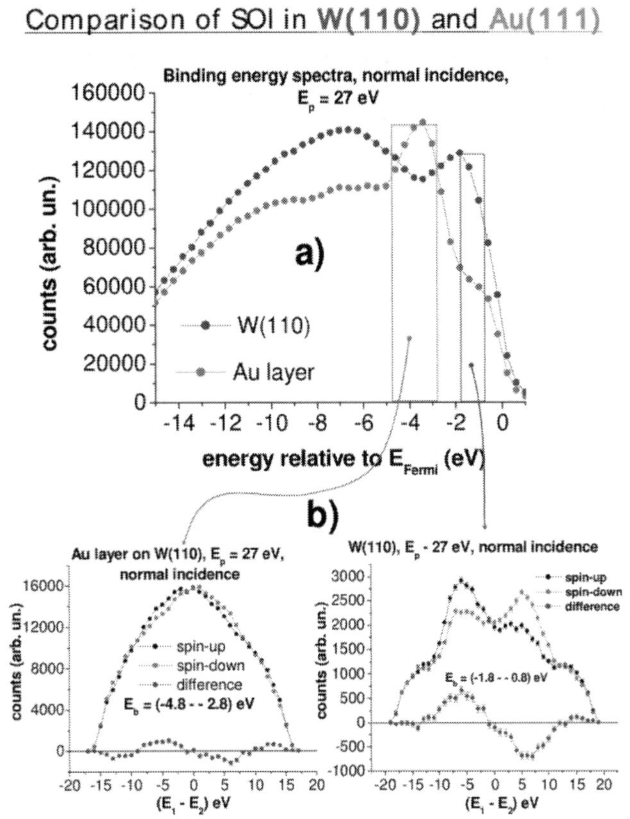

Figure 2. Binding energy distributions for W(110) and Au(111) – a). Corresponding spin-resolved energy sharing distributions – b).

As samples we studied a W(110) surface and thin films of Au grown on this surface.

The W(110) crystal was mounted on a rotatable manipulator with the $[1\bar{1}0]$ direction along the rotational axis and perpendicular to the scattering plane that contains the normal to the sample surface and the detectors (see figure 1).

The tungsten crystal was cleaned in vacuum prior to measurements and to a gold film deposition using a standard procedure [5] including oxygen treatment at 10^{-7} Torr oxygen pressure and 1400 K sample temperature followed by a few high-temperature flashes up to 2300 K. The cleanliness of the surface was monitored by Auger Electron Spectroscopy and Low Energy Electron Diffraction (LEED).

The Au film was evaporated from a small piece of gold wire melted on a V-shaped tungsten filament heated resistively. The layer-by-layer growth of such a film was confirmed by MEED oscillations [6]. The quality of the film was demonstrated by observing quantum well states in such a layer of gold using photoelectron spectroscopy [7]. The thickness of the gold layer (2 to 4 ML in our experiment) was estimated using Auger peak intensity as a function of deposition time as well as

LEED technique and comparison with the previous LEED study of the epitaxy of gold on (110) surface of tungsten [8].

3. Experimental results and discussion

Two high-Z metals were chosen to study spin-orbit interaction in the (e,2e) reaction: i) single crystal of W(110) and ii) thin film of Au deposited on the W(110) substrate. To explain the way the experimental results are presented we recall that the energy conservation law in the (e,2e) reaction implies that $E_b = (E_1 + E_2) - E_0$, where E_b is the binding energy of the valence electron, E_0 is the primary electron energy and E_1 and E_2 are the energies of the two detected electrons. Besides, the momentum conservation for the component parallel to the surface implies that $k_{b\parallel} = k_{1\parallel} + ng_{\parallel} + k_{2\parallel} - k_{0\parallel}$ (indices $_{b,1,2,0}$ have the same meaning as for energies, $g_{\parallel}$ - surface reciprocal lattice vector and n = 0 for energies and geometry used in our experiments). The number of correlated pairs plotted as a function of the x-component of the bound electron momentum is called K_x – distribution. Thus, energy and momentum conservation allow the valence electron involved in the collision to be localized in energy-momentum space in contrast to a single-electron spectroscopy (EELS), where the result of a scattering event is integrated over the unresolved second electron state.

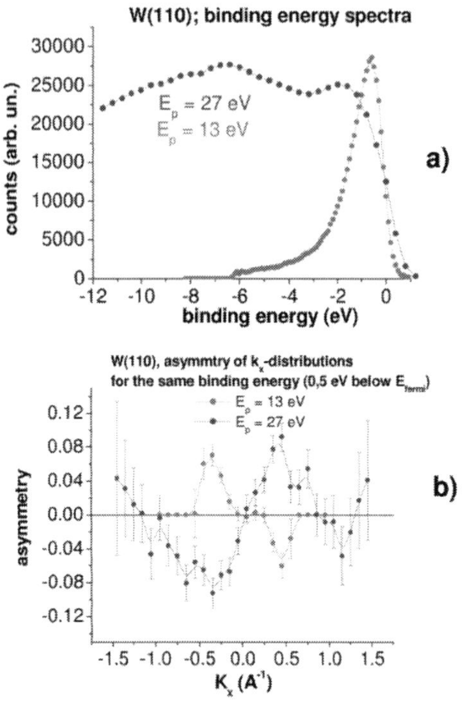

Figure 3. Binding energy spectra for W(111) measured at different primary energies (a) and corresponding asymmetry spectra of K_x - distributions (b).

Our experimental (e,2e) spectra I^+ and I^- measured for two polarizations of the incident beam can be presented in various coordinate systems. Then the asymmetry defined as $A = (I^+ - I^-)/(I^+ + I^-)$ can be calculated. We present in figure 2 the binding energy spectra (a) for tungsten and the gold layer excited by 27 eV spin-polarized primary electrons. For a particular binding energy (band of binding energy) one can plot an energy sharing distribution. These distributions can be constructed for two opposite polarizations of the incident beam and then the asymmetry can be calculated. In such a way the asymmetry of these distributions for various binding energies were calculated. It was found that the most pronounced asymmetry for W(110) is located at about 1 eV below the Fermi level and for Au film it is located at about 4 eV below the Fermi level. The spin-resolved energy sharing distributions and the spin-difference spectra for the tungsten and gold layer are shown in figure 2b. One can see that the sharing distributions for these two samples are different. On the other hand the shapes of the spin-difference curves are quit similar: in both cases the curves change sign at $E_1 - E_2 = 0$. This shape is typical for spin-orbit interaction and consistent with the symmetry consideration. Indeed, the reflection in the R1 symmetry plane (figure 1) reverses the spin of the incident electron (because the spin is an axial vector) and interchanges of the two outgoing electrons. It implies that $I^+(E_1 - E_2) = I^-(E_2 - E_1)$. In other words the spin-asymmetry will remain the same if we mirror reflect the curve in R1 plane and then flip it vertically. This "left-right" symmetry of the curves is typical for spin-orbit-induced asymmetry.

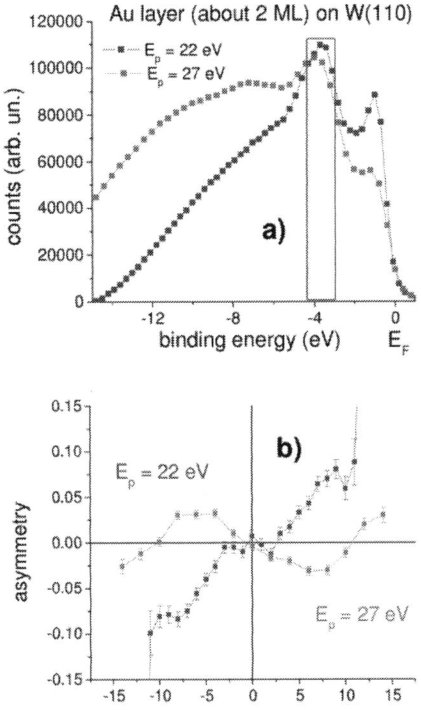

Figure 4. Binding energy spectra of Au(111) film recorded at different primary energies (a) and corresponding asymmetry spectra (b) at binding energy shown by black rectangular in (a).

We note here that in the case of tungsten a substantial density of states below the Fermi level (where the maximum of asymmetry is observed) is formed by d-states, whereas in the Au film the d-states start to contribute to the density of states only at 4 eV below the Fermi level (where the maximum asymmetry is observed for the gold layer). Thus, the energy location of the observed asymmetry indicates that the spin-orbit effect shows up when the d-states of the valence band are involved in the (e,2e) scattering.

Spin-orbit effects in the (e,2e) reaction on W(100) were analyzed in [9]. In this work the collision of an incident electron with a valence electron of the target was treated in a distorted-wave Born approximation formalism with exchange, and the four relevant quasi-one-electron states being solutions of the Dirac equation. To obtain detailed insight into the origin of the individual features of the (e,2e) spectra of W(100) and their spin asymmetry, additional calculations were performed where each of the four relevant wave-functions was modified artificially such that spin-orbit coupling and the elastic scattering amplitudes from the ion-core planes were selectively switched off. The calculated spin asymmetry of the (e,2e) cross-section on W(100) was found to be due to the spin-orbit coupling in the valence electron states mainly. Even though these results are obtained for a W(100) surface the approach and qualitative explanation of the spin-asymmetry in the (e,2e) spectra must be valid for the W(110) surface and thin Au(111) film. Our intension was to check to what extent the spin-asymmetry in the energy sharing distributions does not depend on the primary electron energy and is only determined by the spin-orbit interaction in the valence state of the target.

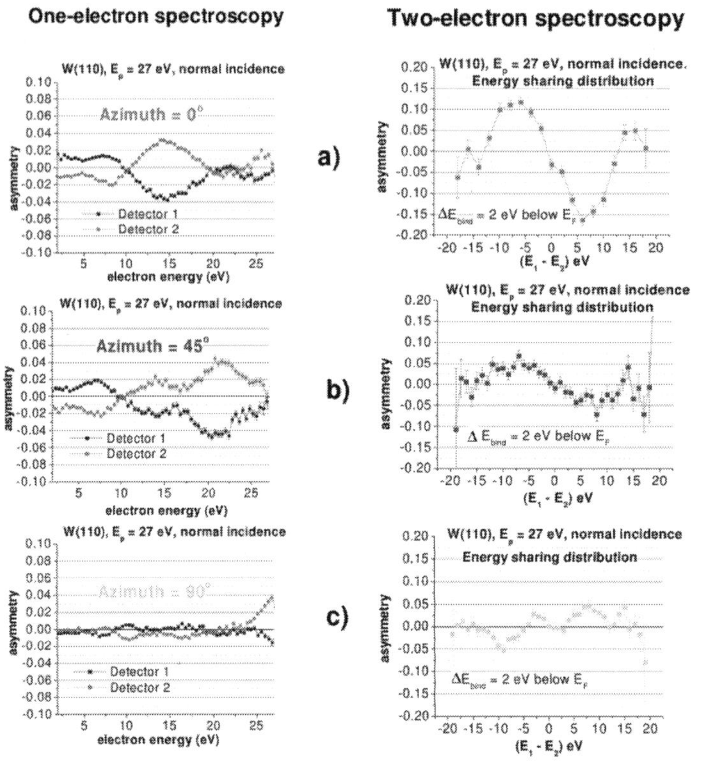

Figure 5. Asymmetries of single-electron spectra (left column) and two-electron spectra (right column) of a W(110) crystal for various azimuth orientations. Azimuth = 0 corresponds to the crystal orientation of figure 1.

We measured (e,2e) spectra on W(110) and Au(111) at various primary energies and analyzed spin-asymmetry in the energy sharing distributions (or K_x – distributions). In the case of tungsten the shape of asymmetry in the energy sharing distributions (at binding energy just below the Fermi level) remains the same in a wide energy range from 32 eV down to 15 eV, although the size of asymmetry changes. However when the primary energy decreased to 13 eV the asymmetry changed the sign (see figure 3). In the case of the Au film the change of the sign in the spin-asymmetry occurs even at higher energy, about 22 eV (see figure 4).

These results indicate that the spin-orbit-induced asymmetry in the (e,2e) spectra of W(110) and Au(111) is not determined by the spin-orbit interaction in the valence state of the target only, but also by a contribution of the orbital momentum of the incident electron.

For the two-fold symmetry surface of W(110) we measured single-electron spectra as well as coincidence spectra for various azimuthal orientation of the surface for primary energy of 27 eV and normal incidence. It turns out that the spin-asymmetry in both single- and two-electron spectra depend on the profile of the potential in the scattering plane (figure 5). The comparison of spectra a) and c) shows that at some electron energies (or combination of the energies in the electron pair) the asymmetry even changes the sign when the azimuth of the sample orientation changes by 90°. The comparative analysis of the asymmetry in the single- and two-electron spectra shows that if the asymmetry is observed in a single-electron spectrum at particular secondary electron energy E_1, then in the corresponding coincidence spectrum the asymmetry shows up in the energy sharing distribution at the combination of energies where one of the electrons has the energy E_1. Thus the spin features of the secondary emission can be linked to the spin features of the energy loss part of the secondary emission spectrum.

4. Conclusions

Experimental results presented in the paper show that the spin-orbit-induced asymmetry in the energy sharing distribution:

- depends on the binding energy of the valence electron involved in the collision;
- sensitive to the anisotropy of the scattering potential;
- depends on the primary electron energy;

The spin-orbit-induced asymmetry observed in a single-electron spectrum (secondary emission spectrum) is observed also in the sharing distribution of the (e,2e) spectrum but with larger value. These findings allow drawing the following qualitative picture of the spin-dependent effects in the (e,2e) scattering from non-magnetic heavy-atom metals. An incident electron shares orbital momentum with the valence electron of the target forming a "quasi-particle" while scattering back to the surface. The "spin-filter" effect [10,11] on the surface potential barrier of the sample contributes to the observed intensity asymmetry in the (e,2e) spectra.

5. Acknowledgments

This research was supported by the Australian Research Council and the University of Western Australia. OMA is grateful to the RFBR (10 – 02 – 00547 - a) for financial support.

References

[1] Kirschner J, Artamonov OM, Terekhov AN, *Phys. Rev. Lett.* 69 (1992) 1711
Kirschner J, Artamonov OM, Samarin SN *Phys. Rev. Lett.* 75 (1995) 2424
[2] Iacobucci S, Marassi L, Camilloni R, Nannarone S, Stefani G, *Phys. Rev.* B51 (1995-II) 10252
[3] Artamonov OM, Samarin SN, Kirschner J *Appl. Phys.* A65 (1997) 535
[4] Samarin SN, Artamonov OM, Waterhouse DK, Kirschner J, Morozov A, Williams JF, *Rev. Sci. Instr.* 74 (2003) 1274
[5] Cortenraad R, Ermolov SN, Semenov VN, Denier van der Gon A. W, Glebovsky VG, Bozhko SI, and Brongersma HH, *J. Crystal Growth* **222** (2001) 154
Zakeri Lori K, Peixoto TRF, Zhang Yu, Prokop J, Kirschner J, *Surface Science* **604** (2010) L1
[6] Knoppe H and Bauer E, *Phys. Rev. B* 48 (1993) 5621
[7] Shikin AM, Rader O, Prudnikova GV, Adamchuk VK, and Gudat W, *Phys. Rev. B* 65 (2002) 075403
[8] Augustus PD and Jones JP, *Surf. Sci.* **64** (1977) 713

[9] Gollisch H, Xiao Yi, Scheunemann T and Feder R, *J. Phys.C: Condens Matter* 11 (1999) 9555
[10] Oepen HP, Hünlich K, and Kirschner J, *Phys. Rev. Lett.* 56 (1986) 496
[11] Samarin S, Artamonov OM, Sergeant AD, and Williams JF, *Surf. Sci.*, 579/2-3 (2005) 166

Journal of Physics: Conference Series **288** (2011) 012016 doi:10.1088/1742-6596/288/1/012016

Auger electron spectra of hydrogenated Si(111)-1×1 surface obtained from Si-$L_{23}VV$ Auger electron Si-2p photoelectron coincidence measurements

Takahiko Yamazaki[1,*], Shogo Hashimoto[1], Takuhiro Kakiuchi[2], Kazuhiko Mase[3,4] and Masatoshi Tanaka[1]

[1] Department of Physics, Faculty of Engineering, Yokohama National University, 79-5 Tokiwadai, Hodogaya-ku, Yokohama 240-8501, Japan

[2] Department of Chemistry, Faculty of Science, Ehime University, 2-5 Bunkyo-cho, Matsuyama 790-8577, Japan

[3] Institute of Materials Structure Science, KEK, 1-1 Oho, Tsukuba 305-0801, Japan

E-mail: mase@post.kek.jp

Abstract. The Si-$L_{23}VV$ Auger electron Si-2p photoelectron coincidence spectra (Si-$L_{23}VV$-Si-2p APECS) of a hydrogenated Si(111)-1×1 surface are measured with an analyzer that is used for electron-electron and electron-ion coincidence measurements of surfaces. The Si-$L_{23}VV$-Si-2p APECS show a sharp peak, a small shoulder, and a broad large peak at electron kinetic energies of about 85, about 80.5, and about 70 eV, respectively. The clarity of the last two peaks is lower in the Si-$L_{23}VV$-Si-2p APECS of a clean Si(111)-7×7 surface. We attribute the three peaks to the Si-$L_{23}V_{3p}V_{3p}$, Si-$L_{23}V_{3p}V_{3s}$, and Si-$L_{23}V_{3s}V_{3s}$ Auger lines, respectively. The three-peak structure indicates that the local density of states of the 3s + 3p band is reduced in the case of the hydrogenated Si(111)-1×1 surface. This finding is consistent with that of a previous study on valence electronic states of hydrogenated Si(111) using ultraviolet photoemission spectroscopy.

1. Introduction

Auger photoelectron coincidence spectroscopy (APECS) has been widely used to study dynamics induced by core-level excitations of surfaces [1–5]. In particular, core-valence-valence APECS is a unique technique to obtain information on local valence electronic structures at a site with a specific chemical environment that causes a certain shift of the core-level photoelectron peak [6, 7]. APECS is

[*] Present address: Fuji Xerox Co. Ltd., Takematsu Center, 1600 Takematsu, Minamiashigara 250-0111, Japan

[4] To whom any correspondence should be addressed.

more surface-sensitive than the conventional synchrotron radiation core-level photoelectron spectroscopy (SR-CL-PES) and Auger electron spectroscopy (AES) because the escape depth (ED_{APECS}) of electrons detected by APECS is given by the equation $\frac{1}{ED_{APECS}} = \frac{1}{\lambda_{Pe}\cos\theta_{Pe}} + \frac{1}{\lambda_{Ae}\cos\theta_{Ae}}$, where λ and θ denote the inelastic mean free path and the emission angles of the detected electrons, respectively [2]. The subscripts Pe and Ae denote photoelectron and Auger electron, respectively. In the case of Si-$L_{23}VV$ Auger electron Si-2*p* photoelectron coincidence spectra (Si-$L_{23}VV$-Si-2*p* APECS) at $h\nu$ = 130 eV, the value of ED_{APECS} is ~1.3 Å [8]. Thus, Si-$L_{23}VV$-Si-2*p* APECS is a unique technique to investigate the valence electronic structures of topmost silicon surfaces [8, 9]. In the present paper, we report a study of a hydrogenated Si(111)-1×1 surface using Si-$L_{23}VV$ Auger electron Si-2*p* photoelectron coincidence measurements.

2. Experimental

We use a coincidence analyzer that has been developed for APECS as well as for electron-ion coincidence (EICO) spectroscopy [10]. It consists of a coaxially symmetric electron energy analyzer (ASMA), double-pass cylindrical mirror electron energy analyzer (DP-CMA), time-of-flight ion mass spectrometer (TOF-MS), magnetic shield, and positioning system. The ASMA was originally proposed by Siegbahn *et al.* [11]; the electron acceptance angle from the surface normal (α) is 59.5° ± 11.5° (solid angle (Ω) is 1.81 sr) and the energy resolution ($E/\Delta E$) is about 55. For the DP-CMA, α and $E/\Delta E$ are 37.5° ± 4° (Ω= 0.43 sr) and ~55, respectively. The ASMA and DP-CMA are positioned such that they share the same axis and the same focus point. The TOF-MS is installed in the DP-CMA.

An *n*-type Si(111) single-crystal wafer with a resistivity of 0.02 Ω cm was used as the sample substrate. A clean Si(111)-7×7 surface was prepared by direct-current heating at ~1100°C for 5 s and annealing at ~900°C for 8 s in an ultrahigh vacuum chamber with a base pressure of about 1.5 × 10^{-8} Pa. A sharp 7×7-pattern was observed by low-energy electron diffraction (LEED). Subsequently, a hydrogenated Si(111)-1×1 surface was prepared by exposing the Si(111)-7×7 surface to atomic hydrogen at a substrate temperature of 350°C. The atomic hydrogen was generated by a hot tungsten filament with a diameter of 0.6 mm (figure 1) that was heated to 1800°C with a direct current of 17 A. Hydrogen gas of 1.3 × 10^{-3} Pa was supplied to the hot filament for 100 s. The distance between the tungsten filament and the sample was about 140 mm. The filament was surrounded by a nipple made of forged 0.2% beryllium copper alloy to suppress outgassing [12]. A sharp 1×1-pattern was observed by LEED (figure 2). The sample was irradiated with *p*-polarized SR at an incident angle of 84° relative to the surface normal at beamline 1C of the SR facility at KEK (Photon Factory, PF). The photon

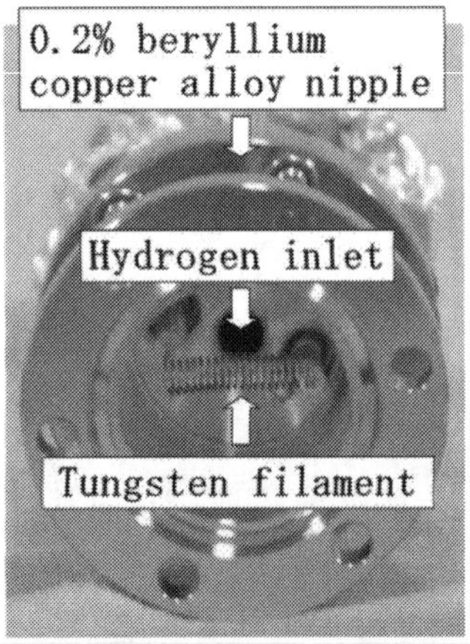

Figure 1. Tungsten filament with a diameter of 0.6 mm that was use to generate atomic hydrogen.

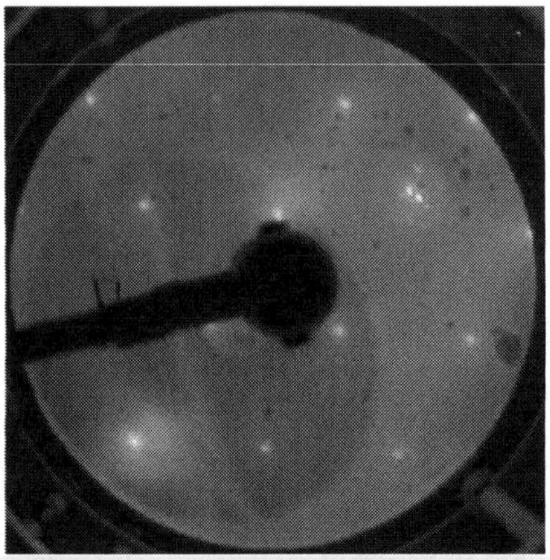

Figure 2. LEED pattern of the hydrogenated Si(111)-1×1 surface.

energy was fixed at 130 eV, with a typical energy resolution ($E/\Delta E$) greater than 2000. The spot size on the Si surface was about 0.1 mm × 4 mm (vertical × horizontal).

3. Results and discussion

Figure 3 shows the Si-$L_{23}VV$ AES and Si-2p PES of the hydrogenated Si(111)-1×1 surface at $h\nu$ = 130 eV. A sharp Si-2p photoelectron peak appears at a kinetic energy of 25.8 eV, while Si-$L_{23}VV$ Auger electron peaks are observed in the kinetic energy range 60–90 eV. Si-2p photoelectrons are used as the trigger for the measurement of the Si-$L_{23}VV$-Si-2p APECS.

The Si-$L_{23}VV$-Si-2p APECS of the hydrogenated Si(111)-1×1 surface (figure 4) show a sharp maximum peak, a small shoulder, and a broad large peak at Auger electron kinetic energies (*AeKE*s) of about 85, about 80.5, and about 70 eV, respectively. We attribute the three peaks to the Si-$L_{23}V_{3p}V_{3p}$, Si-$L_{23}V_{3p}V_{3s}$, and Si-$L_{23}V_{3s}V_{3s}$ Auger lines, respectively. The clarity of the last two peaks is lower in the Si-$L_{23}VV$-Si-2p APECS of the clean Si(111)-7×7 surface (figure 5) [10]. The smooth structure with a sharp peak at the *AeKE* of ~87.5 eV can be explained by the formation of 3s, 3s + 3p, and 3p bands of the Si(111)-7×7 surface [13, 14]. The three-peak structure in the Si-$L_{23}VV$-Si-2p APECS of the hydrogenated Si(111)-1×1 surface indicates that the local density of states of the 3s + 3p band is reduced after hydrogenation. This finding is consistent with that of a previous study on the valence electronic states of hydrogenated Si(111) using ultraviolet photoemission spectroscopy (UPS) [15]; in the study, the valence peak with a binding energy of 9–12 eV, corresponding to the 3s + 3p band, was found to be relatively suppressed. In the present study, the *AeKE* of the maximum Si-$L_{23}V_{3p}V_{3p}$ peak shifts from ~87.5 eV to ~85 eV when Si(111) is hydrogenated. This result is consistent with that of the previous study, which reported the shift of the Si-3p band to higher binding energies upon hydrogenation [15]. The proposed Auger processes are schematically shown in figure 6.

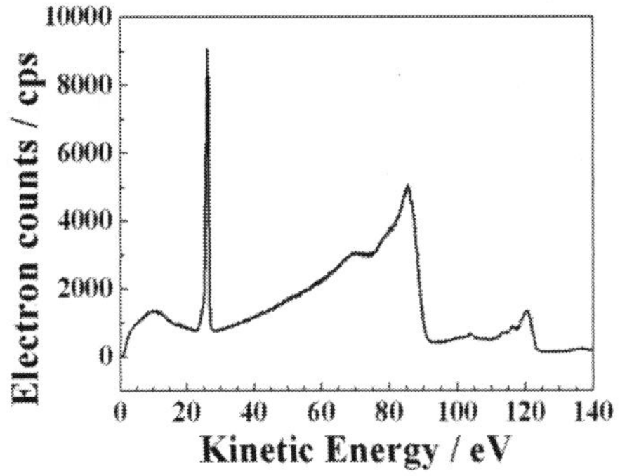

Figure 3. Si-$L_{23}VV$ AES and Si-2p PES of the hydrogenated Si(111)-1×1 surface at $h\nu$ = 130 eV.

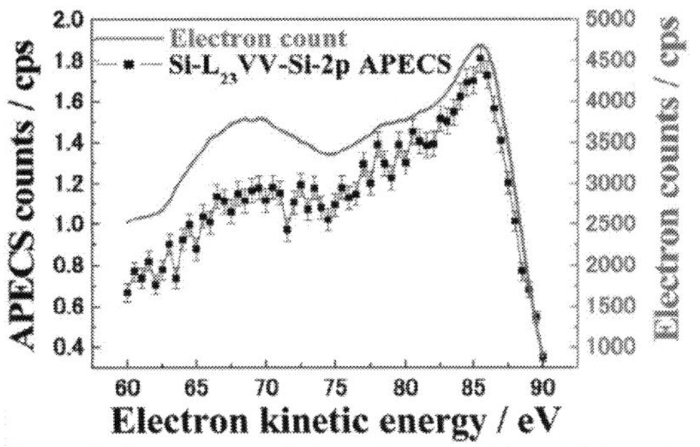

Figure 4. Si-$L_{23}VV$ Auger electron Si-2p photoelectron coincidence spectrum (Si-$L_{23}VV$-Si-2p APECS, •) and the Si-$L_{23}VV$ Auger electron spectrum (Si-$L_{23}VV$ AES, ——) of the hydrogenated Si(111)-1×1 surface at $h\nu$ = 130 eV.

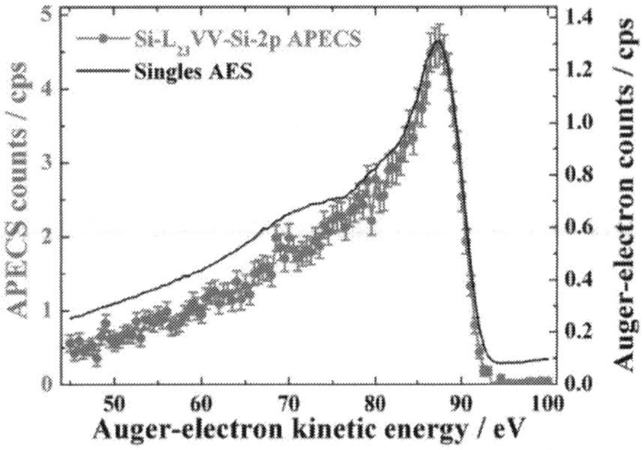

Figure 5. Si-$L_{23}VV$ Auger electron Si-2p photoelectron coincidence spectrum (Si-$L_{23}VV$-Si-2p APECS, ●) and the Si-$L_{23}VV$ Auger electron spectrum (Si-$L_{23}VV$ AES, ——) of a clean Si(111) surface at $h\nu$ = 130 eV [10].

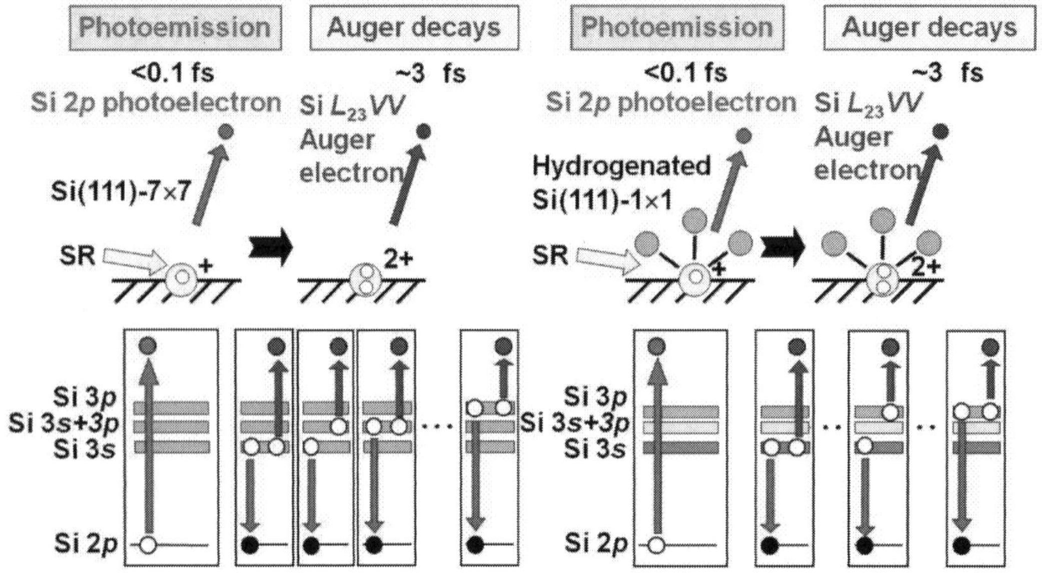

Figure 6. Proposed Si-$L_{23}VV$ Auger processes of the Si(111)-7×7 surface and hydrogenated Si(111)-1×1 surface.

4. Conclusions

Auger photoelectron coincidence spectroscopy (APECS) is used to investigate valence electronic structure of hydrogenated Si(111)-1×1. The Si-$L_{23}VV$-Si-2p APECS of hydrogenated Si(111)-1×1 show a sharp peak, a small shoulder, and a broad large peak at electron kinetic energies of about 85, about 80.5, and about 70 eV, respectively. We attribute the three peaks to the Si-$L_{23}V_{3p}V_{3p}$, Si-$L_{23}V_{3p}V_{3s}$, and Si-$L_{23}V_{3s}V_{3s}$ Auger lines, respectively. The three-peak structure indicates that the local density of states of the $3s + 3p$ band is reduced in the case of the hydrogenated Si(111)-1×1. These findings are consistent with that of a previous study on valence electronic states of hydrogenated Si(111) using ultraviolet photoemission spectroscopy [15].

Acknowledgments

We express our sincere thanks to the members of the Photon Factory for their valuable help during the course of the experiments. This work was partly supported by PRESTO (Structure Function and Measurement Analysis) from the Japan Science and Technology Agency (JST). This work has been performed under the approval of the Photon Factory Program Advisory Committee (PF PAC Nos. 2006S2-002 and 2009G222).

References

[1] Haak H W, Sawatzky G A and Thomas T D 1978 *Phys. Rev. Lett.* **41** 1825

[2] Sawatzky G A 1988 *Auger Electron Spectroscopy* ed C L Briant and R P Messer (San Diego: Academic Press) chapter 5, pp 167-243

[3] Stefani G, Iacobucci S, Ruocco A and Gotter R 2002 *J. Electron Spectrosc. Relat. Phenom.* **127** 1

[4] Berakdar J and Kirschner J ed 2004 *Correlation Spectroscopy of Surfaces, Thin Films, and Nanostructures* (Weinheim: Wiley-VCH)

[5] Riessen G, Wei Z, Dhaka R S, Winkler C, Schumann F O, and Kirschner J 2010 *J. Phys.: Condens. Matter* **22** 092201

[6] Jensen E, Bartynski R A, Weinert M, Hulbert S L, Johnson E D and Garrett R F 1990 *Phys. Rev. B* **41** 12468

[7] Bartynski R A, Yang S, Hulbert S S L, Kao C -C, Weinert M and Zehner D M 1992 *Phys. Rev. Lett.* **68** 2247

[8] Kakiuchi T, Hashimoto S, Fujita N, Tanaka M, Mase K and Nagaoka S 2010 *J. Phys. Soc. Jpn.* **79** 064714

[9] Werner W S M, Smekal W, Störi H, Winter H, Stefani G, Ruocco A, Offi F, Gotter R, Morgante A and Tommasini F 2005 *Phys. Rev. Lett.* **94** 038302

[10] Kakiuchi T, Hashimoto S, Fujita N, Mase K, Tanaka M and Okusawa M 2008 *J. Vac. Soc. Jpn.* **51** 749

[11] Siegbahn K, Kholine N and Golikov G 1997 *Nucl. Instrum. Meth. Phys. Res. A* **384** 563

[12] Watanabe F 2004 *J. Vac. Sci. Technol. A* **22** 181, 739

[13] Pandey K C 1976 *Phys. Rev. B* **14** 1557

[14] Cohen M L 1980 *J. Phys. Soc. Jpn.* **49** Suppl. A 13

[15] Fujiwara K 1982 *Phys. Rev. B* **26** 2036

Journal of Physics: Conference Series **288** (2011) 012017 doi:10.1088/1742-6596/288/1/012017

Elastic scattering and impact ionization by returning electrons induced in a strong laser field

Junliang Xu[1], Yaqiu Liang[1,2], Zhangjin Chen[1] and C. D. Lin[1]

[1] J. R. Macdonald Laboratory, Physics Department, Kansas State University, Manhattan, Kansas 66506-2604, USA

[2]College of Physics, Liaoning University, Shenyang 110036, People's Republic of China

Abstract. We study elastic scattering and impact ionization by returning electrons induced in a strong laser field based on the recently developed quantitative rescattering theory. The high-energy angle-resolved photoelectron spectra of molecules in strong fields are attributed to elastic scattering of the returning electrons with the parent ion. We demonstrate that the high-energy photoelectron spectra can be used to retrieve the position of atoms in a molecule. We also investigate the nonsequential double ionization of an atom in intense laser fields which is partly due to the impact ionization of the parent ion by the returning electron.

1. Introduction

The rescattering model proposed by Corkum [1] and Krause *et al* [2] is one of the most significant advances in strong laser physics. According to this model, an electron that was released earlier in the laser pulse may return to recollide with the target ion. From the recollision, many possible processes can occur. The elastic scattering experienced by the returning electron accounts for the high energy photoelectron spectra, while the nonsequential double ionization (NSDI) is due to inelastic scattering of the returning electron by the parent ion.

High energy photoelectrons for atomic targets have been well studied based on a recently developed quantitative rescattering (QRS) model [3, 4, 5]. It has been demonstrated that the two-dimensional (2D) photoelectron momentum distributions can be separated into a returning wave packet multiplied by laser-free electron elastic differential cross section (DCS). This separation makes the evaluation of high energy photoelectron spectra much easier than the traditional theoretical method, say, the solution of time-dependent Schrödinger equation (TDSE), and much more accurate than the strong field approximation (SFA). More significantly, it also allows one to retrieve atomic structure information [6, 7, 8, 9]. It is well known that the highest returning electron energy is $3.2U_p$, where U_p is the ponderomotive energy. For a typical laser pulse at an intensity of 10^{14} W/cm^2 and wavelength of 800 nm, the returning electron energy is less than 20 eV. The elastic differential cross sections at such low energies are sensitive to the specific ion, and hence carries target structure information.

In this paper, we apply the QRS model to molecular targets. However, the retrieval of molecular structure is not straightforward since accurate evaluation of elastic differential cross sections for laser-free electrons with the molecular ion is extremely difficult, especially for low incident energies. Fortunately, we will show that the independent atom model (IAM) is valid for DCS at large angles for molecular scattering at energies as high as 100 eV, corresponding to

$U_p \sim 30$ eV. In experiments, this high ponderomotive energy can be achieved by using a laser pulse with long wavelength while keeping the intensity as low as possible to avoid saturation.

In another rescattering process, NSDI can occur if the returning electron knocks out another electron in the parent ion directly, similar to the (e, 2e) process in field-free collisions. This process represents a unique example where electron-electron interaction plays an essential role in the presence of a strong laser field. While these phenomena have been understood qualitatively based on the rescattering model, a full quantum mechanical theory for the complete quantitative description of the NSDI, even for helium atom target, is still desirable. In this paper, we apply the QRS model to the NSDI of helium in an intense laser field. The experimental measurements on NSDI which provide the most detailed information are the correlated two-electron momentum spectra. To carry out numerical simulation on the correlated momentum spectra, we first calculate the returning electron wave packet by using QRS. Then by multiplying this wave packet with standard field-free (e, 2e) differential cross sections, we obtain the correlated two-electron momentum spectra. Actually, besides the (e, 2e) process, another mechanism involved in the NSDI is excitation-tunneling, in which the ground state electron in the parent ion is excited to a higher energy bound state, followed by subsequent tunnel ionization. The investigation of excitation-tunneling in NSDI will be reported elsewhere.

Atomic units are used in this paper unless otherwise noted.

2. Theory

2.1. Strong field rescattering

Single ionization of an atom or molecule in a strong laser field can be obtained by solving the TDSE numerically. However, this is not practical due to the high computational demands. An alternative method is to base on the SFA. Although it has been well recognized that the SFA does not treat the target structure accurately, the role of lasers is adequately described by the SFA.

We use linearly polarized laser pulse in which the electric field is written as

$$\mathbf{F}(t) = F_0 a(t) \cos(\omega t + \phi)\hat{z}, \tag{1}$$

where ω is the frequency of the carrier wave and ϕ the carrier envelope phase, with the envelope function $a(t)$ chosen to be

$$a(t) = \cos^2\left(\frac{\pi t}{\tau}\right) \tag{2}$$

for the time interval $(-\tau/2, \tau/2)$ and zero elsewhere. It should be mentioned that in all the calculations the carrier envelope phase is set to zero in this paper.

In SFA, by treating the electron-target interaction as a perturbation, the amplitude of detecting a photoelectron with momentum p at an outgoing angle θ with respect to the polarization is given by

$$f_{\mathrm{SFA}}(\mathbf{p}) = f_1(\mathbf{p}) + f_2(\mathbf{p}), \tag{3}$$

where only the first two terms of the perturbation series are included.

In Eq. (3), the first term $f_1(\mathbf{p})$, called SFA1, is the traditional strong field approximation,

$$f_1(\mathbf{p}) = -i\int_{-\infty}^{\infty} dt\, \langle \chi_{\mathbf{p}}(t)\, | H_i(t) |\, \Psi_0(t) \rangle\,, \tag{4}$$

where Ψ_0 is the ground state wave function, $\chi_{\mathbf{p}}$ the Volkov state, and $H_i(t) = \mathbf{r}\cdot\mathbf{F}(t)$ the atom-field interaction in length gauge.

The second term in Eq. (3), called SFA2, is expressed as

$$f_2(\mathbf{p}) = -\int_{-\infty}^{\infty} dt \int_{t}^{\infty} dt' \int d\mathbf{k} \langle \chi_{\mathbf{p}}(t') | V | \chi_{\mathbf{k}}(t') \rangle \times \langle \chi_{\mathbf{k}}(t) | H_i(t) | \Psi_0(t) \rangle . \quad (5)$$

It consists of three time-ordered steps by the electron: the electron is first tunnel ionized at time t, propagating in the laser field, and rescattered elastically with the parent ion at time t'. The potential V is the electron-ion interaction. The integration over $\mathbf{k}$ is performed by means of saddle point approximation.

The momentum distribution of emission of an electron with energy $E = p^2/2$ in the direction of θ is given by

$$D_{\mathrm{SFA}}(p,\theta) = |f_{\mathrm{SFA}}(\mathbf{p})|^2. \quad (6)$$

The form of f_{SFA} in Eq. (3) allows us to identify the contribution from each individual term. The standard SFA results can be obtained if f_2 is ignored in Eq. (3).

While the SFA1 contributes to the direct ionization, the indirect ionization related to rescattering is attributed to the SFA2. The fatal shortage of SFA lies in the fact that both the long range Coulomb potential and the short range potential from the target are completely ignored when an electron is described by the Volkov state. As a result, the target structure information is missed in SFA2.

The QRS model is set up in the light of improving SFA2 and based on the assumption that photoelectron momentum distribution in Eq. (6) can be written as

$$D_{\mathrm{SFA}}(p,\theta) = W(k_r)\sigma_{\mathrm{B}}(k_r,\theta_r), \quad (7)$$

where $\sigma_{\mathrm{B}}(k_r,\theta_r)$ is the elastic differential cross section in the first-order Born approximation for electron with momentum k_r scattered at an angle θ_r with respect to the incident direction, which is given by

$$\sigma_{\mathrm{B}}(k_r,\theta_r) = |V(\mathbf{q})|^2, \quad (8)$$

where $V(\mathbf{q})$ is the Fourier transform of the potential

$$V(\mathbf{q}) = \frac{1}{2\pi}\int d\mathbf{r}\exp(i\mathbf{q}\cdot\mathbf{r})V(r), \quad (9)$$

and $\mathbf{q}$ is the momentum transfer and its magnitude is $q = 2k_r\sin(\theta_r/2)$. In Eq. (7), $W(k_r)$ is called the returning wave packet which can be interpreted as the momentum distribution of the returning electron beam.

The photoelectron momentum $\mathbf{p}$ and the momentum $\mathbf{k}_r$ of the electron after scattering with the parent ion are related by

$$\mathbf{p} = \mathbf{k}_r - \mathbf{A}_r, \quad (10)$$

where $\mathbf{A}_r$ is the vector potential at the time when electrons return to the origin. In addition, based on the classical simulation [5], the QRS relates

$$k_r = 1.26|A_r|. \quad (11)$$

Figure 1 illustrates the relationship between the photoelectron momentum $\mathbf{p}$ and the momentum $\mathbf{k}_r$ of the rescattered electron for the case in which the returning electron revisits the parent ion along $-\hat{z}$.

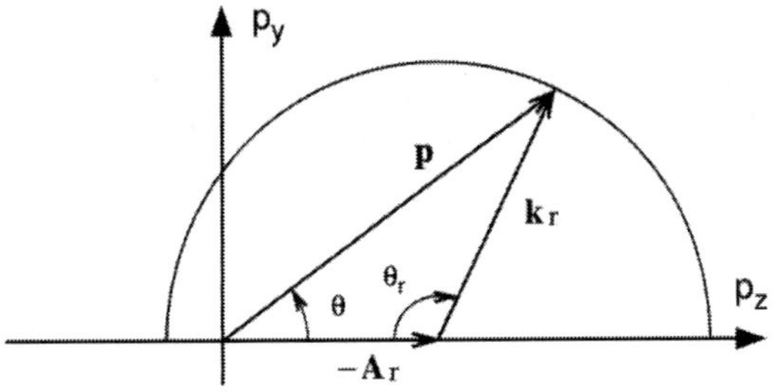

Figure 1. Relationship between the photoelectron momentum $\mathbf{p}$ and the momentum $\mathbf{k}_r$ of the rescattered electron.

The projection of photoelectron momentum in the parallel and perpendicular directions are

$$\begin{aligned} p^{||} &= p\cos\theta = -A_r - k_r\cos\theta_r = -A_r + k_r^{||}, \\ p^{\perp} &= p\sin\theta = k_r\sin\theta_r = k_r^{\perp}. \end{aligned} \quad (12)$$

The good property of the wave packet $W(k_r)$ is that it does not depend on the rescattering angle. This property has been well tested. Recently, Cerkic *et al* [10] also presented a derivation of the factorization formula for the back-rescattering ridge, i.e. electrons returning with maximum kinetic energy of about $3U_p$, and showed that the cross sections in the first-order Born approximation can be factored out.

The great advantage of Eq. (7) is that in the photoelectron spectra, the wave packet depends only on the laser pulse and the DCS only on the target. This allows one to insert the structure information by replacing $\sigma_{\rm B}(k_r, \theta_r)$, calculated within the first-order Born approximation in Eq. (7), by the accurate DCS, $\sigma(k_r, \theta_r)$, i.e.,

$$D_{\rm QRS}(p, \theta) = W(k_r)\sigma(k_r, \theta_r). \quad (13)$$

The key idea of the QRS model is that, one first evaluates the wave packet $W(k_r)$ from Eq. (7) using

$$W(k_r) = D_{\rm SFA}(p, \theta)/\sigma_{\rm B}(k_r, \theta_r) \quad (14)$$

at an arbitrary angle, say $\theta = 170°$, then uses Eq. (13) to obtain the 2D momentum spectra in the whole space. It should be noted that the wave packet only needs to be evaluated at one angle.

2.2. Field-free elastic scattering for atoms

The standard potential scattering theory has been well documented in textbooks, see Ref. [11, 12], for example. Here we address elastic scattering of an electron by a spherical potential

$$V(r) = -\frac{Z}{r} + V_{\rm s}(r), \quad (15)$$

where $V_{\rm s}(r)$ is a short-range potential. For elastic scattering of the returning electron with parent ion, $Z = 1$. For a modified Coulomb potential in Eq. (15), the scattering amplitude for detecting an electron with energy $E = k^2/2$ at an outgoing angle θ with respect to the incident direction is given by

$$f(\theta) = f_{\rm c}(\theta) + \hat{f}(\theta). \quad (16)$$

The first term in Eq. (16) is the scattering amplitude by a pure Coulomb potential

$$f_c(\theta) = -\eta \exp(2i\sigma_0)\frac{\exp\{-i\eta \ln[\sin^2(\theta/2)]\}}{2k\sin^2(\theta/2)}, \tag{17}$$

and the second term is given by

$$\hat{f}(\theta) = \sum_{l=0}^{\infty} \frac{2l+1}{k} \exp(2i\sigma_l)\exp(i\delta_l)\sin\delta_l P_l(\cos\theta), \tag{18}$$

where the $P_l(\cos\theta)$ are Legendre polynomials, δ_l is the phase shift from the short-range potential, and σ_l is the Coulomb phase shift which is given by

$$\sigma_l = \arg[\Gamma(l+1+i\eta)] \tag{19}$$

with $\eta = -Z/k$. The elastic scattering DCS is then given by

$$\sigma(k,\theta) = |f_c(\theta) + \hat{f}(\theta)|^2. \tag{20}$$

2.3. Field-free elastic scattering for molecules
Electron-molecule interaction potential does not have spherical symmetry. Rigorous theory for electron-molecule scattering is quite complicated. Instead, we examine the independent atom model [13, 14, 15] used in the standard diffraction theory. Within IAM, a molecule is modeled as a collection of its component atoms fixed in space, and the electron-molecule interaction potential is assumed to be the sum of the potentials from each atom, which is centered at each atomic site $\mathbf{R}_i$ with i the index of atom.

According to IAM, the total scattering amplitude for a molecule fixed in space is expressed as a superposition of atomic scattering amplitudes f_i, as given in Eq. (16),

$$F(k,\theta,\varphi;\Omega_L) = \sum_i f_i e^{i\mathbf{q}\cdot\mathbf{R}_i}, \tag{21}$$

where $\mathbf{q} = \mathbf{k} - \mathbf{k}_0$ is the momentum transfer vector, and Ω_L is the orientation/alignment angles of the molecule. The incident electron moves along z-axis with momentum $\mathbf{k}_0$, and is scattered into the direction $\mathbf{k} = (k,\theta,\varphi)$. The condition $|\mathbf{k}| = |\mathbf{k}_0|$ tells this is elastic scattering.

Taking the modulus squared of Eq. (21), we have the molecular differential cross section to be the form of

$$I_{\text{tot}}(\theta,\varphi;\Omega_L) = I_A + \sum_{i\neq j} f_i f_j^* e^{i\mathbf{q}\cdot\mathbf{R}_{ij}}, \tag{22}$$

which reveals important features of electron diffraction from molecules. The first term on the right, called atomic term, is the incoherent sum of all atomic scattering cross sections in the molecule

$$I_A = \sum_i |f_i|^2. \tag{23}$$

The second term in Eq. (22) includes all of the interatomic interferences, and is defined as the molecular interference term (MIT). It depends on the molecular geometry, and embeds full molecular structural information, i.e., positions of all atoms in the molecule, in the form of position vector displacement $\mathbf{R}_{ij} = \mathbf{R}_i - \mathbf{R}_j$ between any two atoms i and j.

For gas-phase electron diffraction (GED), molecule samples are in random orientations such that Eq. (22) should be averaged over all orientations. The recorded electron scattering intensity is then given by

$$\langle I_{\text{tot}}\rangle(\theta) = I_A + \sum_{i\neq j} f_i f_j^* \frac{\sin(qR_{ij})}{qR_{ij}}, \tag{24}$$

in which q and R_{ij} are the moduli of $\mathbf{q}$ and $\mathbf{R}_{ij}$, respectively. It is interesting to note that molecular characteristic interference features are still preserved after the angular average, which was pointed out early by Cohen and Fano in 1966 [16]. This is essential to GED experimentalists, allowing them to pull out interatomic distances.

The IAM has been widely applied in stereochemistry to reconstruct molecules. As the molecular structure parameters are determined from the molecular interference term, we can define a molecular contrast factor (MCF) as the ratio of molecular interference term to atomic interference-free term

$$\gamma = \frac{1}{I_A}\sum_{i\neq j} f_i f_j^* \frac{\sin(qR_{ij})}{qR_{ij}}. \tag{25}$$

2.4. Field-free (e, 2e)

We consider electron impact ionization of He^+. Using the prior form, the direct transition amplitude for this (e, 2e) process is expressed by

$$f_{e2e}(\mathbf{k}_1, \mathbf{k}_2) = \langle \Psi^-_{\mathbf{k}_1,\mathbf{k}_2} | V_i | \Psi_{\mathbf{k}_i} \rangle, \tag{26}$$

where V_i is the perturbation interaction,

$$V_i = \frac{1}{|\mathbf{r}_1 - \mathbf{r}_2|} - \frac{Z_N}{r_1} - U_i(r_1), \tag{27}$$

where $\mathbf{r}_1$ and $\mathbf{r}_2$ are the position vectors for the projectile and the bound state electron, respectively. Here, we assume that the He^+ ion is initially in the ground state, and the charge of the nucleus is $Z_N = 2$. In Eq. (27), $U_i(r_1)$ is the initial state distorting potential which is used to calculate the initial state wave function for the projectile.

In (26), $\Psi^-_{\mathbf{k}_1,\mathbf{k}_2}$ is the final state wave function describing the two outgoing electrons with momenta $\mathbf{k}_1$ and $\mathbf{k}_2$, respectively. In this work, we use the BBK model [19] in which the final continuum state is expressed as

$$\begin{aligned}\Psi^-_{\mathbf{k}_1,\mathbf{k}_2}(\mathbf{r}_1, \mathbf{r}_2) = (2\pi)^{-3}\exp(i\mathbf{k}_1\cdot\mathbf{r}_1)\exp(i\mathbf{k}_2\cdot\mathbf{r}_2)\\ \times C(\alpha_1, \mathbf{k}_1, \mathbf{r}_1)C(\alpha_1, \mathbf{k}_2, \mathbf{r}_2)C(\alpha_{12}, \mathbf{k}_{12}, \mathbf{r}_{12}),\end{aligned} \tag{28}$$

where

$$\begin{aligned}\mathbf{k}_{12} = \frac{1}{2}(\mathbf{k}_1 - \mathbf{k}_2), \quad \mathbf{r}_{12} = \mathbf{r}_1 - \mathbf{r}_2,\\ \alpha_1 = -\frac{Z_N}{k_1}, \quad \alpha_2 = -\frac{Z_N}{k_2}, \quad \alpha_{12} = \frac{1}{2k_{12}}.\end{aligned} \tag{29}$$

The Coulomb part of the wave function in Eq. (28) is defined as

$$C(\alpha, \mathbf{k}, \mathbf{r}) = \exp(-\pi\alpha/2)\Gamma(1 - i\alpha)\,{}_1F_1[i\alpha; 1; -i(kr + \mathbf{k}\cdot\mathbf{r})], \tag{30}$$

where Γ is the gamma function and ${}_1F_1$ the confluent hypergeometric function. The Coulomb interaction between the two outgoing electrons included in Eq. (28) can be turned off by setting $\alpha_{12} = 0$.

The initial state $\Psi_{\mathbf{k}_i}$ in (26) is a product of the wave function $\varphi_{\mathbf{k}_i}(\mathbf{r}_1)$ describing the incident electron with momentum $\mathbf{k}_i$ and the ground state wave function of He^+

$$\Psi_{\mathbf{k}_i}(\mathbf{r}_1, \mathbf{r}_2) = \varphi_{\mathbf{k}_i}(\mathbf{r}_1)\phi_{He^+}(\mathbf{r}_2). \tag{31}$$

In principle, $\varphi_{\mathbf{k}_i}(\mathbf{r}_1)$ can be obtained by solving the differential equation

$$\left[-\frac{1}{2}\nabla_1^2 + U_i(r_1) - \frac{1}{2}k_i^2\right]\varphi_{\mathbf{k}_i}(\mathbf{r}_1) = 0, \tag{32}$$

where the distorting potential $U_i(r_1)$ is not determined by the formalism. For the current purpose, we simply choose $U_i(r_1) = 0$ such that a plane wave is obtained to describe the incident electron, i.e.,

$$\varphi_{\mathbf{k}_i}(\mathbf{r}_1) = \frac{1}{(2\pi)^{3/2}}\exp(i\mathbf{k}_i \cdot \mathbf{r}_1). \tag{33}$$

Finally, the triple differential cross section (TDCS) for the electron impact ionization process is given by

$$\frac{d^3\sigma_{\text{e2e}}}{d\Omega_1 d\Omega_2 dE} = (2\pi)^4\frac{k_1k_2}{k_i}\Big[\frac{3}{4}|f_{\text{e2e}}(\mathbf{k}_1,\mathbf{k}_2) - g_{\text{e2e}}(\mathbf{k}_1,\mathbf{k}_2)|^2 + \frac{1}{4}|f_{\text{e2e}}(\mathbf{k}_1,\mathbf{k}_2) + g_{\text{e2e}}(\mathbf{k}_1,\mathbf{k}_2)|^2\Big], \tag{34}$$

where $\Omega_1(\theta_1,\phi_1)$ and $\Omega_2(\theta_2,\phi_2)$ are the solid angles of detection of the two electrons leaving the collision with momenta $\mathbf{k}_1$ and $\mathbf{k}_2$, and $g_{\text{e2e}}(\mathbf{k}_1,\mathbf{k}_2)$ is the exchange amplitude with $g_{\text{e2e}}(\mathbf{k}_1,\mathbf{k}_2) = f_{\text{e2e}}(\mathbf{k}_2,\mathbf{k}_1)$.

The TDCS in Eq. (34) is spin-averaged for the usual (e, 2e) collision. We have used Eq. (34) in our previous QRS calculations for NSDI on helium [20]. However, for ionization of He^+ by the impact of a returning electron induced in laser field, the two electrons in the process started in the singlet ground state, and their singlet coupling is preserved during ionization [21]. Consequently, for laser induced (e, 2e) process, the TDCS should be given by

$$\frac{d^3\sigma_{\text{e2e}}^{\text{Laser}}}{d\Omega_1 d\Omega_2 dE} = (2\pi)^4\frac{k_1k_2}{k_i}|f_{\text{e2e}}(\mathbf{k}_1,\mathbf{k}_2) + g_{\text{e2e}}(\mathbf{k}_1,\mathbf{k}_2)|^2. \tag{35}$$

3. Applications

3.1. Dynamic imaging of molecules with infrared lasers

To image microscopic objects, x-ray diffraction and electron diffraction are the two principal experimental tools for the determination of structure of molecules over many decades. These methods can achieve sub-angstrom spatial resolution, but it is unable to perform time-resolved measurements of molecular dynamics on femtosecond time scale. Although ultrafast electron diffraction (UED) [22, 23] and free-electron x-ray lasers are aiming at improving the temporal resolution, many technical challenges still remain [24].

Alternatively, infared (IR) laser pulse offers the potential for dynamic imaging since the temporal resolution is already in femtoseconds. Moreover, two recent theoretical developments in strong field rescattering physics [4, 5, 25] have made it quite promising for using lasers to image the structure of a molecule. First, using high energy above-threshold-ionization (HATI) spectra, it has been shown that accurate field-free electron-target-ion scattering differential cross sections can be extracted [5, 25]. Second, these HATI electrons are backscattered from the atomic centers in the molecule. The IAM introduced above is adequate to give a good description of those collisions for energies at or above hundreds eV. Such returning energies are much lower than those used in GED, but can be easily obtained with mid-infared (MIR) lasers. At such energies, the corresponding de Broglie wavelength of electrons is less than 1 angstrom, which is short enough to resolve different atoms in a molecule.

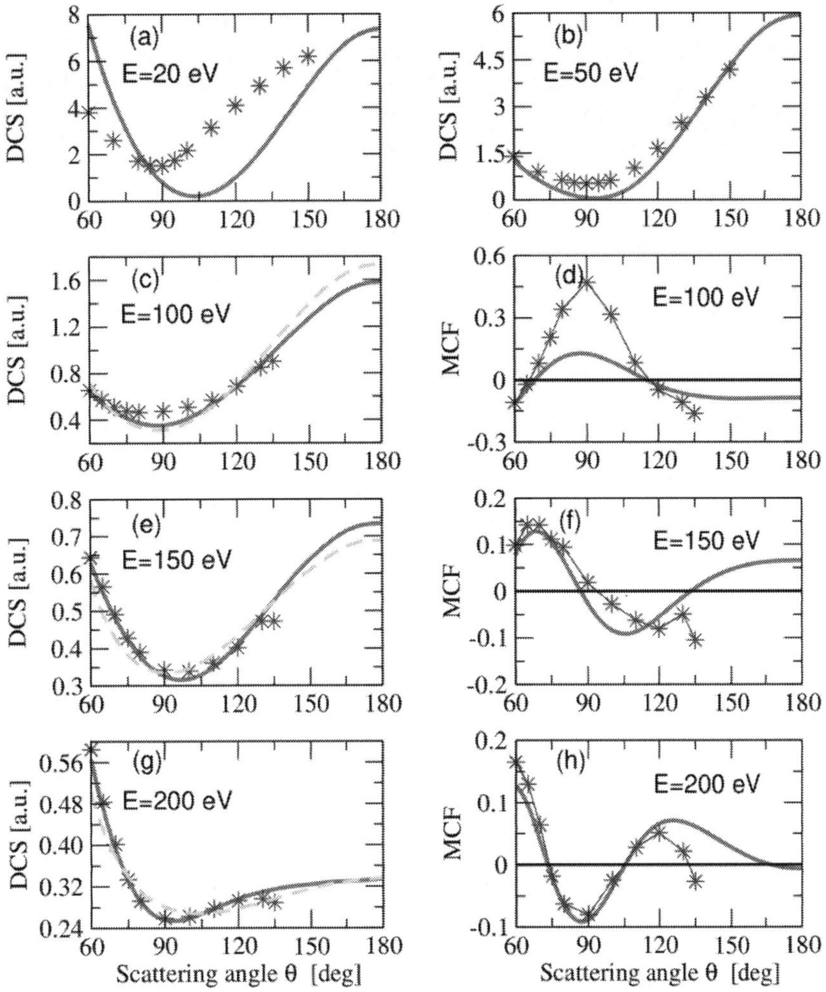

Figure 2. (Color online) (a,b,c,e,g) Elastic differential cross sections for e^--N_2 collisions at incident energies of 20, 50, 100, 150 and 200 eV, respectively. Solid red curves are IAM simulations, asterisk symbols are experimental data from DuBois *et al* (a,b) and Herrmann *et al* (c,e,g). For energies 100 eV and above, atomic terms are also displayed by dashed green lines, around which the total DCS oscillates. (d,f,h) Molecular contrast factor of IAM (solid red curves) compared with experimental data (asterisk symbols) for energies of 100, 150 and 200 eV, respectively.

3.1.1. The validity of IAM In this section, we examine the validity of the simple independent atom model for elastic electron-molecule collisions since no such data are available for molecular ions as target. The molecular DCS's are taken directly from experimental measurements using conventional collision techniques with electron guns.

In Fig. 2, we compare the DCS's from IAM for elastic collision of N_2 with experimental measurements at energies of 20, 50, 100, 150, and 200 eV, respectively. The experimental data are taken from DuBois *et al* [26] and Herrmann *et al* [27]. The IAM calculations are carried out by adopting the known N-N bond length. The nitrogen atomic potential takes the Yukawa form with a damping factor of 1.695 a.u. We only present the data for scattering angles $\theta > 60°$. It can be seen from Fig. 2(a) that, at 20 eV, although the IAM reproduces the similar parabolic angular dependence as in experiment, the dip position in DCS from IAM is shifted at least 20° toward small angles. As shown in Fig. 2(b), much improved agreement between theory and experiment is achieved at 50 eV. Clearly when energy is above 100 eV, the IAM predicts quite reliable DCS's in comparison with the experimental data, as shown in Figs. 2(c,e,g). In the covered angular region, all the DCS's show a broad minimum. This broad minimum is due to the diffraction from individual atoms, i.e., the atomic term. For energies above 100 eV, the atomic terms are also displayed in Figs. 2(c,e,g).

As indicated in Sec.2.3, the broad minimum, the main feature of interference-free atomic term, is of no use for the retrieval of molecular structure information. The molecular interference, i.e., the small oscillations in the total DCS against the smooth atomic DCS, can be clearly seen in Figs. 2(c,e,g). For the purpose of molecular imaging, we are more interested in the molecular interference. To check its accuracy, we subtract the atomic DCS from experimental data and show the difference as the molecular interference term. We renormalize the experimental data to the atomic DCS, such that the mean square of their difference is minimum. In Figs. 2(d,f,h), we display the molecular contrast factors from both experiment and the IAM. Clearly, theoretical MCF's at all energies shows similar behaviors as those from experiment. As the energy increases, the theoretical MCF is almost on top of the experimental one.

The accuracy of DCS predicted with IAM is generally determined by the momentum transfer q. More detailed discussion about this issue can be found in our recent work [18].

Table 1. Retrieved N-N bond length of N_2 from experimental DCS's at 100 eV, 150 eV, and 200 eV, respectively, compared to the known experimental values. Errors are also indicated.

	N-N bond length (Å)
experiment	1.09
100 eV	1.161(+6.5%)
150 eV	1.039(−4.7%)
200 eV	1.091(+0.1%)

3.1.2. Retrieval of molecular structure Once the validity of IAM has been established, we are ready to retrieve molecular geometry information from the experimental data, assuming that the bond lengths and bond angles are not known. Genetic algorithm (GA) is applied here to search for the global structure parameters in the parametric space. Simple molecule N_2 is taken in the first attempt. The experimental data shown in Fig. 2(c,e,g) for energies at 100, 150, and 200 eV are used as the fitting objects, from which the retrieved N-N bond lengths are 1.161, 1.039, and 1.091, respectively, see Table 1. Compared to the experimental bond length of 1.09 Å, the errors are consistent with our analysis in Sec.3.1.1 on both DCS's and MCF's: 6.5% for DCS at 100 eV is the worst, −4.7% for 150 eV shows IAM works better, and 0.1% for 200 eV is the best where the retrieved N-N bond length is almost on top of the known value.

Table 2. Retrieved structural information of C_2H_4 from experimental DCS's at 100 eV and 150 eV, respectively, compared to the known experimental values. Errors are also indicated.

	C-C length (Å)	C-H length (Å)	∠HCH
experiment	1.339	1.087	117.4°
100 eV	1.436(7%)	1.281(18%)	144.4°
150 eV	1.324(-1%)	1.075(-1%)	105.8°

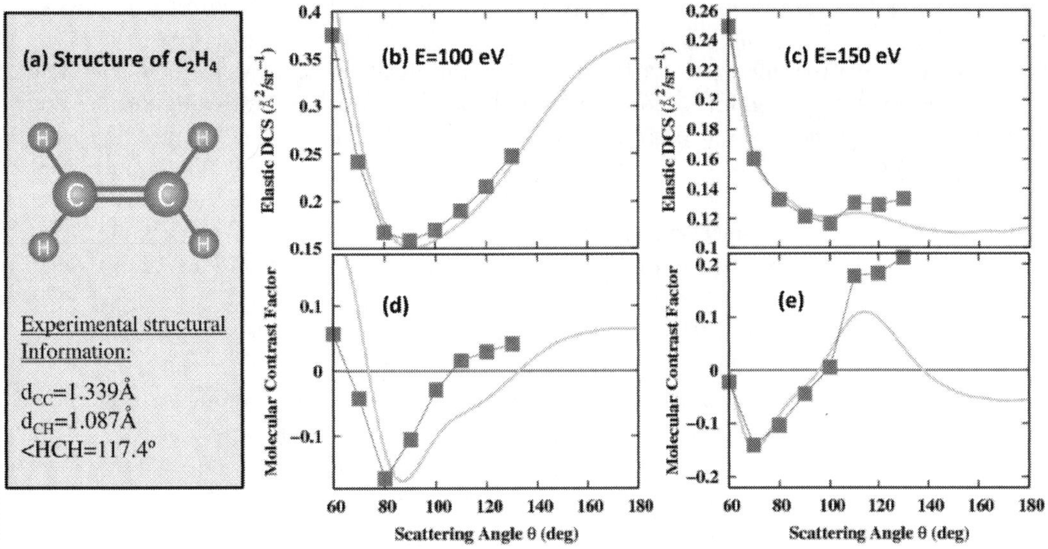

Figure 3. (Color online) (a) Molecular structure of C_2H_4 in ball-and-stick model, with structure information shown at the bottom. (b,c) Comparison between measured DCS's (red solid squares) of C_2H_4 and corresponding IAM simulations (green full lines) for electron energies at 100 and 150 eV. (d,e) Same as (b,c) but for comparison of molecular contrast factors.

We take ethylene molecule as another example, whose structure is shown in Fig. 3(a). C_2H_4 is a planar molecule and has D_{2h} symmetry. The structural information, the lengths of the double C-C bond and the C-H bond, and the angle ∠HCH, are shown at the bottom. In Figs. 3(b,c) the DCS's from IAM simulation are presented in comparison with experimental data at 100 and 150 eV, which are taken from Brescansin *et al* [28]. In both cases, theory matches with experiment quite well. Similar to N_2, we also show the molecular contrast factors, see Figs. 3(e,f). The overall agreement looks very good, except that the last 3 data points appear to be too high. Fitting to experimental DCS's with GA, the retrieved results are listed in Table 2. At 100 eV, the bond lengths and angles are not well recovered, and the error for C-H bond length is quite big; nevertheless, the C-C bond length is still acceptable. However, the three parameters are all accurately retrieved from DCS at 150 eV. Considering that the structure determination of C_2H_4 is a challenge for GED, as the light atom H has a very small scattering cross section, such accuracy is quite encouraging.

3.1.3. HATI spectrum The checking we have done in the above section shows that the IAM is an adequate theory in the description of electron-molecule collisions at large scattering angles for energies on the order of 100 eV and more, and that it is feasible to apply IAM for molecular structure determination from the DCS of electron-molecule collisions. Based on such tests, we believe that our proposal to image transient molecules with intense infared laser pulses are quite promising. In considering the DCS extracted from the HATI spectra, Eq. (24) has been modified into

$$\langle I_{\rm tot}\rangle(\theta) = (\sum_i |f_i(\theta)|^2)\int N(\Omega_L)d\Omega_L + \sum_{i\neq j} f_i(\theta)f_j^*(\theta)\int e^{i\mathbf{q}\cdot\mathbf{R}_{ij}}N(\Omega_L)d\Omega_L, \tag{36}$$

where $N(\Omega_L)$ is the ionization rate for each orientation angle Ω_L calculated using MO-ADK theory [29, 30].

Whether we have to refine our proposal, i.e., the application of IAM to molecular ions, we have to evaluate the situation until HATI spectra from MIR lasers become available. For energies of interest here, however, we did observe that the DCS's at large angles for atomic target and atomic ion target are quite close to each other. So far, experiments [31, 32, 33, 34] have reported HATI spectra for N_2, O_2, and CO_2 using 800 nm Ti-Sapphire lasers where the returning electron energies are below 50 eV. At such low energies, the IAM is not expected to work well. Therefore, HATI spectra for molecules measured in laser pulses at longer wavelengths are highly desirable.

3.2. Nonsequential double ionization of helium due to the (e, 2e) processes in strong laser fields
Nonsequential double ionization of atoms by linearly polarized laser pulses has drawn considerable interest for over two decades. Experimentally, measurements on the full momentum vectors of the two outgoing electrons have been carried out since 2000's [35, 36, 37, 38, 39, 40, 41, 42]. So far, such experiments have been performed only for 800 nm lasers, and often only the longitudinal momentum distributions, with respect to laser's polarization axis, of the two electrons have been presented. These data have attracted intensive theoretical research including the "S-matrix theory" [43, 44, 45, 46] based on the strong field approximation, and the pure numerical approaches by solving classical equation of motion [47] or the TDSE [40, 48].

In this section, we apply the QRS model to NSDI due to (e, 2e) collisions by the returning electrons induced in a strong laser field.

3.2.1. Quantitative rescattering theory for laser-induced (e, 2e) processes The QRS model has already been successfully applied to obtain the total nonsequential double ionization of Ar atoms by intense laser pulses [49]. The purpose of the present work is to simulate the correlated momentum spectra measured by Staudte *et al* [40].

For the laser-induced (e, 2e) process, the incident electron is actually the recollision electron, which is first tunnel ionized in the laser field and then driven back to recollide with the parent ion. Therefore the momentum $\mathbf{k}_i$ of the incident electron is

$$k_i = k_r. \tag{37}$$

In the correlated two-electron momentum spectra induced in a strong laser field, the experimental measurements were taken for the momentum components along the laser polarization of the two electrons against each other such that the data were integrated over the other components. Correspondingly, to compare with experiment, we integrate the TDCS, for laser-free (e, 2e) process at a given incident energy $E_i = k_i^2/2$, over ϕ_2 and E_2

$$Y_{E_i}^{\mathrm{e,2e}}(k_1^{||}, k_2^{||}) = \frac{4\pi}{k_1 k_2}\int_0^{E_{\max}} dE_2 \int_0^{\pi} d\phi_2 \frac{d^3\sigma_{\mathrm{e2e}}^{\mathrm{Laser}}}{d\Omega_1 d\Omega_2 dE_2}\Big|_{\phi_1=0}, \tag{38}$$

where $E_{\max} = E_i - I_p$. Due to the cylindrical symmetry, we set $\phi_1 = 0$ in Eq. (38). In the actual calculations, the integration over ϕ_2 is performed only from 0 to π since the TDCS is symmetric about the plane containing $\mathbf{k}_1$ and $\mathbf{k}_i$.

By shifting the momenta of the two outgoing electrons using the relation in Eq. (10), we obtain the correlated momentum spectra at an incident energy of E_i for NSDI,

$$D_{E_i}^{\mathrm{e,2e}}(p_1^{||}, p_2^{||}) \equiv D_{E_i}^{\mathrm{e,2e}}(k_1^{||} - A_r, k_2^{||} - A_r) = Y_{E_i}^{\mathrm{e,2e}}(k_1^{||}, k_2^{||}). \tag{39}$$

Similar to Eq. (13), the final distribution can be evaluated by

$$D_{\mathrm{e,2e}}(p_1^{||}, p_2^{||}) = \int_{I_p}^{\infty} dE_i D_{E_i}^{\mathrm{e,2e}}(p_1^{||}, p_2^{||})\bar{W}(E_i), \tag{40}$$

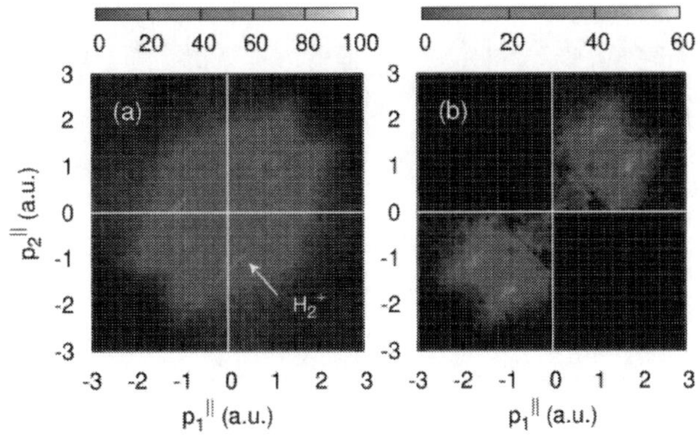

Figure 4. (Color online) (a) Experimental correlated momentum spectra along the polarization direction for double ionization of helium at 800 nm, 4.5×10^{14} W/cm^2. The indicated lines are H_2^+ background from cold H_2 in the gas jet. (b) Same as (a), only the (e, 2e) portion is retained. The original data are from Staudte *et al* [40].

where $\bar{W}(E_i)$ is the volume-integrated wave packets [5], i.e., the wave packet integrated over the volume distribution of the laser pulse at peak intensity I_0.

3.2.2. Correlated momentum spectra due to (e, 2e) processes in strong laser field Now we apply the QRS model to simulate the correlated two-electron momentum spectra of NSDI on helium in a 800 nm, 4.5×10^{14} W/cm^2 laser field measured by Staudte *et al* [40], in which a fingerlike structure in the correlated electron momentum distribution was observed. The measured correlated momentum spectra from Staudte *et al* are shown in Fig. 4(a).

It has been well recognized that NSDI consists of both (e, 2e) and excitation-tunneling. To compare with the experimental data due to (e, 2e) process only, one needs to separate these two mechanisms. The separation of (e, 2e) and excitation-tunneling can be carried out according to the analysis of the recollision mechanisms in NSDI by Feuerstein *et al* [36]. Based on the analysis, the (e, 2e) events are located in the first and third quadrants only, while the excitation-tunneling events are distributed symmetrically in all four quadrants. Following Feuerstein *et al* [36], by subtracting the corresponding number of excitation-tunneling events from the measured number of all events in the first and third quadrants, we obtain the correlated momentum distributions for the "pure" (e, 2e) process, which are displayed in Fig. 4(b). It should be noted that the original experiment spectra include contamination from the H_2^+ background as indicted by the lines in Fig. 4(a). This background should be removed to ensure that the spectra in the second quadrant are from excitation-tunneling only.

The simplest theoretical model used in the (e, 2e) TDCS calculations in this work is called P-CC in which a plane wave (P) is used to describe the incident electron and two Coulomb waves (C) are used for the outgoing electrons. The calculated P-CC correlated momentum spectra for two outgoing electrons in the recollisional (e, 2e) process on He^+ at an incident energy of 85 eV is displayed in Fig. 5(a). The momentum spectra shows no minimum along the diagonal of equal momenta.

The primary weakness of the P-CC model lies in the fact that the final state electron-electron correlation is contained only in the first order. This interaction is taken into account in the well-known BBK model [19] in which two Coulomb waves are used to describe the two outgoing electrons and an additional Coulomb interaction is employed for the interaction between the two outgoing electrons. In this paper, this model is referred to as P-CCC. The P-CCC result for the

Journal of Physics: Conference Series **288** (2011) 012017 doi:10.1088/1742-6596/288/1/012017

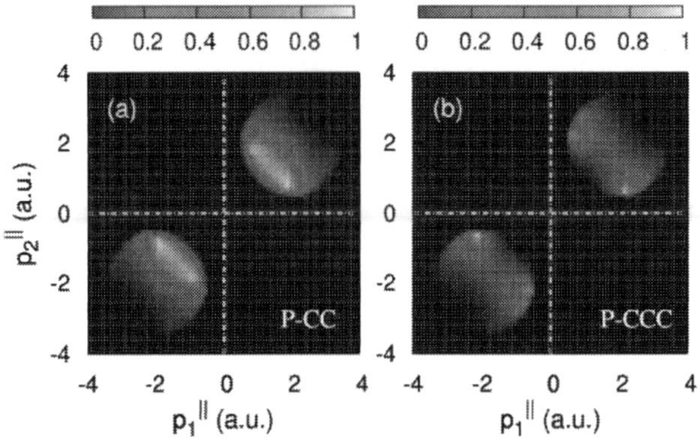

Figure 5. (Color online) Theoretical normalized correlated momentum spectra for two outgoing electrons in the recollisional (e, 2e) process on He^+ at an incident electron energy of 85 eV. Shown are the momentum components along the polarization direction. The theoretical models used in the simulation are (a) P-CC and (b) P-CCC. See text for detail.

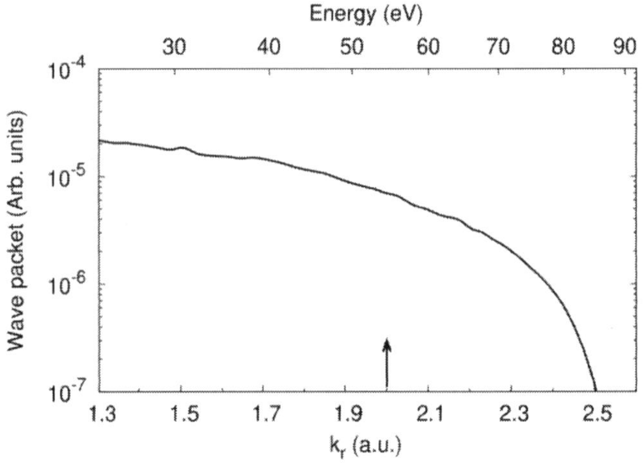

Figure 6. Volume integrated wave packet against the momentum of the recollision electron (the bottom horizontal axis) for 45 fs (FWHM) laser pulses at peak intensity of 4.5×10^{14} W/cm^2 with wavelength of 800 nm. The arrow indicates the minimum momentum of the recollision electron to ionize the parent ion He^+. The corresponding energy of the returning electron is also used as the top horizontal axis.

same collision parameters is shown in Fig. 5(b). One can see that the fingerlike structure, which was observed in experiment [40], is clearly reproduced by the P-CCC model. The significant discrepancy in the correlated momentum spectra between P-CC and P-CCC indicates that the Coulomb interaction plays a crucial role in the (e, 2e) process at least for the energy regime considered here.

To obtain correlated electron momentum spectra that can be compared to experimental data, the integration over all possible incident energies weighted by the returning wave packet should be performed using Eq. (40). For the 800 nm laser at peak intensity of 4.5×10^{14} W/cm^2 used by Staudte *et al*, the maximum energy of the returning electron is $E_i^{\max} = 3.17U_p = 85.6$ eV. The returning electron wave packet is shown in Fig. 6.

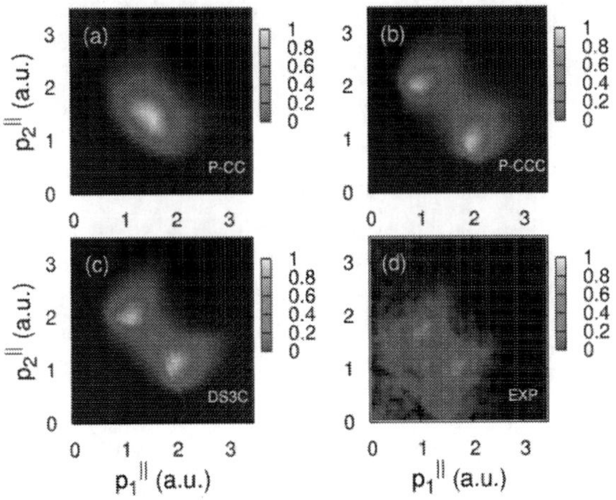

Figure 7. (Color online) Momentum spectra for two out-going electrons in the recollisional (e, 2e) process on He^+ at 800 nm, 4.5×10^{14} W/cm^2. Shown are the momentum components along the polarization direction. The theoretical models used are: (a) P-CC, (b) P-CCC, and (c) DS3C, respectively. (d) Experiment.

The volume-integrated correlated momentum spectra of P-CC and P-CCC are plotted in Figs. 7(a) and (b). Since the spectra in the third quadrant is symmetric with respect to the first quadrant, only the first quadrant spectra are shown. One can see that the final correlated momentum spectra of P-CC and P-CCC are slightly different from the results for fixed incident returning energy shown in Fig. 5 while the main features remain. Again, different from the P-CC results, the P-CCC predicts a fingerlike structure which is in better agreement with the experimental measurements shown in Fig. 7(d).

It has already been recognized that the BBK model is not in agreement with the measurements at low energies because the interaction between each pair of charges is not screened by the presence of the third charge. This deficiency was first corrected by Berakdar and Briggs [50], by introducing an effective Sommerfeld parameter for the case in which the two outgoing electrons have equal energies. Such a modification represents a dynamic screening (DS) of the three-body Coulomb interactions and hence the improved BBK model is called DS3C. The DS3C model was later generalized by Chen *et al* [51] for any geometry and energy sharing. In this work, we also performed the DS3C calculations and the final correlated momentum spectra are shown in Fig. 7(c). A comparison of P-DS3C with P-CCC shows that the two separated peaks in the spectra of P-CCC is connected by an almost evenly distributed bridge. This indicates the effect of screening included in the P-DS3C model, which suppresses the strong electron-electron Coulomb repulsion.

4. Conclusion

In conclusion, we have studied the elastic scattering of the laser-induced returning electrons with the parent molecular ion. From high energy photoelectron spectra, laser-free scattering DCS can be extracted. Using traditional electron-molecule collision data, we show that IAM gives reasonably good predictions on the scattering DCS for large scattering angles when electron energy is above 100 eV, and the bond lengths of N_2 and C_2H_4 are successfully retrieved. Based on the QRS and IAM, the feasibility of self-imaging of transient molecules with intense infrared laser pulses has been established.

The NSDI on helium in an intense laser field due to the (e, 2e) process has also been studied. We obtain the correlated momentum spectra for the two outgoing electrons by multiplying the returning electron wave packet with the differential cross sections for laser-free (e, 2e) reactions. It turns out that the post-collision interaction between the two outgoing electrons is crucial in reproducing the finger-like structure in the correlated momentum space observed in the experiment.

Acknowledgment

We thank Dr. A. Staudte for providing experimental data in digital form that allowed us to separate (e, 2e) from excitation-tunneling. This work was supported in part by Chemical Sciences, Geosciences and Biosciences Division, Office of Basic Energy Sciences, Office of Science, US Department of Energy. Y Liang was also supported by financial aids from China Scholarship Council.

References

[1] Corkum P B 1993 *Phys. Rev. Lett.* **71** 1994
[2] Krause J L, Schafer K J and Kulander K C 1992 *Phys. Rev. Lett.* **68** 3535
[3] Chen Zhangjin, Morishita Toru, Le Anh-Thu and Lin C D 2007 *Phys. Rev. A* **76** 043402
[4] Morishita Toru, Le Anh-Thu, Chen Zhangjin and Lin C D 2008 *Phys. Rev. Lett.* **100** 013903
[5] Chen Zhangjin, Le Anh-Thu, Morishita Toru and Lin C D 2009 *Phys. Rev. A* **79** 033409
[6] Okunishi M, Morishita T, Prümper G, Shimada K, Lin C D, Watanabe S and Ueda K 2008 *Phys. Rev. Lett.* **100** 143001
[7] Ray D, Ulrich B, Bocharova I, Maharjan C, Ranitovic P, Gramkow B, Magrakvelidze M, De S, Litvinyuk I V, Le A T, Morishita T, Lin C D, Paulus G G and Cocke C L, 2008 *Phys. Rev. Lett.* **100** 143002
[8] Chen Zhangjin, Le Anh-Thu, Morishita Toru and Lin C D 2009 *J. Phys. B: At. Mol. Opt. Phys.* **42** 061001
[9] Xu Junliang, Zhou Hsiao-Ling, Chen Zhangjin and Lin C D 2009 *Phys. Rev. A* **79** 052508
[10] Čerkić A, Hasović E, Milošević D B and Becker W 2009 *Phys. Rev. A* **79** 033413
[11] Schiff L I 1968 *Quantum Mechanics* 3rd Edition, McGraw-Hill, New York, p. 145
[12] Joachain C J 1983 *Quantum Collision Theory* 3rd Edition, North-Holland, Amsterdam, p. 144
[13] Lothar Schäfer 1976 *Appllied Spectroscopy* **30** 123
[14] McCaffrey P D, Dewhurst J K, Rankin D W H, Mawhorter R J and Sharma S 2008 *J. Chem. Phys.* **128** 204304
[15] Hinchley S, Wann D A and Rankin D W H 2005 *Int. J. Quant. Chem.* **101** 878
[16] Cohen H and Fano U 1966 *Phys. Rev.* **150** 30
[17] Reckenthaeler P, Centurion M, Fuβ W, Trushin S A, Krausz F and Fill E E 2009 *Phys. Rev. Lett.* **102** 213001
[18] Xu Junliang, Chen Zhangjin, Le Anh-Thu and Lin C D 2010 *Phys. Rev. A* **82** 033403
[19] Brauner M, Briggs J S and Klar H 1989 *J. Phys. B: At. Mol. Opt. Phys.* **22** 2265
[20] Chen Zhangjin, Liang Yaqiu and Lin C D 2010 *Phys. Rev. Lett.* **104** 253201
[21] Yudin G L and Ivanov M Y 2001 *Phys. Rev. A* **64**, 035401
[22] Zewail A H and Thomas J M 2009 *4D Electron Microscopy: Imaging in Space and Time* (London: Imperial College Press)
[23] Siwick B J, Dwyer J R, Jordan R E and Miller D 2003 *Science* **302** 1382
[24] Ivanov M Y and Marangos J P 2007 *J. Mod. Opt.* **54** 899
[25] Lin C D, Le A T, Chen Z, Morishita T and Lucchese R R 2010 *J. Phys. B: At. Mol. Opt. Phys.* **43** 122001
[26] DuBois R D and Rudd M E 1976 *J. Phys. B: At. Mol. Opt. Phys.* **9** 2657
[27] Herrmann D, Jost K, Kessler J and Fink M 1976 *J. Chem. Phys.* **64** 1
[28] Brescansin L M, Iga I and Lee M T 2004 *J. Phys. B: At. Mol. Opt. Phys.* **37** 471
[29] Tong X M, Zhao Z X and Lin C D 2002 *Phys. Rev. A* **66** 033402
[30] Zhao S F, Jin C, Le A T, Jiang T F and Lin C D 2010 *Phys. Rev. A* **81** 033423
[31] Cornaggia C 2009 *J. Phys. B: At. Mol. Opt. Phys.* **42** 161002
[32] Okunishi M, Itaya R, Shimada K, Prümper G, Ueda K, Busuladžić M, Gazibegović-Busuladžić A, Milošević D B and Becker W 2008 *J. Phys. B: At. Mol. Opt. Phys.* **41** 21004
[33] Busuladžić M, Gazibegović-Busuladžić A, Milošević D B and Becker W 2008 *Phys. Rev. Lett.* **100** 203003
[34] Meckel M, Comtois D, Zeidler D, Staudte A, Pavičić D, Bandulet H C, Pépin H, Kieffer J C, Dörner R, Villeneuve D M and Corkum P B 2008 *Science* **320** 1478

[35] Weber Th, Giessen H, Weckenbrock M, Urbasch G, Staudte A, Spielberger L, Jagutzki O, Mergel V, Vollmer M and Dörner R 2000 *Nature* (London) **405** 658

[36] Feuerstein B, Moshammer R, Fischer D, Dorn A, Schröter C D, Deipenwisch J, Lopez-Urrutia J R Crespo, Höhr C, Neumayer P, Ullrich J, Rottke H, Trump C, Wittmann M, Korn G and Sandner W 2001 *Phys. Rev. Lett.* **87** 043003

[37] Eremina E, Liu X, Rottke H, Sandner W, Dreischuh A, Lindner F, Grasbon F, Paulus G G, Walther H, Moshammer R, Feuerstein B and Ullrich J 2003 *J. Phys. B: At. Mol. Opt. Phys.* **36** 3269

[38] Moshammer R, Ullrich J, Feuerstein B, Fischer D, Dorn A, Schröter C D, López-Urrutia J R Crespo, Höhr C, Rottke H, Trump C, Wittmann M, Korn G, Hoffmann K and Sandner W 2003 *J. Phys. B: At. Mol. Opt. Phys.* **36** L113

[39] Weckenbrock M, Zeidler D, Staudte A, Weber Th, Schöffler M, Meckel M, Kammer S, Smolarski M, Jagutzki O, Bhardwaj V R, Rayner D M, Villeneuve D M, Corkum P B and Dörner R 2004 *Phys. Rev. Lett.* **92** 213002

[40] Staudte A, Ruiz C, Schöffler M, Schössler S, Zeidler D, Weber Th, Meckel M, Villeneuve D M, Corkum P B, Becker A and Dörner R 2007 *Phys. Rev. Lett.* **99** 263002

[41] Rudenko A, Jesus V L B de, Ergler Th, Zrost K, Feuerstein B, Schröter C D, Moshammer R and Ullrich J 2007 *Phys. Rev. Lett.* **99** 263003

[42] Liu Yunquan, Tschuch S, Rudenko A, Dürr M, Siegel M, Morgner U, Moshammer R and Ullrich J 2008 *Phys. Rev. Lett.* **101** 053001

[43] Becker A and Faisal F H M 2002 *Phys. Rev. Lett.* **89** 193003

[44] Faria C Figueira de Morisson, Schomerus H, Liu X and Becker W 2004 *Phys. Rev. A* **69** 043405

[45] Faria C Figueira de Morisson, Liu X and Becker W 2004 *Phys. Rev. A* **69** 021402(R)

[46] Becker A and Faisal F H M 2005 *J. Phys. B: At. Mol. Opt. Phys.* **38** R1

[47] Haan S L, Smith Z S, Shomsky K N and Plantinga P W 2009 *J. Phys. B: At. Mol. Opt. Phys.* **42** 134009

[48] Parker J S, Doherty B J S, Taylor K T, Schultz K D, Blaga C I and DiMauro L F 2006 *Phys. Rev. Lett.* **96** 133001

[49] Micheau Samuel, Chen Zhangjin, Le Anh-The and Lin C D 2009 *Phys. Rev. A* **79** 013417

[50] Berakdar J and Briggs J S 1994 *Phys. Rev. Lett.* **72** 3799

[51] Chen Zhangjin, Shi Qicun, Zhang Suimeng, Chen Ji and Xu Kezun 1997 *Phys. Rev. A* **56** R2514

Journal of Physics: Conference Series **288** (2011) 012018 doi:10.1088/1742-6596/288/1/012018

Infrared laser assisted photoabsorption of Ne atoms

X. M. Tong[1,2], **and N. Toshima**[1]

[1] Institute of Materials Science, Graduate School of Pure and Applied Sciences, University of Tsukuba, 1-1-1 Tennodai, Tsukuba, Ibaraki 305-8573, Japan

[2] Center for Computational Sciences, University of Tsukuba, 1-1-1 Tennodai, Tsukuba, Ibaraki 305-8577, Japan

E-mail: tong@ims.tsukuba.ac.jp

Abstract. We proposed recently a theoretical method to study the infrared (IR) laser assisted photoabsorption cross sections over a broad energy range by a single calculation. We apply this method to study the IR laser assisted photoabsorption of Ne atoms near the first ionization threshold. The simulation results show that the IR field modifies the photoabsorption cross sections and the energy structures. This photoabsorption cross section is an important quantity to analyze the IR-laser assisted dynamical processes by an attosecond pulse, a pulse train or a free electron laser.

1. Introduction

Understanding and controlling natural phenomena is one of the main goals of the physical researches. Photoabsorption spectra have been used to study atomic structures and to identify the constitutes in a material for a long time [1]. As the laser technology advances, one can control even the photoabsorption of atoms in a very short time scale with the assistance of a moderately intense infrared (IR) laser field by a non-coherent light source from the synchrotron radiations [2, 3, 4] or by a coherent light source from the attosecond XUV source [5, 6]. The IR field modifies the atomic structure and the photoabsorption processes, for example, the IR field induced transparency for x rays [7, 8]. For most of the moderately intense IR field assisted processes, the IR field only modifies the energy structures of the excited or continuum states without perturbing the ground state wave function. Recently, using the IR assisted photoexcitation process, Swodoba *et al.* [9] could even investigate the IR field induced energy shift of a specified excited state.

Stimulated by those recent developments, we developed a general theoretical method to study the photoabsorption process controlled by a periodic external field and apply it here to study the IR assisted photoabsortion process of Ne atoms. The details of the theoretical method can be found in our previous work [10]. We brief some key equations here for the completeness of the paper.

2. Theoretical Method

When an atom is placed in a periodic external field, its motion is governed by the following time-dependent Schrödinger equation (atomic units $\hbar = m = e = 1$ are used unless stated otherwise):

$$i\frac{\partial \Psi(t)}{\partial t} = H(t)\Psi(t) \tag{1}$$

with $H(t) = H_0 + V(t)$ where H_0 is the atomic external field-free Hamiltonian and $V(t)$ is periodic in T as $V(t+T) = V(t)$. Based on the Floquet theory [11] , the solution of the above equation can be generally expressed as

$$\Psi_\alpha(t) = e^{-i\epsilon_\alpha t} \sum_{n=-\infty}^{\infty} e^{-in\omega t}\phi_{\alpha,n}, \tag{2}$$

where ϵ_α is the quasi-energy, $\phi_{\alpha,n}$ is the time-independent wave function of the nth Fourier component of the α Floquet state, and $\omega = \frac{2\pi}{T}$ is the frequency of the periodic external field. If the periodic field is strong enough to modify the structures of the excited and continuum states but weak enough not to perturb the ground state, the photoabsorption cross section from the ground state to the Floquet states can be expressed as

$$\sigma(\omega_p) = \frac{4\pi\omega_p}{c}\mathrm{Im}\sum_{\alpha,n} \frac{|\langle\Psi_g|\mathbf{d}|\phi_{\alpha n}\rangle|^2}{(\epsilon_\alpha + n\omega) - (\epsilon_g + \omega_p) - i\eta}, \tag{3}$$

where η is an infinitesimal positive number to represent adiabatical switching on of the probe light, ω_p is the frequency of the probe light, $\mathbf{d}$ is the dipole operator, Ψ_g is the ground state wave function, and ϵ_g is the ground state energy. If there is no periodic external field, there is only one non-zero Fourier component for a given Floquet state and the above equation turns to the standard one for photoabsorption without the periodic field [1]. In the traditional procedure, we obtain the time-dependent wave function $\Psi_\alpha(t)$ solving the time-independent Floquet equation. The computational load increases cubically as the number of the Floquet blocks increases and the applicability of the method to real systems is limited. If many continuum states are involved, this is the case for the IR laser assisted photoabsorption process, the problem becomes even more difficult and more complicated.

We developed a theoretical method to study the photoabsorption process in an oscillating magnetic field for a non-degenerate system [12]. We extend the method to a general case in which both bound and continuum degenerate states are involved. For such a purpose, we define the time-dependent Hamiltonian as $H(t,\tau) = H_0 + V(t+\tau)$ with τ a time shift which corresponds to an initial phase $\delta = \frac{2\pi\tau}{T}$ and set the initial wave function $|\Psi(t=0,\tau)\rangle = \mathbf{d}|\Psi_g\rangle$. The time-dependent wave function $\Psi(t,\tau)$ is obtained from Eq. (1) using the split-operator method with the generalized pseudospectral grid in the energy representation [13] for a given $H(t,\tau)$. Defining a generalized autocorrelation function as

$$C(t) = \frac{1}{T}\int_0^T \langle\Psi(t=0,\tau)|\Psi(t,\tau)\rangle d\tau, \tag{4}$$

the photoabsorption cross section can be represented as

$$\sigma(\omega_p) = \frac{4\pi\omega_p}{c}\mathrm{Im}\int_0^\infty iC(t)e^{i(\omega_p+\epsilon_g)t-\eta t}dt. \tag{5}$$

In the practical simulation, we confine the numerical calculation in a finite space and have to filter out the wave function in the outer region using an energy dependent optical potential [14, 15]. Generally speaking, the quasi-energy ϵ_α in Eq. (2) is a complex number with the

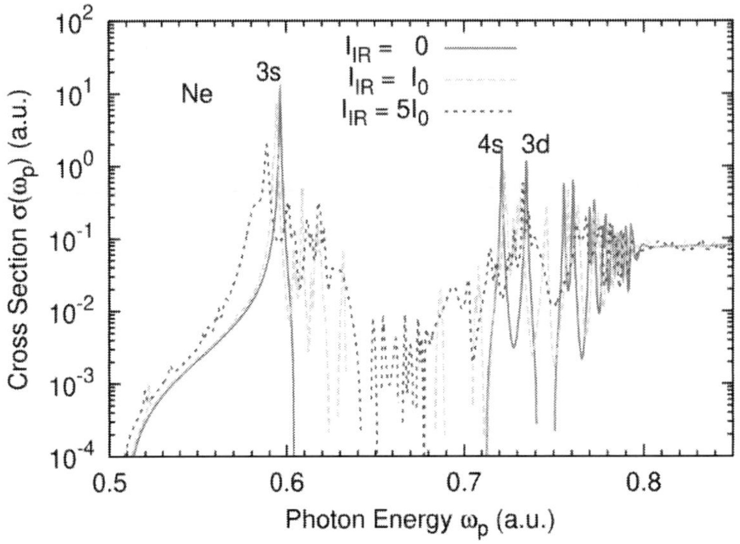

Figure 1. The IR assisted photoabsorption cross sections of Ne at three IR intensities: 0 (solid line), I_0 (dashed line) and $5I_0$ (dotted line), respectively with $I_0 = 10^{12}$ W/cm^2.

imaginary part representing the width of the Floquet state. To avoid the divergency at $t \to -\infty$ in Eq. (5), we adiabatically switch off the probe field instead of adiabatically switching it on.

The equivalence of Eqs. (3) and (5) has been proved in our previous work [10]. We focus on how to apply this general theoretical method to study the IR assisted photoabsorption cross sections of Ne atoms at different intensities for an 800 nm IR laser. The atomic Hamiltonian can be written as $H_0 = -\frac{1}{2}\nabla^2 + V_{eff}(r)$ with a model potential $V_{eff}(r)$ [16] describing the electron parent-core interaction. The interaction of the IR field with the active electron is written as $V(t) = -zE_0\cos(\omega t + \delta)$, where z is the electron z coordinate, E_0 is the IR field strength and $\delta = \frac{2\pi\tau}{T}$ is the initial phase. We assume that both the IR laser and the probe light are linearly polarized along the z direction. Similarly to the He case [10], we propagate the wave function for 256 o.c. (o.c. denotes the IR optical cycle).

3. Results and Discussion

Figure 1 shows the IR assisted photoabsorption cross sections at 3 different IR intensities. Without the IR field, the photoabsorption spectra are divided into two parts, photoexcitation below the ionization threshold and photoionization above the threshold. The photoexcitation consists of many sharp resonances and the energy widths of the resonances are very narrow. For photoionization, the spectra change smoothly as a function of the photon energy and they are insensitive to the propagation time. As we switch on the IR laser, the atomic structures are described by the Floquet states and there are many sidebands for a bound state. For the photoionization far above the ionization threshold, the spectra depend on the IR intensity weakly. There are some structures above the ionization threshold and they are attributed to the sidebands of the bound states. A moderate intense IR laser does not modify the occupied orbits significantly due to the electron binding energies are much larger than the electron laser field interaction. For excited states, the binding energies are comparable to the electron laser field interaction so the excited energy structures are sensitive to the applied IR intensity. The

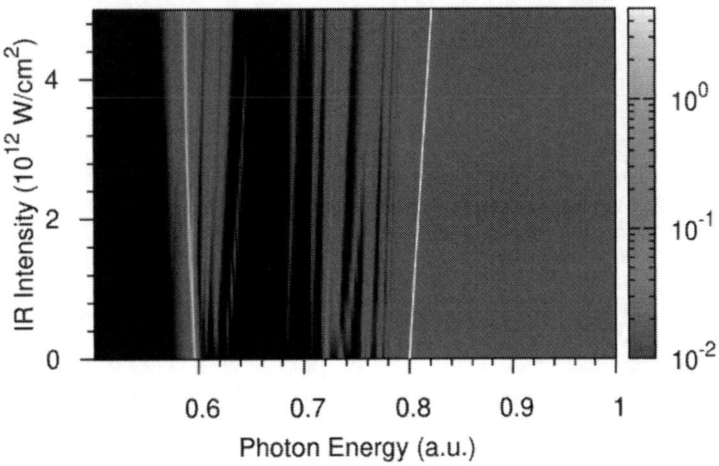

Figure 2. The IR assisted photoabsorption cross sections of Ne as a function of the IR intensity. The solid line shows the ionization threshold due to the ponderomotive shift.

strengths of the sideband contribution increase as the IR intensity increases. The ionization potential also increases as the IR intensity increases owing to the ponderomotive shift [17, 18].

Now let us focus on the bound state energy shifts, which can be seen more clearly in Fig. 2. The ionization threshold increases linearly as the IR intensity increases and it is well represented by $I_p + U_p$ with I_p atomic ionization potential and U_p the ponderomotive energy of a free electron in the laser field. The ponderomotive shift mainly affects the highly excited states and the energy shifts are well represented by the ponderomative energy. This is the reason why the above threshold ionization peaks from the highly excited states (so called Freeman resonances [19]) do not depend on the IR intensity. For the lower bound states, such as the $3s$ state, the shift moves to the opposite direction. From the energy difference, we can classify the origin of the sharp peak above the ionization threshold. As the IR intensity increases, we do not see a peak structure clearly for the highly excited states because the width of the highly excited states in the IR field is comparable to or even larger than the energy separation between the different excited states. Note that in the present non-relativistic calculation, we did not take into account the fine structure splitting. To compare with high precision experiment, we need to split the calculated spectra into two parts and shift the photon energy to fit the two ionization thresholds as we did before [20].

To summarize, we investigated the IR assisted photoabsorption processes of Ne atoms using a recently developed theoretical method. The IR field modifies the photoabsorption cross section as well as the energy structures. For the highly excited states, the IR induced energy shifts can be well explained by ponderomotive shift. For the lowest excited state, the IR induced energy shifts move to opposite direction significantly. The calculated cross sections also provide us a guide to investigate how to control the photoabsorption process by tuning the energy range of attosecond XUV source.

Acknowledgments
This work was supported by Grant-in-Aids for Scientific Research (C) from the Japan Society for the Promotion of Science.

References

[1] Fano U and Cooper J W 1968 *Rev. Mod. Phys.* **40** 441–507
[2] Hertlein M P, Adaniya H, Amini J, Bressler C, Feinberg B, Kaiser M, Neumann N, Prior M H and Belkacem A 2006 *Phys. Rev. A* **73** 062715
[3] Young L, Arms D A, Dufresne E M, Dunford R W, Ederer D L, Höhr C, Kanter E P, Krässig B, Landahl E C, Peterson E R, Rudati J, Santra R and Southworth S H 2006 *Phys. Rev. Lett.* **97** 083601
[4] Southworth S H, Arms D A, Dufresne E M, Dunford R W, Ederer D L, Höhr C, Kanter E P, Krässig B, Landahl E C, Peterson E R, Rudati J, Santra R, Walko D A and Young L 2007 *Phys. Rev. A* **76** 043421
[5] Johnsson P, Mauritsson J, Remetter T, L'Huillier A and Schafer K J 2007 *Phys. Rev. Lett.* **99** 233001
[6] Ranitovic P, Tong X M, Gramkow B, De S, DePaola B, Singh K P, Cao W, Magrakvelidze M, Ray D, Bocharova I, Mashiko H, Sandhu A, Gagnon E, Murnane M M, Kapteyn H, Litvinyuk I and Cocke C L 2010 *New J. Phys.* **12** 013008
[7] Harris S E, Field J E and Imamoğlu A 1990 *Phys. Rev. Lett.* **64** 1107–1110
[8] Buth C, Santra R and Young L 2007 *Phys. Rev. Lett.* **98** 253001
[9] Swoboda M, Fordell T, Klünder K, Dahlström J M, Miranda M, Buth C, Schafer K J, Mauritsson J, L'Huillier A and Gisselbrecht M 2010 *Phys. Rev. Lett.* **104** 103003
[10] Tong X M and Toshima N 2010 *Phys. Rev. A* **81** 063403
[11] Chu S I and Telnov D A 2004 *Phys. Rep.* **390** 1–131
[12] Shirahama T, Tong X M, Hino K I and Toshima N 2009 *Phys. Rev. A* **80** 043414
[13] Tong X M and Chu S I 1997 *Chem. Phys.* **217** 119–130
[14] Perie J and Jolicard G 1993 *J. Phys. B* **26** 4491–4502
[15] Tong X M, Hino K and Toshima N 2006 *Phys. Rev. Lett.* **97** 243202
[16] Tong X M and Lin C D 2005 *J. Phys. B:* **38** 2593–2600
[17] Davidson M D, Wals J, Muller H G and van Linden van den Heuvell H B 1993 *Phys. Rev. Lett.* **71** 2192–2195
[18] Toma E S, Muller H G, Paul P M, Breger P, Cheret M, Agostini P, Le Blanc C, Mullot G and Cheriaux G 2000 *Phys. Rev. A* **62** 061801
[19] Freeman R R, Bucksbaum P H, Milchberg H, Darack S, Schumacher D and Geusic M E 1987 *Phys. Rev. Lett.* **59** 1092–1095
[20] Domondon A T and Tong X M 2002 *Phys. Rev. A* **65** 032718

Journal of Physics: Conference Series **288** (2011) 012019 doi:10.1088/1742-6596/288/1/012019

Ultra-intense X-ray Induced Non-linear Processes in Molecular Nitrogen

L. Fang[1], M. Hoener[1,2] and N. Berrah[1]

[1]Physics Department, Western Michigan University, Kalamazoo, MI 49008, USA
[2]Advanced Light Source, Lawrence Berkeley National Laboratory, Berkeley, CA 94720, USA

E-mail: lifang@slac.stanford.edu, nora.berrah@wmich.edu

Abstract.
Sequential multiple photoionization of the prototypical molecule N_2 is studied with femtosecond time resolution using the Linac Coherent Light Source (LCLS). A detailed picture of intense x-ray induced ionization and dissociation dynamics is revealed, including a molecular mechanism of frustrated absorption that suppresses the formation of high charge states at short pulse durations. The inverse scaling of the average target charge state with x-ray peak brightness has possible implications for single pulse imaging applications. Moreover, we investigate the creation of double K-shell holes in N_2 molecules via sequential absorption of two photons on a timescale shorter than the core-hole lifetime by using intense x-ray pulses from the LCLS. The production and decay of these states is characterized by photoelectron spectroscopy and Auger electron spectroscopy. In molecules, two types of double core holes are expected, the first with two core holes on the same N atom, and the second with one core hole on each N atom. We report the first direct observations of the former type of core hole in a molecule, in good agreement with theory, and provide an experimental upper bound for the relative contribution of the latter type.

1. Introduction

The development of femtosecond light sources covering the infrared to the extreme ultraviolet (EUV) has led to a revolution in time-resolved studies of physical and chemical processes [1]. However, even with the development of high-harmonic generation sources in the EUV and soft-x-ray regions, the relatively long wavelengths of these sources preclude the direct imaging of the motion of individual atoms. As a result, there has been a tremendous interest in "fourth generation" accelerator-based free electron lasers (FEL's) [2, 3, 4, 5], which offer an unprecedented combination of spatial and temporal resolution. The Free Electron Laser in Hamburg (FLASH) [2, 3] pushed the wavelengths of these sources into the soft x-ray regime, while the recently commissioned Linac Coherent Light Source (LCLS) now accesses the hard x-ray regime [4, 5]. The intense, short pulse, x-ray FELs will allow the exploration of novel states of atoms, molecules and clusters, with potential impact on applications ranging from single-pulse imaging of biomolecules to high energy density materials [6, 7, 8, 9, 10, 11]. In this proceeding, we present (1) the first study of the response of a molecular system, the prototypical molecule N_2, to ultraintense femtosecond duration laser pulses at a wavelength of 1.1 nm (1100 eV). We observe intense x-ray induced ionization and dissociation dynamics leading to various charge states up to fully-stripped N^{7+} ions. A molecular mechanism of frustrated absorption that suppresses the formation of high charge states at short pulse durations is revealed. We show that this phenomenon can be explained by a simple rate equation model that accounts for molecular valence electron dynamics during the sequence of ionization events. (2) The first experimental attempt to characterize the creating double core hole (DCH) states of a molecule, N_2, by sequential two-photon absorption from an x-ray FEL, the LCLS. The production and decay of these states are characterized by using photoelectron spectroscopy and Auger-electron spectroscopy. The experimental results are interpreted with the help of *ab initio* calculations [12, 13, 14, 15, 16] of the singly, doubly, and triply ionized states and their corresponding Auger spectra [17]. These results will serve as a basis for understanding double- and multiple- core-hole states in more complex molecules.

The experiment was conducted using the LCLS Atomic Molecular and Optical (AMO) physics instrument [18]. We used a photon energy of 1000 and 1100 ($\pm$15) eV and pulse durations of 280 fs, 80 fs, $\sim$7 fs and $\sim$4 fs [19, 20]. The pulse duration of 280 fs and 80 fs were produced with electron bunches with 250 pC and the pulse duration of 7 fs and 4 fs were obtained with 20 pC. Unless otherwise noted, the pulse energies quoted here are nominal values measured upstream from the beamline optics. These values are reduced by $\sim$65 - 85% in the interaction region due to various photon beam transport losses [20]. The photon beam was focused by Kirkpatrick-Baez (KB) mirrors to an area equal to about $2\,\mu m^2$. The details of the apparatus and x-ray facility are described in Refs [19, 21, 22]. The LCLS had a repetition rate of 30 Hz. It took about 5 minutes to measure one of the ion-time of flight spectra and about 2 hours to obtain one of the photoelectron spectra presented in this work.

2. Multiple photon ionization and intense x-ray induced frustrated photon absorption [21]

The x-ray photons used in this study interact predominantly with the N_2 1s electrons, representing an entirely different regime than previous FEL-based experiments in the EUV [23]. Absorption of the first photon produces a single 1s inner shell vacancy in N_2. The core hole state undergoes Auger decay into N_2^{2+} with two valence-shell holes within 6.4 fs [24]. The subsequent decay of ${N_2}^{2+}$ occurs on a range of timescales; most dissociates to N^+ + N^+ and N^{2+} + N on a femtosecond timescale, but some of the ${N_2}^{2+}$ is metastable [25, 26, 27]. The high peak brightness of the LCLS opens the possibility of DCHs by sequential two-photon absorption within the core-hole lifetime [19]. These DCH states of N_2 can have both holes on a single N atom, or a single hole on each atom. DCH states decay via Auger processes on timescales that are shorter than the corresponding SCH lifetimes, accelerating multiple ionization. Cyclic photon absorption and core-hole decay sequences produce ions of increasing charge states, eventually reaching the extreme case where all electrons are removed.

If the x-ray pulse duration is shorter than the Auger decay time, absorption will effectively terminate after the ejection of the two K-shell electrons in atomic N, or after the ejection of four K-shell electrons in N_2 (assuming that absorption by the other shells is negligible). In contrast, if the x-ray pulse is long relative to the Auger decay time, *sequential multiple photoionization* sets in. One or more additional photons from the same laser pulse can be absorbed by the products of the Auger decay of N_2^+ (dominantly N, N^+, and N^{2+}). For these fragments, a second photoelectron-Auger (PA) cycle will result in N^{2+}, N^{3+}, and N^{4+} ion charge states, respectively. The Auger relaxation timescale for this second PA cycle will be longer than that of the first cycle because there are fewer valence electrons available that can participate in the Auger decay [36]. The N^{2+} and N^{3+} fragments can undergo a third and still slower PA cycle, producing N^{4+} and N^{5+} ions. These ions can absorb additional photons, but now Auger decay is no longer possible due to the lack of valence electrons. Absorption of five or more photons will result in the production of fully stripped N^{7+} ions. A time-of-flight (TOF) mass spectrometer [18] was used to analyze ions produced in the interaction of the sample with the LCLS pulses as shown

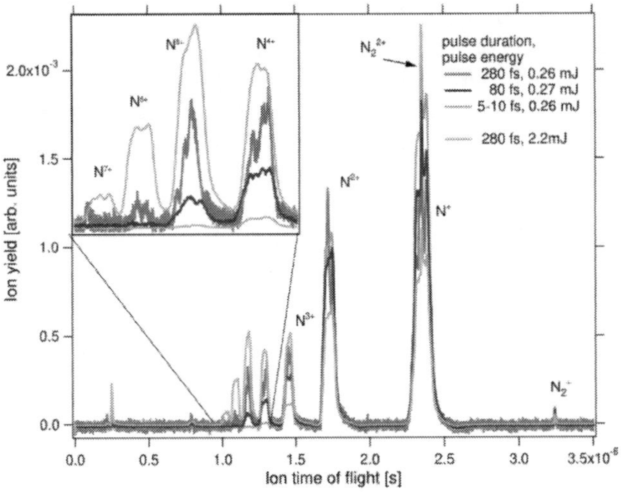

Figure 1. Ion time of flight mass spectra of N_2 recorded at 1100 eV at three different pulse durations and similar pulse energies (see text for details).

in Fig. 1. The relative ion yields have been corrected for the spectrometer-specific acceptance efficiency. The "bare" ions above were indeed observed for long (280 fs) pulses with high pulse energies (2.2 mJ), represented by the gray curve in Fig. 1.

The tremendous interest in single-pulse imaging and structure determination has raised questions about radiation damage and the ability to extract information before the sample is degenerated or destroyed. Previous calculations and experiments have focused on the idea of "inertial delay" of the Coulomb explosion during scattering from a pulse that is significantly shorter than the disassembly timescale for the object [6, 28, 29, 7]. We examine another effect that might have an even larger impact than inertia based distortion limitations: PA cascades induced by sufficiently short x-ray pulses are subject to frustrated absorption. As a result, the effective total photoionization cross section is reduced and Coulomb-repulsion effects in the sample during exposure to an imaging light pulse might be minimized.

Figure 1 shows ion time of flight spectra for three different pulse durations ($\sim$7 fs, 80 fs and 280 fs) at similar pulse energies (0.26 mJ). Over the range of parameters investigated, longer pulses generally result in the observation of the highest charge states (red curve) whereas shorter pulses lead almost exclusively to lower charge states. Highly charged N^{5+} and N^{6+} ions are clearly observed for 80 fs and 280 fs pulse lengths. In contrast, the $\sim$7 fs spectrum shows only charge states up to N^{4+}. The intensity of the N^{3+}, N^{4+}, and N^{5+} peaks decreases monotonically from 280 fs pulse duration to $\sim$7 fs pulse duration. This indicates that molecular N_2 and its fragments cannot absorb photons as efficiently for short pulses as for longer pulses. The effect shown in Fig. 1 can be explained by frustrated absorption after the removal of K-shell electrons with a short pulse. The finite Auger relaxation time determines how fast the K-shell can be replenished with electrons, and thereby limits the number of photons that can be absorbed in a given time interval. This leads to a reduction of multiple ionization events with increasing peak power and therefore a suppression of the production of higher charge states for short pulses.

To support the interpretation of our measurements within the concept of frustrated absorption, we have developed a theoretical description of the interaction of intense x-rays with nitrogen molecules based on a parameterized rate equation model. As shown in the following, these x-ray induced chemical dynamics are of fundamental importance to understand charge formation in samples that consist of more than a single atom. The model presented here is based on a complete set of atomic ionization cross sections and relaxation rates. The effects of valence electron dynamics on the measured partial ion yields are modeled by including valence charge exchange between different atoms. Experimental parameters such as the pulse shape and the light intensity distribution across the focal volume have been included.

Fig. 2a-d show the measured charge state distributions for four different x-ray pulse durations along with the results of the parameterized rate equation model. The average charge states $q_{av}(\tau)$ shown in Fig. 2e are derived by $q_{av}(\tau) = \sum_{i=1}^{7} N_i(\tau)q_i / \sum_{i=1}^{7} N_i(\tau)$ (1). Here, N_i are the relative abundances of the charge states q_i detected in the partial ion yield measurement. The frustrated absorption effect is clearly seen as a drop of the average charge towards short pulse durations and it is well reproduced by the model calculations. The brown band in Fig. 2e indicates the variation of the model calculations when adjusting the pulse energy between 17 % and 21 % of the nominal pulse energy (0.26 mJ). The good agreement of the band with the measured data indicates that the fundamental physics of the frustrated absorption effect is captured in the model. The extent of the theory band is a measure for the precision with which the pulse energy at the sample can be determined within the rate equation model. The effect of frustrated absorption on the average charge state is well reproduced by the parameterized rate equation model based entirely on available spectroscopic data, *ab-initio* calculations, and experimental parameters provided by LCLS.

To demonstrate the importance of molecular valence charge dynamics in the modeling of charge formation, the results of a rate equation model that is entirely based on an independent atom picture are included in Fig. 2 a-d as dashed lines. By comparing the atomic and molecular model, it shows that the inclusion of molecular valence electron dynamics in the theoretical description is instrumental to describe charge formation, and therefore damage formation, on the microscopic scale.

Beyond the fundamental importance of modeling a new regime of x-ray induced chemical dynamics, the ability to describe atom-specific transient charge formation in molecular samples will be vital to model damage mechanisms induced by the LCLS with an Angström spatial resolution and femtosecond time resolution. Fig. 3 provides a demonstration of how the findings described above can be employed to estimate charge-induced distortion effects in single-shot imaging of molecular samples. These distortion effects are directly related to the Coulomb repulsion between charged atoms within the molecular complex and thus to the time-dependent charging of the sample. In contrast to the asymptotic charge distributions that define the partial ion yields described in Fig. 2, only those charges that are present at the moment that the photons pass through the sample are of interest for imaging applications. An important benchmark for the achievable image quality is therefore the photon flux weighted charge as defined by $Q(\tau) = \int_{-\infty}^{+\infty} q(t)I_\tau(t)dt / \int_{-\infty}^{+\infty} I_\tau(t)dt$ (2). Here, q(t) is the transient average charge in the molecule

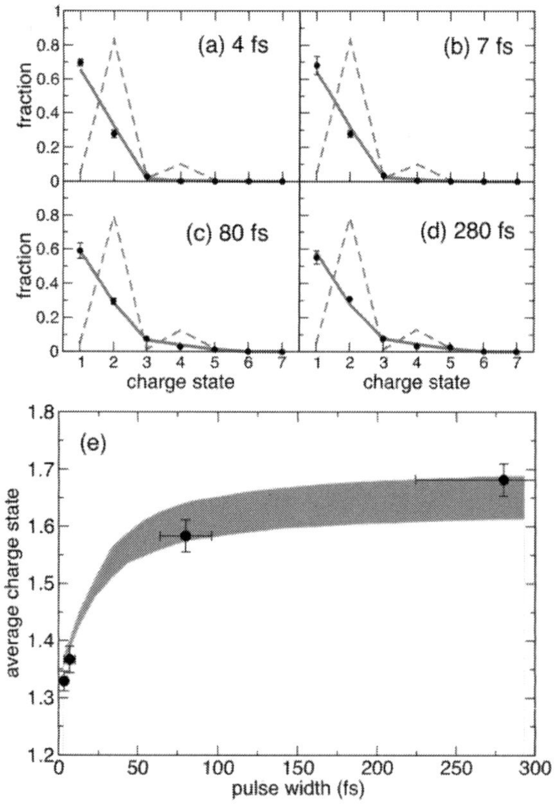

Figure 2. (a)-(d) Charge state distributions and (e) average charge states of N^{n+} ions generated in the photoionization and relaxation of N_2 molecules using x-ray pulses with different pulse durations but a constant pulse energy. The results of a rate equation model are shown as red lines (a)-(d) and brown band (e), respectively. The results of an isolated atom based rate equation model for a pulse energy of 49 μJ have been included as dashed lines in (a)-(d). (See text for details.)

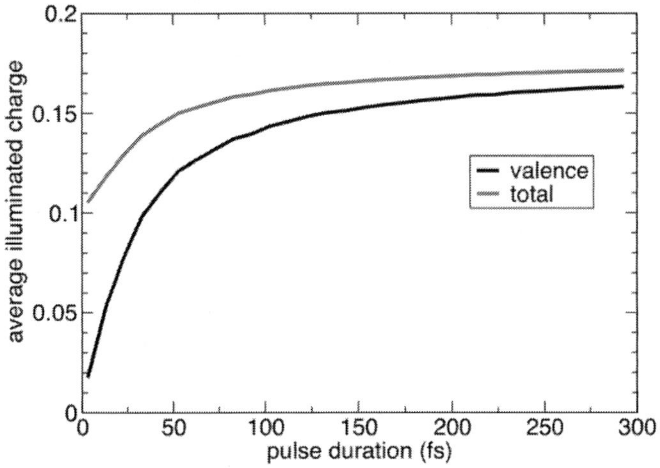

Figure 3. Model calculation of the illuminated charge per atom as a function of pulse length (the pulse energy is 49 μJ).

and $I_\tau(t)$ is the time dependent intensity of the x-ray pulse with a pulse envelope width τ. The illuminated charge $Q(\tau)$ represents the average charge that is encountered by an imaging photon while it traverses the sample. $Q(\tau)$ is significantly more affected by frustrated absorption than the asymptotic average charge $q_{av}(\tau)$ (Fig. 2 and 3). The illuminated valence charge is of particular interest since valence charges are predominantly responsible for distortion effects in the regime of low average charge. Core electrons contribute only little to molecular bonding. In addition, Coulomb repulsion between core holes that are shielded by a number of valence electrons is expected to be significantly smaller than the repulsion between valence charges. Thus, the results presented in Fig. 3 suggest that the relatively modest frustrated absorption effect presented in Fig. 2e might reduce the Coulomb forces that are ultimately responsible for target distortion during imaging applications by about one order of magnitude.

3. Double core hole formation and decays [30]

One intriguing possibility enabled by intense x-ray sources is the ability to produce atoms and molecules with multiple electron vacancies in core orbitals through the sequential absorption of multiple photons on a timescale faster than Auger decay, and it has been suggested that DCHs could provide the basis for richer, sensitive, spectroscopies than conventional inner-shell photoelectron spectroscopy [31, 32]. The different atomic sites in molecules introduce multiple possibilities for the DCH configurations, e.g., DCHs with both vacancies on a single site (DCHSS) and DCHs with single vacancies on two different sites (DCHTS). For such states, the presence and location of the first hole is predicted to affect the energy to produce the second hole, as well as the decay mechanisms and fragmentation patterns of the resulting DCH states [31]. The magnitude of the energy shifts of these states will provide unique spectral signatures as well as new information on the chemical environment of the core holes [33, 32]. While the intensity of conventional x-ray sources is too low to produce observable DCHs through sequential absorption processes, DCH states have been observed as a result of electron correlation in single-photon absorption. Unfortunately, these processes have a very low yield [34, 35], and only DCHSS states have been observed in this manner. At the intensities of the new x-ray FELs, photoabsorption should compete effectively with Auger decay [19, 21], allowing the production of both types of DCH states through sequential two-photon processes.

Figure 4 shows photoelectron spectra recorded at 1.0 keV and 1.1 keV photon energy and the theoretically predicted electron energies for the relevant photo-processes. The electron energies in the 1 keV spectrum have been shifted by 100 eV to higher kinetic energies. The labels indicate the energies of photoelectrons resulting from a number of different processes based on the calculations [17, 36]. These include : (1) Photoelectrons from single core-shell ionization of molecular N_2, N_2^+, and N_2^{2+}, where the initial state electron vacancies are located in the valence shells (dot-dashed blue lines). (2) Photoelectrons from core-shell ionization of atomic N^{m+} (initial state with m valence holes) (dashed black lines). (3) Photoelectrons from core-shell ionization of atomic $N^{m+}(1s)$, with a single core hole and the remaining holes in the valence shells (dotted green lines); the processes responsible for these photoelectrons result in atomic DCH production. (4) Photoelectrons from core-shell ionization of $N_2^+(1s)$ with a single hole in the 1s orbital of one N atom, where the second electron is removed either from the same atom, i.e., the DCHSS process (purple solid line), or from the other N atom i.e., the DCHTS process (red solid line). Most of the spectral features are the result of multiple photon absorption by a single molecule. In particular, processes (3) and (4) present sequential two-photon inner-shell double ionization events on timescales that are fast compared to Auger decay. For the higher charge states, the reduced number of valence electrons increases the Auger lifetime, and thus increases the probability of producing atomic DCH states. The peak identification is particularly clear for $N^{4+}(1s)$, $N^{5+}(1s)$, and $N^{6+}(1s)$ ionization, because no other photoelectrons are expected at energies below ~550 eV, and the observed peaks are in good agreement with the predicted line positions [36].

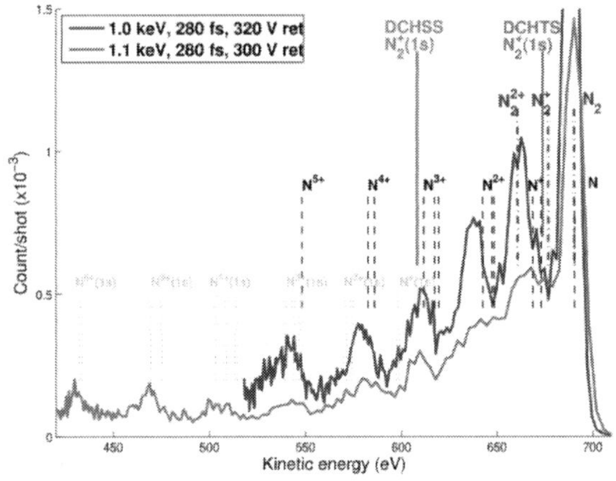

Figure 4. Photoelectron spectra recorded at 1.1 keV and 1.0 keV. The electron energies in the 1 keV spectrum have been shifted by 100 eV to higher kinetic energies. Vertical markers indicate the calculated photoelectron line positions for various molecular [17] and atomic states of N_2 [36]. (See text for details).

Fig. 5 shows the corresponding Auger spectra for N_2, where the solid curves are the experimental data and the dashed curves are our theoretical results. Above ~400 eV, no Auger electrons of N_2 have been observed with conventional x-ray sources. In contrast, the new spectra shown in Fig. 5 exhibit two new features lying 50-80 eV above the main SCH Auger peaks and with kinetic energies of 413 eV and 442 eV. These signals are clearly associated with Auger processes since the kinetic energies of the peaks do not vary with the photon energy. Contributions from photoelectrons observed in Fig. 4 are strongly suppressed in Fig. 5 due to the selected electron emission direction perpendicular to the light polarization. Our calculation shown in Fig. 5 (yellow shaded curve) predicts peaks at ~410 eV, which are produced by the Auger decay of the main DCHSS state of N_2^{2+} involving mainly $2\sigma_u$ and $1\pi_u$ electrons. In light of the expected underestimate in theory of 2-4 eV due to partially unaccounted electron relaxation in the final states [37], the prediction agrees very well with the observed peak position. We conclude that the newly observed Auger electron signal at ~410 eV kinetic energy stems from the decay of DCHSS states and is associated with so-called Auger hypersatellites [34, 35]. We estimate the DCHSS signal intensity to be ~1% of the main Auger peak signal between 355 eV- 370 eV (±5eV). Experiments have also been performed at the LCLS to characterize the angular distribution of these hypersatellite electrons [38].

Although calculations of the Auger decay of DCHSS shakeup states have not been performed

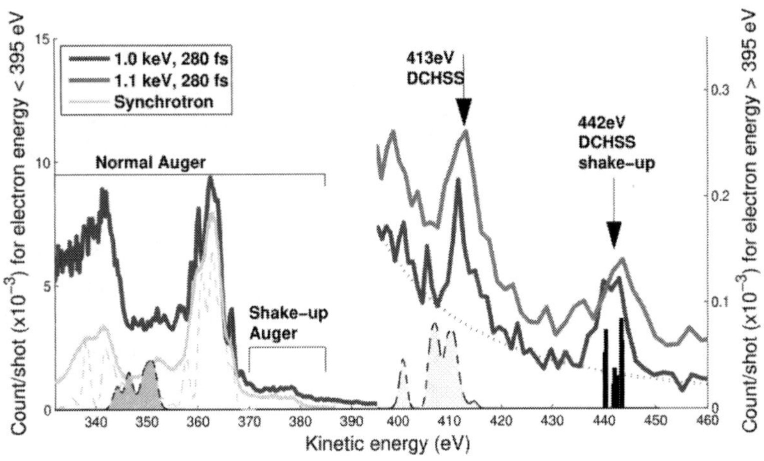

Figure 5. Auger spectra from the LCLS and synchrotron experiments, and from theoretical calculations. Thick solid curves: Auger spectra recorded at 1.0 keV and 1.1 keV. The 1.1 keV spectrum is offset vertically for clarity. Solid green curve: Auger spectrum recorded with synchrotron source [39]. Dashed curves: calculated Auger spectra (scaled) of various initial states, including SCH (green), DCHTS (grey shaded curve), DCHSS (yellow shaded curve). Black stems indicate predicted energies of DCHSS shake-up Auger electrons. Dotted curve: spline fit to background. (See text for details).

explicitly, the energy splitting between the DCHSS main Auger peak and the DCHSS shake-up Auger peaks can be estimated by the energy difference between the calculated DCHSS photoline and the DCHSS shake-up photoline. The predicted energies of the DCHSS shake-up Auger electrons are indicated by black stem lines in Fig. 5. The good agreement of the calculated positions with the peak at 442 eV suggests that this peak might correspond to the Auger decay of the DCHSS shake-up states.

The detection and assignment of electrons associated with the production of the uniquely molecular DCHTS states requires improved energy resolution, which is achieved by increasing the retardation voltage in the photoelectron spectrometer to 480 V, as shown in Fig. 6. The spectrum is analyzed by taking into account the contributions from single-photon shake-up/off (SUO) processes as previously determined by Svensson et al. [40] and Kaneyasu et al. [41], and convoluting them with the experimental energy resolution of the current study (blue shaded area in Fig. 6). To disentangle the major contributions generated by multiple photon ionization, a nonlinear least-squares fit is performed based on five spectral components with identical peak shapes and independent amplitudes and center energies (green and red shaded areas).

Within the model of identical, apparatus-limited peak shapes, the fit result is globally unique as determined by a Monte-Carlo based sampling of the available fit starting parameter space by $\sim 10^5$ independent fit cycles. One spectral component is consistently found exactly at the position of the expected DCHTS photoline (574 eV). As indicated by the red area, the estimate of the total multiple photon ionization signal at 574 eV varies between zero and 8% of the main photoline intensity. Given the uncertainty in the contributions from overlapping spectral components, we can provide only an upper bound for the DCHTS signal rather than an absolute value. The upper limit of the DCHTS contribution is mainly defined by the minimum contribution of the $N^+(1s^2 2s^1 2p^3)$ ionization signal. This leads to an upper bound for DCHTS contributions in the photoelectron spectrum of 4% relative to the intensity of the $N_2(1s^{-1})$ main photoline. This value is in qualitative agreement with the $\sim$1% DCHSS signal described above and an estimated ratio of $\sim$1.65 between DCHTS and DCHSS signal intensities. An independent estimate of

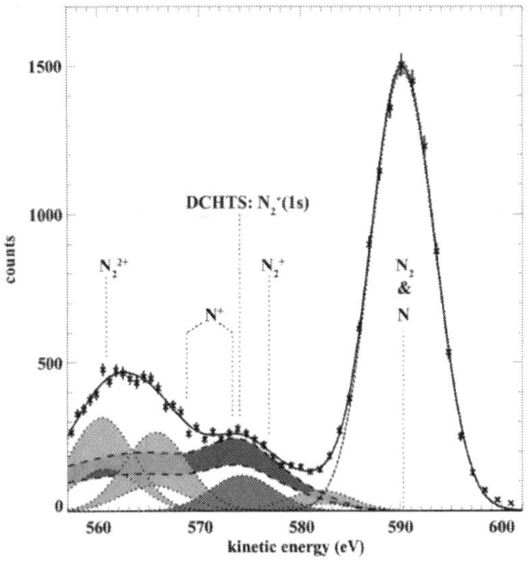

Figure 6. Photoelectron spectrum (crosses with error bars) recorded at 1.0 keV photon energy and 480 V spectrometer retardation voltage. The blue, green, and red shaded areas mark the correlated uncertainties for various spectral components (see text). Solid black curve: constant sum of all components. Vertical dashed lines: calculated line positions. The red shaded area allows for an upper limit estimate of DCHTS contributions.

the intensity of the DCHTS process by examination of the Auger spectrum is not possible at this stage. The nature of these states marked by a single core hole per atom leads to Auger electron energies that are embedded in the SCH Auger spectrum, as indicated by the grey shaded structure near 350 eV kinetic energy in Fig. 5.

4. Conclusion

In summary, the results presented here highlight some of the new frontiers in molecular physics that can be explored with high-intensity x-ray FELs. We present a microscopic picture of charge formation on a femtosecond time-scale and frustrated absorption in molecules that includes the decisive x-ray induced valence electron dynamics. The inclusion of molecular valence charge dynamics is fundamental for a correct description of frustrated absorption and time-dependent distortion effects in molecules. The present results are a significant step towards understanding ultra-intense x-ray induced physical processes and chemical dynamics in molecules. The predicted suppression of Coulomb repulsion effects in ultrashort x-ray pulses by one order of magnitude might have far reaching implications for single shot imaging applications, which are amongst the most promising fields of research at 4th Generation Light Sources. We also present the first experimental results on the characterization of DCH states produced by sequential two-photon absorption in molecules. The observation of DCHSS states is supported by photoelectron and Auger-electron spectra, as well as by theoretical calculations. In contrast, the observation of the uniquely molecular DCHTS states remains ambiguous, and only an upper bound to their production can be determined. Higher intensity (afforded by improved beam transport optics), improved stability of photon energy, higher electron energy resolution, and detailed studies of the intensity dependence of the electron spectra will thus be required to fulfill the potential of DCH spectroscopies.

5. Acknowledgments

This work was supported by the U.S. Department of Energy, Office of Science, Basic Energy Sciences, Division of Chemical Sciences, Geosciences, and Biosciences. This work was conducted in collaboration with O. Kornilov, O. Gessner, S.T. Pratt, F. Tarantelli, M. Gühr, E.P. Kanter, C. Blaga, C. Bostedt, J.D. Bozek, P.H. Bucksbaum, C. Buth, M. Chen, R. Coffee, J. Cryan, L. DiMauro, M. Glownia, E. Hosler, E. Kukk, S.R. Leone, B. McFarland, M. Messerschmidt, B. Murphy, V. Petrovic and D. Rolles. We thank the LCLS staff for their assistance, and A. Kivimäki, T. Jahnke and R. Dörner for providing us with unpublished data. We thank

Robin Santra for fruitful discussion. We thank R. Dörner, P. Emma, J. Frisch and H. Zhiron for their assistance. We thank B. Krässig for his eTOF models. FT thanks italian FIRB and PRIN grants. CBu was supported by NSF grants. MH thanks the Alexander von Humboldt foundation for the Feodor Lynen fellowship. The LCLS is funded by DOE-BES.

6. References

[1] Krausz F and Ivanov M 2009 *Rev. Mod. Phys.* **81** 163
[2] Ackermann W *et al.* 2007 *Nat. Photonics* **1** 336
[3] Gutt G *et al.* 2009 *Phys. Rev. B* **79** 212406
[4] Feldhaus J, Arthur J and Hastings J 2005 *J. Phys. B* **38** S799
[5] Emma P *et al.* 2004 *Phys. Rev. Lett.* **92** 074801 [Online] 2009. http://lcls.slac.stanford.edu/
[6] Neutze R *et al.* 2000 *Nature* **406** 752
[7] Chapman H N *et al.* 2006 *Nat. Phys.* **2** 839
[8] Lindl J D *et al.* 2004 *Phys. Plasmas* **11** 339
[9] Dyer G, Bernstein A, Cho B and Osterholz J 2008 *Phys. Plasmas* **101** 015002
[10] Berrah N *et al.* 2010 *J. Mod. Opt.* **57** 1015
[11] Bostedt C *et al.* 2008 *Phys. Rev. Lett.* **100** 133401
[12] Schirmer J and Barth A 1984 *Z. Phys. A* **317** 267
[13] Tarantelli A and Cederbaum L 1992 *Phys. Rev. A* **46** 81
[14] Tarantelli F 2006 *Chem. Phys.* **329** 11
[15] Dunning T H 1989 *J. Chem. Phys* **90** 1007
[16] Woon D E and Dunning T H 1995 *J. Chem. Phys* **103** 4572
[17] Tarantelli F, Sgamellotti A and Cederbaum L S 1991 *J. Chem. Phys* **94** 523
[18] Ding Y *et al.* 2009 *Phys. Rev. Lett.* **102** 254801
[19] Young L *et al.* 2010 *Nature* **466** 56
[20] Emma P *et al.* 2010 *Nature Photonics* (to be published)
[21] Hoener M *et al.* 2010 *Phys. Rev. Lett.* **104** 253002
[22] Bozek J D 2009 *Eur. Phys. J. Special Topics* **169** 129
[23] Jiang Y H *et al.* 2009 *Phys. Rev. Lett.* **102** 123002
[24] Kempgens B *et al.* 1996 *J. Phys. B* **29** 5389
[25] Moddeman W E *et al.* 1971 *J. Chem. Phys* **55** 2317
[26] Eberhardt W and J Stöhr 1983 *Phys. Rev. Lett.* **51** 2370
[27] Suzuki I and Saito N 1989 *J. Chem. Phys* **91** 5324
[28] Bogan M J *et al.* 2008 *Nano Lett.* **8** 310
[29] Marchesini S *et al.* 2008 *Nat. Photonics* **2** 560
[30] Fang L *et al.* 2010 *Phys. Rev. Lett.* **105** 083005
[31] Cederbaum L S, Tarantelli F and Sgamellotti A 1986 *J. Chem. Phys* **85** 6513
[32] Santra R, Kryzhevoi N V and Cederbaum L S 2009 *Phys. Rev. Lett.* **103** 013002
[33] Tashiro M *et al.* 2010 *J. Chem. Phys* **132** 184302
[34] Kanter E P, Dunford R W, B Krässig and Southworth S H 1999 *Phys. Rev. Lett.* **83** 508
[35] Southworth S H *et al.* 2003 *Phys. Rev. A* **67** 062712
[36] Chen M H, Laiman E, Crasemann B, Aoyagi M and Mark H 1979 *Phys. Rev. A* **19** 2253
[37] Schulte H D, Cederbaum L S and Tarantelli F 1996 *J. Phys. B* **105** 11108
[38] Cryan J *et al.* 2010 *Phys. Rev. Lett.* **105** 083004
[39] Kivimäki A unpublished data from Elletra synchrotron light source
[40] Svensson S *et al.* 1992 *J. Phys. B* **25** 135
[41] Kaneyasu T *et al.* 2008 *J. Phys. B* **41** 135101

Journal of Physics: Conference Series **288** (2011) 012020 doi:10.1088/1742-6596/288/1/012020

Application of a VMI spectrometer to near-threshold photoionization with synchrotron radiation

P. O'Keeffe[1], P. Bolognesi[1], A. Mihelic[2], R. Richter[3], A. Moise[3], E. Ovcharenko[1,4], G.C. King [5] and L. Avaldi[1,*]

[1] CNR Istituto di Metodologie Inorganiche e dei Plasmi, Area della Ricerca di Roma 1, CP10,00016 Monterotondo Scalo, Italy
[2] Jožef Stefan Institute, Jamova cesta 39, SI-1000 Ljubljana, Slovenia
[3] Sincrotrone Trieste, Area Science Park, I-34149 Basovizza (Trieste), Italy
[4] Institute of Electron Physics, Uzghorod, Ukraine
[5] Dept. of Physics and Astronomy, University of Manchester, Manchester, England

E-mail: Lorenzo.Avaldi@imip.cnr.it

Abstract. A new developed velocity map imaging spectrometer has been used to study the photoionization of atoms near threshold. The application of the spectrometer to the measurement of the angular distributions of the photoelectrons emitted in the photoionization of the Ne $2p_{3/2}$ state between the 2p spin orbit thresholds and of polarised Ne atoms are presented.

1. Introduction

Measurements of the *photoelectron angular distributions* (PADs) in the photoionization of atoms and molecules have long been known as a valuable tool to characterize the structure of the continuum and bound electronic states, to provide information on photoionization dynamics and to test theoretical models. Since the early experiments in the 1930s [1], PADs have been used to perform complete photoionization experiments, to observe effects due to site specific emission and alignment of fixed-in-space molecules, and to probe chirality and time-dependent dynamics [2 and references therein]. Renewed interest in the measurement of PADs has been triggered by the development of new highly efficient imaging detectors [3] and the availability of new sources, such as third generation synchrotron radiation facilities, fast and intense lasers for multi-photon ionisation and, more recently, by the possibility to combine these two photon sources to tackle processes previously not accessible [4].

The near threshold photoionization offers the opportunity to study different interesting processes, which are illustrated in the case of Ne in figure 1. In case (a) the ionisation occurs in the region between the two spin-orbit thresholds. This region is populated by the $2p^5(^2P_{1/2})$ *nl* Rydberg states that can decay to the Ne^+ $2p_{3/2}$ continuum and therefore affect the cross section and angular distribution of the phototoelectrons.

*To whom any correspondence should be addressed.

In case (b) synchrotron radiation (SR) is used to excite the neutral atoms and then an IR laser ionizes the polarised atom. In this work a novel VMI spectrometer specialized to detect low energy electrons has been used to investigate these two processes which characterize the near threshold photoionization of Ne.

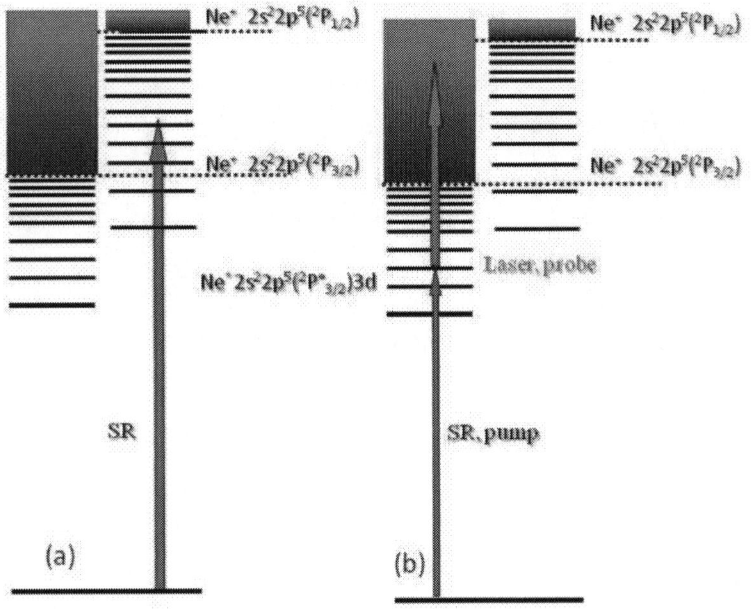

Figure.1. Scheme of the Ne energy levels near the 2p thresholds and of the ionizing processes that can be achieved combining VUV-SR and IR photon sources.

2. Experiment and data analysis

The experiments have been performed at the Gasphase photoemission beamline at the Elettra storage ring (Trieste, Italy). The layout, characteristics and performances of the beamline have been described previously [5] and will be not repeated here. Recently a branch line of the GasPhase beamline has been equipped with a mode-locked Ti:Sapphire oscillator (Tsunami, Spectra Physics) [6] modified to allow operation at a frequency of 83.3MHz, which corresponds to exactly 1/6 of the storage ring frequency. The laser is operated in the picosecond configuration with a measured pulse width of 15 ps, in order to achieve a better temporal overlap with the synchrotron pulses (about 60 ps).

The present VMI spectrometer design (see figure 2) is based on a classical three element system (20 mm aperture, 15 mm spacing) and a 190 mm time of flight tube, at the end of which a set of two MCPs is placed. The spectrometer is equipped with a position sensitive detector [7] developed by the Instrumentation and Detector Laboratory at Elettra. The UHV section of the detector consists of two 40 mm MCPs mounted in a Chevron configuration, and a cross delay line detector for X,Y position identification. The spatial resolution is about 100 μm in one direction and 150 μm in the other. The X,Y start and stop signals from the two ends of each delay line of the detector are immediately amplified at the air side of the UHV flange. The amplified pulses are sent to constant fraction discriminators and then fed into time-to-digital converters (TDCs). The TDCs are controlled by a mother board connected via an external SCSI cable to a PCI Interface board inside a PC. Finally a Labview code manages the data acquisition, analysis and on-line graphical presentation.

The atoms or molecules under investigation are introduced into the chamber via a continuous supersonic beam formed by expanding 1-2 bars of gas through a 50 μm nozzle. This nozzle is mounted on an XYZ-manipulator in order to allow the nozzle to be centered on a 0.5 mm skimmer, the tip of which is placed at a distance of about 12 cm from the interaction zone. In the two colour experiments the synchrotron and laser beams are focussed and overlapped (both spatially and temporally) at the center of the chamber in a counter propagating geometry. Optimization of the overlap of the two light beams with the supersonic jet was achieved by monitoring a double resonant two-colour signal and adjusting the position of the laser beam and of the entire experimental chamber along all three axes.

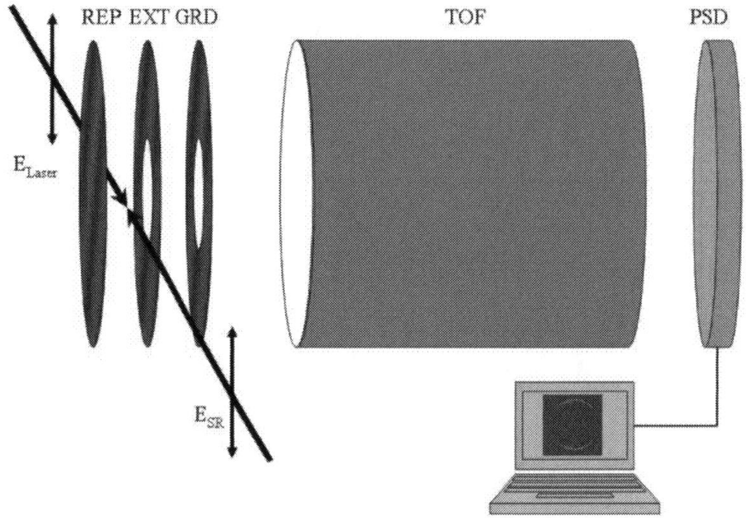

Figure 2. Schematic representation of the VMI spectrometer where REP, EXT and GRD are the repeller, extractor and ground electrodes of the set-up, respectively. TOF represents the field-free time of flight region and PSD is the position sensitive detector (figure not on scale). The directions of the polarizations of the light sources are also shown.

The chamber is lined with a 2mm thick μ-metal shield. This allows the measurement of the angular distribution of photoelectrons with kinetic energy as low as 13 meV with a typical energy resolution of 6% when the detector is used in the high resolution mode [8].The detector can be operated also in a coincidence mode to improve signal to noise in pump-probe experiments either by using a gate to count electrons only when the laser is present (down to a 27 ps gate) or by recording coincidences with the ion formed in the ionization process [8]. Typical acquisition times of an image were of the order of 5 minutes for one photon and 40 minutes for singly resonant two-photon ionisation processes.

The VMI technique works by projecting the 3D distribution of electrons released in the ionization process onto the 2D position sensitive detector. It is then possible to recover the PAD and kinetic energy distribution (KED) by performing an inverse Abel transformation or other inversion methods [9,10,11]. The data analysis in this work was performed using the pBasex method introduced by Garcia *et al.* [11] combined with a new Windows type graphical user interface which has extensive image manipulation features [12].

The reported anisotropy parameters are determined using a weighted average of all points contained within the full width half maximum of the peak in the kinetic energy distribution (as suggested in ref. [11]). The errors are then determined based on the spread of the anisotropy parameters extracted by varying the centre point of the image and other parameters related to the image processing.

3.Results

In one-photon ionisation of randomly oriented targets by fully linearly polarized radiation the photoelectron angular distribution is represented by the double differential cross section

$$\frac{d^2\sigma(E,\theta)}{dEd\Omega} = \frac{\sigma_0}{4\pi}[1+\beta P_2(\cos\theta)] \qquad (1)$$

where σ_0 is the total photoionization cross section, θ the angle of the photoelectron with respect to the polarization direction of the incident radiation, $P_2(\cos\theta)$ is the second order Legendre polynomial and β the asymmetry parameter. The β parameter holds information on the photoionization dynamics, because it depends on the radial matrix elements, $R_{l\pm1}$, and their relative phase, δ_{l-1}-δ_{l+1}, where $l\pm1$ are the final state partial waves of the free photoelectron in the continuum. The interference between the $l\pm1$ partial waves determines the variation of β with energy in its allowed range (-1,2).

In the two-photon ionisation when the two photon beams have parallel polarization, the angular distributions are of the form

$$\frac{d^2\sigma(E,\theta)}{dEd\Omega} = \frac{\sigma_0}{4\pi}[1+\beta_2 P_2(\cos\theta)+\beta_4 P_4(\cos\theta)] \qquad (2)$$

where both the β_2 and β_4 anisotropy parameters are combinations of $R_{l\pm1}$, and δ_{l-1}-δ_{l+1}. Thus an experimental determination of β_2 and β_4 allows the extraction of at least two (even though not all) independent quantities governing the photoionization, i.e. the ratio $\chi=R_{l+1}(j)/R_{l-1}(j)$ of the matrix elements and their relative phase.

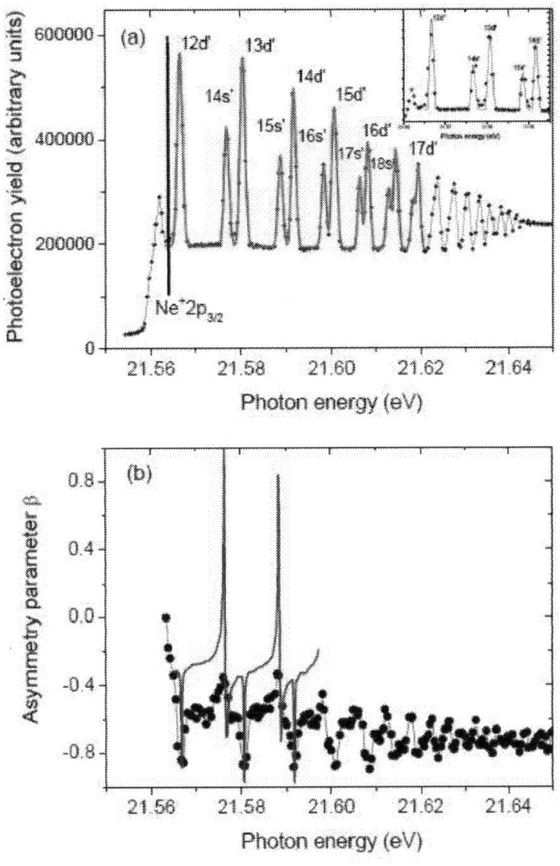

Figure3. The photoelectron electron yield (a) and the β_2 anisotropy parameter (b) of Ne $2p_{3/2}$ The full line in (a) is a fit to the experimental data, while in (b) is the theoretical prediction by [16]. In the inset of figure (a) the experimental spectrum is compared with the theoretical results of ref. [16] convoluted with the experimental resolution.

3.1 The photoionization of Ne $2p_{3/2}$ between the 2p spin orbit thresholds

The high members of the nd' and ns' Rydberg series converging to the $^2P_{1/2}$ ionic states have been studied in several photoabsorption and photoionization experiments [13]. The small energy splitting

between the two fine structure components (97 meV) has made photoelectron investigation difficult and only Caldwell and Krause [14] reported a measurement of the photoelectron spectra and angular distribution over a region of about 30 meV, which includes the 12-14d′ and 14-15s′ resonances. In figure 3 the results of the present measurements are reported. The overall energy resolution and efficiency of the present experiment allow to distinguish clearly the two Rydberg series up to the 18s′ and 17d′. From there on, the two series merge up to the ionization continuum. As for the asymmetry parameter we observe that the autoionizing resonances produce sharp variations superimposed on a β value which is slowly decreasing towards a value of about -0.7 near the $^2P_{1/2}$ ionic threshold. The behaviour of the asymmetry parameter is consistent with the measurements by Southworth et al [15] who observed a minimum value of -0.6 at about 22.5 eV and a rise towards threshold. Caldwell and Krause [13] quoted a value of -0.17±0.15 in the region between the 13d′ and 15s′ resonances, which is higher of the ones measured here. The resonances result in β values less negative at the position of the ns′ peaks, while the opposite occurs at the nd′ peaks.
Radjević and Talman [16] calculated the photoionisation parameter in the region up to 21.95eV using the relativistic random phase approximation together with the relativistic multichannel quantum defect theory. Their spectrum convoluted with the present experimental resolution is compared with the experiments in the inset of figure 3a , while the calculated asymmetry parameters are compared with the experiments in figure 3b. A good agreement in the relative position and intensity of the peaks in the absorption spectrum is observed; the calculated asymmetry parameters agree in shape but not in absolute value, the theoretical parameters being more positive than the experimental ones.

3.2 The photoionization of the polarized Ne atoms

In this experiment the synchrotron radiation was set to 20.04 eV, in order to excite the Ne $2p^6$ ($^1S_0$)$\rightarrow 2p^5(^2P^0_{3/2})3d[3/2]_1$ transition, which is characterized by a quite large oscillator strength ($1.86(\pm 0.09)\mathrm{x}10^{-2}$ [17]), and then three laser wavelengths between 775 and 805 nm have been chosen to ionize the excited state. Thus all the measurements are within 90 meV from the Ne^+ $2p^{-1}$ ($^2P_{3/2}$) ionization threshold, i.e. below the Ne^+ $2p^{-1}$ ($^2P_{1/2}$) ionization threshold. The photon energies were chosen not to lie close to or on-resonance with the resonances in the autoionization region. The measured β_2 and β_4 values are reported in Table 1 and figure 4.

Table 1. The β_2 and β_4 anisotropy parameters for the photoionization of Ne $2p3d[3/2]_1$ intermediate state.

Photoelectron kinetic energy (meV)	β_2	β_4
20	0.84 ± 0.06	0.42 ± 0.02
45	0.87 ± 0.03	0.40 ± 0.02
70	0.93 ± 0.08	0.39 ± 0.04

The β_2 values slowly increase for photon energies moving away from threshold, while β_4 stays almost constant in the region investigated. Despite the small variation of the asymmetry parameters both the observed images and the 3D angular distributions reconstructed from the asymmetry parameters display clear differences, as shown in figure 4. To our best knowledge the only calculation of the anisotropy parameters for the photoionization of a Ne excited state is the one due to Chang et al [18] for Ne $2p^5 3p$ while no data exists for the Ne$2p^5nd$ states. Thus we have calculated the angular distributions using a single-channel quantum defect theory [19]. In this method the interaction between the channels with different angular momenta of the atomic core is neglected. The exact energies of the ground and $2p^5nl$ (l=p, f) excited states converging to the ionization threshold have been taken from the NIST database [20]. In addition, bound-bound dipole matrix elements for the transitions to the aforementioned $2p^5nl$ states have been calculated with GRASP Dirac-Fock codes [21]. The oscillator strength distribution has been extrapolated from the discrete part of the spectrum across the threshold

[22], and the values of the transition matrix elements have been determined from the extrapolated values. The resonances in the autoionization region have been treated as purely bound in the extrapolation procedure. A good agreement between calculations and experiment is found in the region investigated.

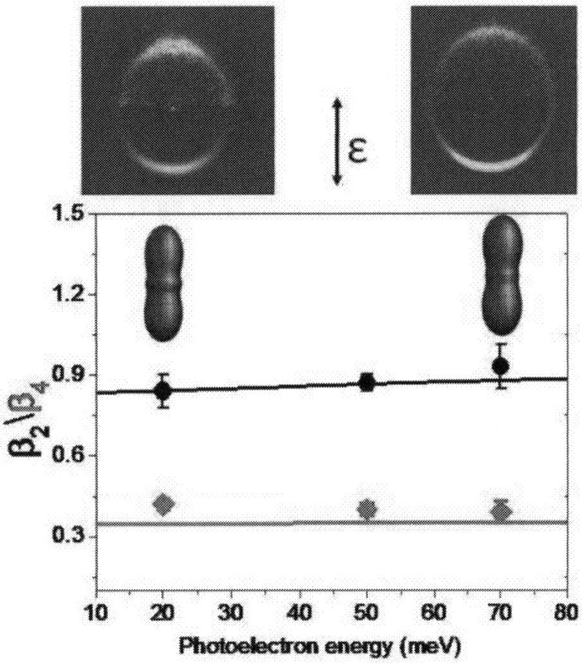

Figure 4. The β_2 and β_4 anisotropy parameters for the photoionization of the Ne $2p^5 3d$ $[3/2]_1$ excited state. The full line are the prediction of a single channel calculation (see text). On the top of the figure the maps of the angular distribution are represented. The upper part of each map contains the raw image, while the lower part shows the pBasex inverted image. The angular distributions are depicted graphically as surfaces in the insets where the distance from the origin to the surface is proportional to the probability that the electron is emitted in that direction.

4. Conclusions

A novel experimental set-up using a VMI spectrometer in conjunction with a VUV synchrotron radiation and IR laser source has been used to measure the photoelectron angular distribution in near threshold photoionization of Ne. In the case of the direct photoionization of the Ne $2p_{3/2}$ state between the two spin orbit thresholds the efficiency and energy resolution of the spectrometer has allowed to extend the investigation of autoionisation processes to higher member of the Rydberg series.

The velocity map imaging detection scheme used for the measurement of the photoelectron angular distributions in the pump-probe experiments allowed the extraction of β_2 and β_4 and therefore provides more information on the ionization process than geometries in which fixed 1D-detectors are used and only the polarization of the ionizing light is changed [23]. The good agreement between the experiment and the calculations indicate that a single channel quantum defect model which accounts for the dependence of the transition matrix elements on the channel quantum numbers of the continuum wavefunction is accurate enough to describe the two-colour Ne $2p_{3/2}$ photoionization near threshold.

5. Acknowledgments
This research was partly supported by a Marie Curie Reintegration Grant within the 7th European Community Framework Programme. E. Ovcharenko acknowledges the support of the Elettra-ICTP Program for supporting his participation to the measurements at Elettra.

References

[1] Lawrence EO and Chaffee MA 1930 *Phys. Rev.* **36** 1099
[2] Reid KL 2003 *Annu. Rev. Phys. Chem.* **54** 397
[3] Eppink AT and Parker JB 1997 *Rev. Sci. Instrum.* **68** 3477
[4] Meyer M *et al* 1987 *Phys. Rev. Lett.* **59** 2963
[5] Blyth RR *et al* 1999 J. *Electron Spectrosc. Relat. Phenom.* **101-103** 959
[6] Moise A, Alagia M, Banchi L, Ferianis M, Prince KC and Richter R 2008 *Nucl. Instr. Meth. A* **588** 502
[7] Cautero G, Sergo R, Stebel L, Lacovig P, Pittana P, Predonzani M and Carrato S 2008 *Nucl. Instr. Meth. A* **595** 447
[8] O'Keeffe P. *et al* 2010 *Rev. Sci. Instrum. in preparation*
[9] Vrakking MJJ 2001 *Rev. Sci. Instrum.* **72** 4084
[10] Dribinski V, Ossadtchi A, Mandelstam VA, and Reisler H 2002 *Rev. Sci. Instrum.* **73** 2634
[11] Garcia GA, Nahon L and Powis I 2004 *Rev. Sci. Instrum.* **75** 4989
[12] Program available from the authors on request.
[13] Radler K and Berkowitz J 1979 *J. Chem. Phys.* **70** 216; Baing MA and Connerade JP 1984 *J. Phys. B: At. Mol. Opt. Phys.* **17** 1785; Ito K, Ueda K, Namioka T, Yoshino K and Morioka Y 1988 *J. Opt. Soc. Am.* B **5** 2006
[14] Caldwell D and Krause M O 1990 *J. Phys. B: At. Mol. Opt. Phys.* **23** 2233
[15] Soutworth SH, Parr AC, Hardis JE, Dehmer JL and Holland DMP 1986 *Nucl. Instr. Meth. A* **246** 782
[16] Radoiević V and Talman JD 1990 *J. Phys. B: At. Mol. Opt. Phys.* **23** 2241
[17] Chan WF, Cooper G, Guo X and Brion CE 1992 *Phys. Rev.* **A** 45 1420
[18] Chang TN and Kim YS 1982 *Phys. Rev.* A **26** 2728
[19] O'Keeffe P et al. 2010 *Phys. Rev. A* submitted
[20] NIST Atomic Spectra Database, URL http://www.nist.gov/physlab/data/asd.cfm.
[21] Dyall KG, Grant I., Johnson CT, Parpia FA and Plummer EP 1989 *Comput. Phys. Commun.* **55** 425
[22] Fano U and Cooper JW 1968 *Rev. Mod. Phys.* **40** 441
[23] Mitsuke K,. Hikosaka Y, and Iwasaki K 2000 *J. Phys. B: At. Mol. Opt. Phys.* **33** 391

Journal of Physics: Conference Series **288** (2011) 012021 doi:10.1088/1742-6596/288/1/012021

Dissociative photoionization of perfluorocyclobutane and *cis*-1,1,2,2,3,4-hexafluorocyclobutane

K Okada[1], T Nakashima[1], M Sakai[1], A Suemitsu[1], C Huang[2,4], H Yagi[2,5], H Katayanagi[2,3], K Mitsuke[2,3] and K Tabayashi[1,6]

[1] Department of Chemistry, Graduate School of Science, Hiroshima University, Higashi-Hiroshima 739-8526, Japan
[2] Institute for Molecular Science, Okazaki 444-8585, Japan
[3] Graduate University for Advanced Studies, Okazaki 444-8585, Japan

E-mail: okadak@sci.hiroshima-u.ac.jp

Abstract. Ion yield spectra produced by the photoionization of perfluorocyclobutane (*c*-C_4F_8) and *cis*-1,1,2,2,3,4-hexafluorocyclobutane (*cis*-*c*-$C_4H_2F_6$) have been measured in the photon energy range of 25–170 eV. For the *c*-C_4F_8 photoionization, the $C_3F_5^+$ and $C_2F_4^+$ fragments are most abundant at all the energies, but the yields decline with the energy, giving at 170 eV about 30% of the yields at 30 eV. The $C_3H_2F_3^+$ and $C_2HF_3^+$ ions are the most dominant fragments for the *cis*-*c*-$C_4H_2F_6$ photoionization, with their yield curves similar to the $C_3F_5^+$ and $C_2F_4^+$ counterparts for *c*-C_4F_8, respectively. Photoelectron–photoion–photoion coincidence spectra were also acquired to study the breakdown pathways.

1. Introduction

Molecular photoionization and ionic fragmentation processes are of fundamental importance in the upper-atmospheric chemistry and plasma physics. Perfluorocyclobutane (*c*-C_4F_8, FC-c318) is extensively used as a reagent for dry etching of semiconductors. However, the use of the gas in industry has atmospheric implications for global warming because *c*-C_4F_8 has a high global warming potential (GWP) of 8700 [1]. The estimated atmospheric lifetime depends on the sink mechanism: Ravishankara *et al* [2] reported that the lifetime of this molecule in the atmosphere is 3200 years. Later, Morris *et al* [3] argued that it is reduced to 1400 years if considering the atmospheric removal of *c*-C_4F_8 by electron interactions. The knowledge of its chemical processes in the upper atmosphere is eagerly required to evaluate the GWP of *c*-C_4F_8.

The photochemical process of *c*-C_4F_8 has not been fully established yet. Tuckett's group [4,5] reported the threshold photoelectron and threshold-photoelectron–photoion coincidence (TPEPICO) spectra of *c*-C_4F_8 in the photon energy range of 10–27 eV. Ion yield curves and branching ratios were

[4] Present address: Department of Chemistry, University of Arizona, Tucson, AZ 85721, USA
[5] Present address: Department of Materials Science and Biotechnology, Graduate School of Science and Engineering, Ehime University, Matsuyama 790-8577, Japan
[6] Present address: Hiroshima Synchrotron Radiation Center, Hiroshima University, Higashi-Hiroshima 739-0046, Japan

determined for the $C_3F_5^+$, $C_2F_4^+$, CF_3^+, CF_2^+ and CF^+ fragment ions. More recently, Limão-Vieira *et al* [6] studied the electron spectroscopy of *c*-C_4F_8 and derive a precise value of 12.291 eV for the vertical ionization energy. As far as we are aware, however, no study has been reported on the ion yield curves above 27 eV. We therefore measured in this study the yield spectra of the fragment ions produced by the photoionization of *c*-C_4F_8 in the 25–170 eV region. The region covers the multiply-charged molecular states, because the appearance energy for the *c*-$C_4F_8^{2+}$ ion is estimated as ~ 34 eV according to the rule of thumb given in ref. [7].

In recent years, significant effort has been devoted to develop the most promising alternative chemicals to perfluorocarbons (PFCs) for etching and chamber cleaning processes. While a practical use in semiconductor device fabrication requires good etching characteristics of the alternatives comparable to the PFCs as well as lower cost, the alternatives have short atmospheric lifetime and low GWP. Thus, we also carried out the same measurements for *cis*-1,1,2,2,3,4-hexafluorocyclobutane (*cis*-*c*-$C_4H_2F_6$, HFC-c336ee), a candidate for replacing *c*-C_4F_8, to provide the photochemical data.

In addition, ion pairs originating from charge separation of the multiply-charged molecular ions were detected with a photoelectron–photoion–photoion coincidence (PEPIPICO) technique to obtain information on the breakdown pathways. A demonstration of this technique for the *c*-$C_4F_8^{2+}$ dissociation was given by Hsieh and Eland [8], but was illustrated only for the reaction *c*-$C_4F_8^{2+} \rightarrow CF_3^+ + C_2F_4^+ + CF$ at 30.4 nm; The PEPIPICO spectrum itself has already been provided in an earlier, separate paper [9]. A more detailed investigation is necessary for discussing the breakdown pathways.

The next section describes our experimental setup and methods. We then discuss the ion yields obtained by the dissociative photoionization of *c*-C_4F_8 and *cis*-*c*-$C_4H_2F_6$. Several breakdown pathways are deduced from their PEPIPICO spectra. Finally, we summarize the conclusions of this study.

2. Experimental

The experiments have been performed on the beamline BL2B at the UVSOR facility, Institute for Molecular Science in Japan. The experimental setup has been described in previous papers [10,11], except a minor change for the purpose of this study: A thickness monitor was removed and a needle for gas inlet was installed. Synchrotron radiation was irradiated at right angles to the effusive beam of the sample gas. Fragment ions produced in the ionization region were extracted toward a time-of-flight (TOF) spectrometer by a pulsed electric field applied at its extractor plate. A microchannel plate (MCP) operated in pulse-counting mode served for the detection of the ions. The pulse and ion-detection signals were fed into the start and stop inputs of a multi-stop time-to-digital converter (FAST ComTec, model P7888), respectively, to record the TOF signals. Partial ion yield (PIY) spectra were obtained by measuring a series of TOF spectra while scanning the photon energy. The spectra were obtained over the energy range of 25–170 eV which can be covered by three gratings installed on the beamline [12]. Ion signal counts were integrated over the respective TOF mass peaks. The integrated ion counts were normalized with the fluxes of synchrotron radiation at the each photon energy.

Breakdown pathways of the multiply-charged molecular ions were also investigated by applying a PEPIPICO technique for detecting a pair of fragment ions resulting from the same ionization event. In this measurement mode continuous electrostatic filed was applied over the ionization region. Emitted electrons were detected with another MCP which was set in the opposite direction to the ion TOF tube. The start of the multi-stop time-to-digital converter was triggered by the electron signals instead of the electric pulses.

The chemicals *c*-C_4F_8 and *cis*-*c*-$C_4H_2F_6$ for this study were purchased from SynQuest Laboratories, Inc. The sample of *c*-C_4F_8 was used as provided while that of *cis*-*c*-$C_4H_2F_6$ was carefully degassed under vacuum by repeated freeze–pump–thaw cycles before use. The pressure in the chamber during the measurements was kept at ~ 2×10^{-6} Torr (1 Torr = 133.2 Pa).

3. Results and discussion

TOF spectra were measured to study the yield of fragment ions formed by the VUV photoabsorption.

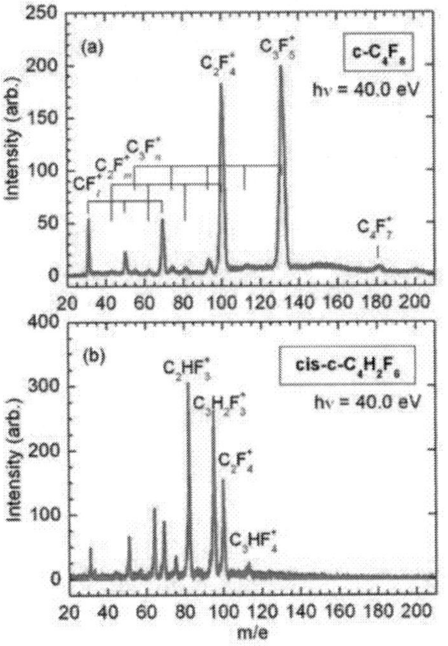

Figure 1. Typical mass spectra obtained by the photoionization of (a) c-C_4F_8 and (b) cis-c-$C_4H_2F_6$ measured at 40.0 eV. Virtually no peaks are seen in the range $m/e \leq 20$.

Figure 1a displays a typical mass spectrum measured at 40.0 eV. The abscissa has been converted from the flight time of ions to the mass/charge ratio. A variety of fragment ions were detected, i.e., CF_l^+ (l = 1–3), $C_2F_m^+$ (m = 1–4), $C_3F_n^+$ (n = 1–5) and $C_4F_7^+$. This is in accordance with the results by Eland *et al* (figure 3a of ref. [9]). However, it is slightly different from the report by Jarvis *et al* [4], in which $C_4F_7^+$ could not be detected by the measurement of TPEPICO spectra acquired at photon energies of 10–27 eV. The production of $C_4F_7^+$ is reasonable as observed in this study, if we consider the fragmentation of other perfluorocarbon compounds, for example, perfluoro-2-butene (2-C_4F_8) [4].

Figure 2 gives PIY curves of the fragment ions. A reasonable agreement is obtained between the curves in the regions which two gratings can cover. While the $C_3F_5^+$ and $C_2F_4^+$ fragments are most

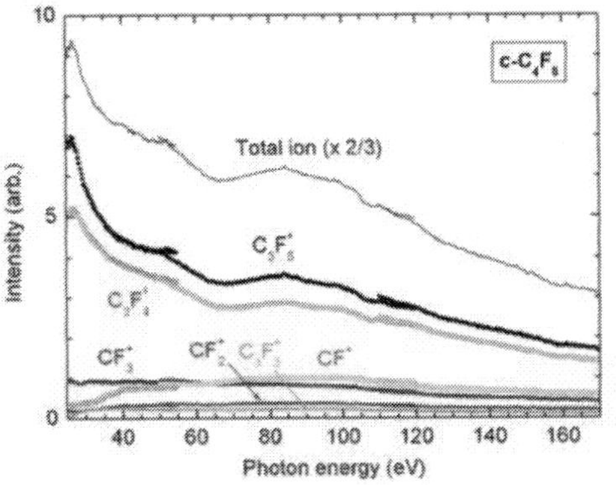

Figure 2. Partial ion yield curves of the fragment ions produced by the photoionization of c-C_4F_8. The total ion yield is also shown and is scaled by two-thirds.

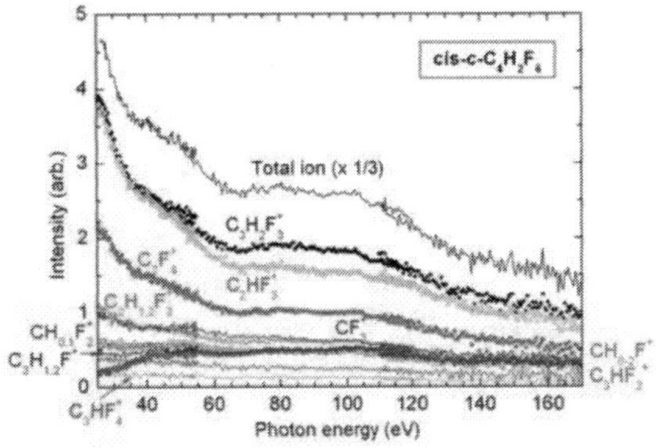

Figure 3. Partial ion yield curves of the fragment ions produced by the photoionization of cis-c-$C_4H_2F_6$. The total ion yield is also shown and is scaled by a third.

abundant at all the energies studied here, the yield of these ions decreases greatly with the photon energy. The yield at 170 eV becomes about 30% of that at 30 eV. The yield of CF_3^+ gives an approximately constant value up to 55 eV and decreases monotonically at the higher energies. The yield of CF^+ increases with the photon energy up to ~ 90 eV and then gradually decreases. Table 1 summarizes the branching ratio of the fragment ions at selected excitation energies. The ions are listed in order of abundance at 30.0 eV. Relative increase of the CF^+ and $C_3F_2^+$ ion yields is found with the energy. This indicates that at higher energies the larger $C_3F_5^+$ and $C_2F_4^+$ fragments further decompose to emit F atoms or CF/CF^+ fragments.

Similar yield curves were observed for the photofragmentation of *cis*-*c*-$C_4H_2F_6$. Figure 1b shows a typical mass spectrum measured at 40.0 eV. A variety of fragment ions were detected, such as CH_kF^+ (k = 0–2), $CH_{0,1}F_2^+$, $C_2H_{1,2}F_2^+$, CF_3^+, $C_2HF_3^+$, $C_3H_2F_3^+$ and $C_2F_4^+$. The peak of $C_3HF_4^+$ is discernible, but that of $C_3F_5^+$ is not. This infers that the C–H bond in the CHF group is not scissile. Figure 3 exhibits the PIY curves of the fragment ions. The $C_3H_2F_3^+$ and $C_2HF_3^+$ fragments are most abundant at all the photon energies and their yields decline with the energy, just like the yield curves of the $C_3F_5^+$ and $C_2F_4^+$ fragments from *c*-C_4F_8, respectively. Yield curves are highly similar between CF_3^+ from *c*-C_4F_8 and CHF_2^+ from *cis*-*c*-$C_4H_2F_6$. In the tail of the intense $C_3H_2F_3^+$ peak the $C_3HF_3^+$ component might exist, but its relative contribution is small, judging from the peak shape. Careful determination was made in this way for the higher mass peaks $m/e \geq 50$.

Table 2 outlines the branching fraction of the fragment ions of *cis*-*c*-$C_4H_2F_6$ at selected excitation energies. Increase in the fraction of the $CH_{0-2}F^+$ and $C_3H_{1,2}F^+$ ions is found, showing that at higher energies the abundant $C_3H_2F_3^+$, $C_2HF_3^+$ and $C_2F_4^+$ fragments further decompose to form these ions. The approximately constant value of 17 for the PIY ratio between $C_3H_2F_3^+$ and $C_3HF_4^+$ reveals that the loss of a CF_2 group is about 17 times easier than that of a CHF group irrespective to the photon energy, if we assume that the CHF/CF_2 dissociation is initiated by removal of a fluorine atom. This fact also explains the lower branching fraction of the $C_3H_2F_3^+$ and $C_3HF_4^+$ fragments compared to the branching fraction of $C_3F_5^+$ for *c*-C_4F_8. Reversely, the total branching fraction for $C_2H_kF_{4-k}^+$ (k = 0–2) production is larger for *cis*-*c*-$C_4H_2F_6$, being 1.2 times as much as the branching fraction of $C_2F_4^+$ from *c*-C_4F_8.

In addition, PEPIPICO spectra were acquired to study the breakdown pathways of the multiply-charged *c*-$C_4F_8^{n+}$ and *cis*-*c*-$C_4H_2F_6^{n+}$ ($n \geq 2$) molecular ions. Figure 4 depicts a typical spectrum obtained for *c*-C_4F_8 at the excitation energy of 45.0 eV. The flight time of the ions arrived at the ion detector first was taken as the abscissa, while that of the ions detected afterwards as the ordinate. Several coincidence islands can be found in the spectrum, along with ridges arising from false coincidences with background noises. They are $C_3F_5^+/CF_2^+$, $C_3F_4^+/CF_2^+$, $C_2F_4^+/CF_2^+$, $C_3F_3^+/CF_2^+$, CF_3^+/CF_2^+, $C_2F_4^+/CF_3^+$, $C_3F_3^+/CF_3^+$, $C_2F_4^+/CF^+$, CF_3^+/CF^+ and CF_2^+/CF^+ fragment pairs. The following breakdown pathways can be identified from these pairs:

$$c\text{-}C_4F_8^{2+} \rightarrow C_3F_5^+ + CF_2^+ + F \quad (1)$$

Table 1. Branching fraction (%) of the fragment ions formed by the dissociative photo-ionization of *c*-C_4F_8.

Ion	Photon energy (eV)					
	30.0	40.0	70.0	100.0	130.0	170.0
$C_3F_5^+$	46.2	40.6	37.8	37.2	37.5	36.3
$C_2F_4^+$	37.5	35.0	31.1	30.8	31.5	30.0
CF_3^+	7.2	8.2	9.6	8.9	8.3	8.8
CF^+	3.1	6.4	10.4	11.3	11.2	12.6
CF_2^+	1.7	3.1	3.9	4.1	3.9	4.2
$C_3F_3^+$	1.7	2.3	2.3	2.3	2.2	2.5
$C_4F_7^+$	1.1	0.8	0.6	0.8	0.8	0.6
C_3F^+	0.2	0.4	0.9	0.9	1.0	1.1
$C_3F_2^+$	0.2	1.0	1.2	1.3	1.4	1.6

Table 2. Branching fraction (%) of the fragment ions formed by the dissociative photoionization of *cis*-*c*-$C_4H_2F_6$.

Ion	Photon energy (eV)					
	30.0	40.0	70.0	100.0	130.0	170.0
$C_3H_2F_3^+$	26.3	24.3	23.2	23.2	23.3	22.4
$C_2HF_3^+$	25.6	23.7	20.4	19.5	20.8	18.7
$C_2F_4^+$	15.1	13.6	12.7	12.9	12.8	11.3
$C_2H_{1,2}F_2^+$	7.5	7.7	7.5	7.6	8.1	7.3
CF_3^+	7.3	7.4	8.5	7.9	8.0	10.1
$CH_{0,1}F_2^+$	4.7	5.3	6.2	6.3	5.4	6.3
$C_3H_{1,2}F^+$	3.5	4.5	5.8	6.5	6.1	6.2
$C_3HF_2^+$	2.8	3.2	3.2	3.0	3.2	4.2
$CH_{0-2}F^+$	2.0	3.8	5.7	6.4	6.6	6.5
$C_3HF_4^+$	1.5	1.7	1.6	1.4	1.3	1.2

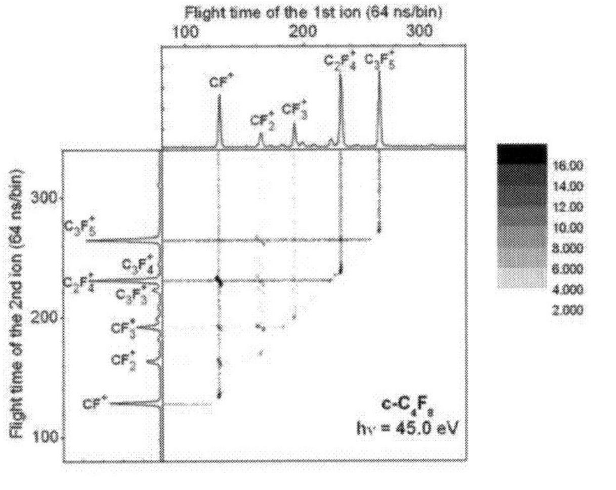

Figure 4. Overview of a typical PEPIPICO map for *c*-C_4F_8 measured at 45.0 eV. The top and left panels indicate the corresponding TOF spectrum.

c-$C_4F_8^{2+} \rightarrow C_3F_4^+ + CF_2^+ + 2F$ (2)
c-$C_4F_8^{2+} \rightarrow C_3F_3^+ + CF_3^+ + 2F$ (3)
c-$C_4F_8^{2+} \rightarrow C_3F_3^+ + CF_2^+ + 3F$ (4)
c-$C_4F_8^{2+} \rightarrow C_2F_4^+ + CF_2^+ + CF_2$ (or CF + F) (5)
c-$C_4F_8^{2+} \rightarrow C_2F_4^+ + CF_3^+ + CF$ (6a)
c-$C_4F_8^{2+} \rightarrow C_2F_4^+ + CF^+ + CF_3$ (6b)
c-$C_4F_8^{2+} \rightarrow CF_3^+ + CF^+ + C_2F_4$ (or its fission products). (6c)

Reaction channel 6a has already been discussed by Hsieh and Eland [8]. We presume that in reactions 1–4 the charge separation follows fluorine atom fission(s), i.e., deferred charge separation.

Similar spectra were obtained for *cis*-*c*-$C_4H_2F_6$ as well. Coincidence islands identified are $C_2HF_3^+/CHF_2^+$, $C_3HF_2^+/CF_2^+$, CF_3^+/CF_2^+, $C_2HF_3^+/C_2HF_2^+$, $C_2HF_3^+/CF^+$, CF_3^+/CF^+, $C_2H_2F_2^+/CF^+$ and CHF_2^+/CF^+ fragment pairs at 45.0 eV. The pairs to the $C_3H_2F_3^+$ ion are left unassigned due to a relatively high ridge appearing in our spectra. The following breakdown pathways can be identified:

cis-*c*-$C_4H_2F_6^{2+} \rightarrow C_2HF_3^+ + C_2HF_2^+ + F$ (7)
cis-*c*-$C_4H_2F_6^{2+} \rightarrow C_3HF_2^+ + CF_2^+ + 2F + H$ (8)
cis-*c*-$C_4H_2F_6^{2+} \rightarrow C_2H_2F_2^+ + CF^+ + CF_3$ (or its fission fragments) (9)

$cis\text{-}c\text{-}C_4H_2F_6^{2+} \rightarrow C_2HF_3^+ + CHF_2^+ + CF$ (10a)
$cis\text{-}c\text{-}C_4H_2F_6^{2+} \rightarrow C_2HF_3^+ + CF^+ + CHF_2$ (10b)
$cis\text{-}c\text{-}C_4H_2F_6^{2+} \rightarrow CHF_2^+ + CF^+ + C_2HF_3$. (10c)

If one formally replaces the Hs with Fs, reactions 8, 9 and 10a–c correspond to reactions 4, 6b and 6a–c for the *c*-C_4F_8 fragmentation, respectively. The *c*-C_4F_8 counterpart for reaction 7 is hardly seen due to the low yield of $C_2F_3^+$, implying that the $C_2F_3^+$ ion is more reactive than $C_2HF_2^+$. Thermochemical stability of the ions is not the factor because the heat of formation is 812 kJ/mol for $C_2F_3^+$ and 916 kJ/mol for $C_2HF_2^+$ [13].

Analyzing the slope of the islands allows us to deduce the fission dynamics [14,15] and is now in progress. The results will be published elsewhere.

4. Summary

Dissociative photoionization of *c*-C_4F_8 in the energy range of 25–170 eV produces CF_l^+ (l = 1–3), $C_2F_m^+$ (m = 1–4), $C_3F_n^+$ (n = 1–5) and $C_4F_7^+$ as fragment ions. The $C_3F_5^+$ and $C_2F_4^+$ ions are most abundant at all the energies studied here, but their yields decrease significantly with the photon energy: The yield at 170 eV becomes about 30% of that at 30 eV. Similar yield curves were obtained for the *cis*-*c*-$C_4H_2F_6$ photoionization. The most dominant fragments are $C_3H_2F_3^+$ and $C_2HF_3^+$, giving a total yield of 52% at 30 eV. The PIY curves of $C_3H_2F_3^+$ and $C_2HF_3^+$ closely resemble those of $C_3F_5^+$ and $C_2F_4^+$ produced by the *c*-C_4F_8 photoionization, respectively. Several breakdown pathways were also identified by applying a PEPIPICO technique. Comparison of the pathways between both molecules implies that the $C_2F_3^+$ ion is more reactive than $C_2HF_2^+$.

Acknowledgments

The authors are grateful to the staff of the UVSOR facility for stable operation of the storage ring. This study was supported in part by the Joint Studies Program of the Institute for Molecular Science. KO also acknowledges the financial support from the Japan Society for the Promotion of Science through its Grant-in-Aid for Scientific Research.

References

[1] Forster P, Ramaswamy V, Artaxo P, Berntsen T, Betts R, Fahey D W, Haywood J, Lean J, Lowe D C, Myhre G, Nganga J, Prinn R, Raga G, Schulz M and Van Dorland R 2007 *Changes in Atmospheric Constituents and in Radiative Forcing* in: *Climate Change 2007: The Physical Science Basis. Contribution of Working Group I to the Fourth Assessment Report of the Intergovernmental Panel on Climate Change*, ed S Solomon *et al* (Cambridge University Press: New York)
[2] Ravishankara A R, Solomon S, Turnipseed A A and Warren R F 1993 *Science* **259** 194
[3] Morris R A, Miller T M, Viggiano A A, Paulson J F, Solomon S and Reid G 1995 *J. Geophys. Res.* **100** 1287
[4] Jarvis G K, Boyle K J, Mayhew C A and Tuckett R P 1998 *J. Phys. Chem.* A **102** 3230
[5] Parkes M A, Ali S, Tuckett R P, Mikhailov V A and Mayhew C A 2006 *Phys. Chem. Chem. Phys.* **8** 3643
[6] Limão-Vieira P, Vasekova E, Giuliani A, Lourenço J M C, Santos P M, Duflot D, Hoffmann S V, Mason N J, Delwiche J and Hubin-Franskin M-J 2007 *Phys. Rev.* A **76** 032509
[7] Tsai B P and Eland J H D 1980 *Int. J. Mass Spectrom. Ion Phys.* **36** 143
[8] Hsieh S and Eland J H D 1995 *Rapid Commun. Mass Spectrom.* **9** 1261
[9] Eland J H D, Coles L A and Bountra H 1989 *Int. J. Mass Spectrom. Ion Processes* **89** 265
[10] Mori T, Kou J, Ono M, Haruyama Y, Kubozono Y and Mitsuke K 2003 *Rev. Sci. Instrum.* **74** 3769
[11] Mitsuke K, Katayanagi H, Kafle B P, Huang C, Yagi H, Prodhan Md S I and Kubozono Y 2007 *J. Phys. Chem.* A **111** 8336
[12] Ono M, Yoshida H, Hattori H and Mitsuke K 2001 *Nucl. Instr. Meth.* A **467–468** 577

[13] Anicich V G, Bowers M T, O'Malley R M and Jennings K R 1973 *Int. J. Mass Spectrom. Ion Phys.* **11** 99
[14] Eland J H D, Wort F S and Royds R N 1986 *J. Electron Spectrosc. Relat. Phenom.* **41** 297
[15] Simon M, Lebrun T, Martins R, de Souza G G B, Nenner I, Lavollee M and Morin P 1993 *J. Phys. Chem.* **97** 5228

Journal of Physics: Conference Series **288** (2011) 012022 doi:10.1088/1742-6596/288/1/012022

Site- and geometry-specific CH···O interaction in small acetaldehyde clusters studied with core-electron excitation spectroscopy in the carbon *K*-edge region

K Tabayashi[1,*], Y Tsutsumi[1], M Chohda[1], O Takahashi[1], Y Tamenori[2], I Higuchi[2], I H Suzuki[3], S Nagaoka[4], T Gejo[5], and K Honma[5]

[1]*Hiroshima University, Higashi-Hiroshima 739-8526, Japan*
[2]*JASRI/SPring-8, Hyogo 679-5198, Japan*
[3]*Photon Factory (PF)/KEK, Tsukuba 305-0801, Japan*
[4]*Ehime University, Matsuyama 790-8577, Japan*
[5]*University of Hyogo, Hyogo 678-1297, Japan*

Abstract. Site-dependent C1*s*-core excitation spectra of small acetaldehyde (AAL) clusters have been studied under the cluster regime of beam conditions. The excitation spectra of the clusters were generated by monitoring the partial-ion-yields of cluster-origin fragments. Comparison of the cluster bands with the monomer bands revealed that the band intensities of C1*s* core-to-Rydberg transitions come to be strongly excitation-site dependent upon cluster formation showing significant reduction of $C1s_{HCO}$→Ryd bands relative to the monomer bands. Computer modeling of X-ray absorption spectra using a density functional theory calculation demonstrated that the intensity reduction is the outcome of the site- and geometry-specific CH···O interaction where hydrogen-bonding association between the aldehyde (HCO) sites of different AAL molecules takes place exclusively within the clusters. As the representatives of small AAL clusters ($n \geq 3$), two types of molecular complexes (a planar cyclic trimer and a tetramer with a stack of two dimers) were derived to rationalize the site-dependence of the experimental C1*s* excitation spectra. Spectral simulation of the above complexes could reproduce the experimental site-dependence of the core-to-Rydberg band-features upon clusterization, indicative of significant contribution of these clusters under the present beam-stagnation condition.

1. Introduction

Inner-shell excitation spectra of small hydrogen bonding (HB) clusters have been studied in the beam condition [1-6], and they provided new spectroscopic findings on the local HB interaction within the molecular clusters as well as the fragmentation mechanism upon H-bond cluster formation. Formic-acid (FA) has two oxygen sites available for strong H-bond formation and forms a stable cyclic-dimer

* To whom any correspondence should be addressed.
Present address: Hiroshima Synchrotron Radiation Center, Hiroshima University, Higashi-Hiroshima 739-0046, Japan
E-mail: kiyotaba@hiroshima-u.ac.jp

configuration in the gas phase, with doubly bridged (OH···O) H-bonds with a binding energy of ~15 kcal/mol [1,7]. The configuration is exclusively the most stable unit-structure among the molecular complexes and provides a major source of building blocks for the FA cluster formation. The comparison of the cluster excitation-bands with those of molecular FA showed that the first resonance $O1s_{CO}\rightarrow\pi^*_{CO}$ band shifts upwards (to a higher energy) by ~0.31 eV whereas the second $O1s_{OH}\rightarrow\pi^*_{CO}$ band shifts downwards by ~0.82 eV [1]. The π^*_{CO} orbital is actually insensitive to the H-bonding due to the out-of-plane character, the band shifts of $O1s_{CO/OH}\rightarrow\pi^*_{CO}$ are strongly attributable to the change in core electron binding energies (CEBEs) of these oxygen atoms upon H-bond formation. In contrast, the antibonding σ^*_{OH} orbital is directly related to the HB to give higher excitation energies and reduced band-intensities relative to the monomer band. The band-structures of FA clusters have been well supported by a recent density functional theory (DFT) calculation [7].

Acetaldehyde (AAL) has a relatively high dipole moment, larger than those of typical polar solvents such as water and methanol [8]. Quantum chemical calculation shows that the charge distribution of AAL is mostly localized on the carbonyl C^+O^- site that actually defines the molecular dipole moment. Although AAL has both hydrogen donor C sites and lone-pair acceptor O site within the molecule, strong H-bonds such as the O–H···O are not expected in neutral cluster species due to a lower H-bond donor strength of C-H [9]. It is thus natural to postulate that the overall structures of small AAL clusters are actually determined and that the weak CH···O interactions are largely influenced, by the strong multipole interactions. Governed by such electrostatic interactions, it can be assumed to constitute nearly anti-parallel CO of dimer configurations within AAL clusters.

Recently, we have studied $O1s_{HCO}\rightarrow\pi^*_{CO}$ excitation of AAL clusters [5,6] in the smallest cluster regime of beam conditions. The intense $O1s_{HCO}\rightarrow\pi^*_{CO}$ band of clusters was found to shift to a high energy by 0.15 eV relative to the monomer band. Computer modeling of X-ray absorption spectra based on the DFT calculation was carried out to rationalize the experimental results. The simulation of the most stable dimer with a non-planar T-shaped geometry could reproduce the experimental cluster-band shift. The band shift was interpreted as being both due to the HOMO-LUMO (n–π^*) interaction within the complex and the stabilization upon dimerization. The CH···O interaction was also identified by the appearance of vibrationally blue-shifting hydrogen bonding [10] in the AAL complex.

Here, we have extended inner-shell excitation studies to small AAL clusters in the Carbon *K*-edge region, and the effects of such intermolecular interactions on the core-excitation spectra have been examined. C1*s* site-specific band probing of AAL clusters is promising since AAL has two different CH sites (the methyl and the aldehyde (carbonyl) carbon sites) available for CH···O interaction and their distinct core-to-Rydberg bands detectable in the Carbon *K*-edge region. The band-structures of core-to-Rydberg transitions observed in small AAL clusters were also compared with those predicted by the DFT calculation.

2. Experiments and theoretical calculations

The experiments were performed using a cluster-photochemistry apparatus [3,11] mounted to a high-resolution plane-grating monochromator at the soft X-ray beamline BL27SU in the SPring-8 facility. Cluster beams were generated in a beam source chamber by supersonic expansion of gaseous mixtures of 4.9–13.5% AAL/He via a ϕ 50 μm nozzle, under the stagnation pressures of $P_0 = 0.11$–0.3 MPa. The mixtures were prepared by bubbling helium gas through the liquid AAL sample in a cylinder. The stagnation pressures were kept constant with immersing the sample cylinder in a thermostated refrigerant. The beams were collimated with a ϕ 1.0mm skimmer and directly introduced into a main chamber of the apparatus [11], where it was crossed with a monochromatized soft X-ray beam at the ionization region of a double-field type TOF spectrometer. For comparison with the inner-shell excitation measurements of AAL clusters, effusive (monomeric) gas was prepared by direct introduction [1] of pure AAL through a 1/16 in. (o.d.) stainless-steel tube into the ionization region. Photoelectron-photoion coincidence (PEPICO) signals were recorded using a fast multihit digitizer with a time resolution typically set to 4 ns. The photon energy was calibrated using the $C1s\rightarrow 3p/4p\pi$

Rydberg transitions of CO_2 at 294.938/296.364 eV [12]. The resolution of the monochromator was $E/\Delta E$ = 5000–7500 at the energy corresponding to the carbon *K*-edge. The details of the procedure have been described elsewhere [1].

Acetaldehyde (AAL: CH_3CHO) was supplied by Aldrich at a purity of ≥ 99.5% and helium by Nihon-Sanso at ≥ 99.99%. They were used without further purification.

In order to examine the molecular configurations and intermolecular interactions within AAL clusters and support the analysis of core-excitation spectra, ab initio molecular orbital (MO) and DFT calculations were carried out for AAL molecule and its complexes. Initial geometries of the small complex were generally optimized using the Gaussian 03 program [13] at the MP2/cc-pVTZ level of approximation. The vibrational frequencies were calculated at the same level to confirm the stationary structures and correct the zero-point vibrational energies (ZPVEs). The frequencies were further scaled [14] with multiplying by 0.9832 after the MP2 calculations. The basis set superposition error (BSSE) was corrected using the procedures proposed by Boys and Bernardi [15]. The molecular complexes larger than the dimer were optimized at the MP2/cc-pVDZ level and their stabilization energies were estimated without BSSE correction.

The calculation of core-excited states was performed with the StoBe-deMon code [16], based on the solution of the Kohn–Sham DFT equations. The theoretical X-ray absorption spectra (XAS) were generated by the transition potential (TP) method combined with a double basis set technique. The orbital energies and oscillator strengths were calculated to obtain the transition energies and absorption intensities in the theoretical XA spectrum. The TP method describes most of the relaxation effects on the core-excitation states and provides a single set of orthogonal orbitals for the spectrum calculation. In determining the absolute energy positions of the XA spectrum, the ΔKohn–Sham calculations [17] that allow full relaxation of core-hole ionization states were performed to derive the ionization energy (IP). The relativistic correction of 0.08 eV [18,19] was added to the IP of the C1*s* electrons. Electron interactions of C and O atoms without any core-electron excitation (core-hole) were described with effective core potentials. The ionized center was described using the IGLO-III basis of Kutzelnigg *et al.* [20], whereas (6311/311/1) and (311/1) basis sets were used for other heavy atoms and hydrogen atoms. The spectrum was generated by a Gaussian convolution of discrete transition lines by adjusting their broadenings to simulate the experimental spectrum. All the DFT calculations were performed using the gradient-corrected exchange and correlation functionals developed by Perdew and Wang [21,22]. Further details of the calculations have been given elsewhere [17,23].

3. Results and Discussion

3.1. Core excitation measurements of AAL molecule and clusters in the carbon K-edge region

3.1.1. Total-ion-yield (TIY) spectra of AAL molecule and clusters

Prior to the measurements of AAL clusters, the total-ion-yield (TIY) of AAL was observed under the effusive condition. Figure 1 shows a typical overview TIY spectrum of AAL molecule recorded in the 285–295 eV energy range that covers carbon *K*-edge transitions. Publications on the X-ray absorption and excitation spectra of free AAL molecule are available in the ISEELS [24] and NEXAFS [25] measurements. The carbon *K*-edge excitation of AAL involves the electron transitions from the aldehyde (carbonyl) HC=O and methyl CH_3 carbon sites. An intense core-to-π^*_{CO} valence band with excitation from $C1s_{HCO}$ site occurs at 286.1 eV; moderate and/or weak Rydberg (3*s* and 3*p*) transitions from $C1s_{CH3}$ site exist at 287–289 eV whereas those from $C1s_{HCO}$ site appear in the 290–292 eV range. The Rydberg transitions of AAL from both C1s sites are fairly well resolved amongst the carbonyl compounds. The CEBEs of $C1s_{CH3}$ and $C1s_{HCO}$ have been determined to be 291.35 and 294.00 eV in the X-ray photoelectron spectroscopy [26]. Based on the previous band assignments, we reassigned the C1*s* excitation bands from the TIY as listed in table 1, where low-lying Rydberg transitions relevant to the present study are indicated. In our measurement, the band energies are in closer agreement (with an average discrepancy of 0.06 eV) with the results of Prince *et al.* [25].

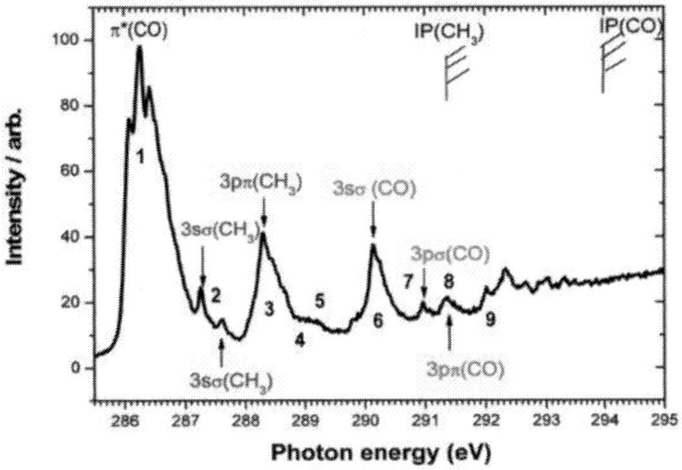

Figure 1. TIY spectrum of free acetaldehyde (AAL) molecule observed with core-excitation in the carbon *K*-edge region. Prominent valence and lower-Rydberg excited states are indicated. The band assignment is given in table 1.

Table 1. Representative band-assignment of free AAL molecule in the carbon *K*-edge region.

Band	Photon energy E / eV			Assignment	Quantum
No.	Present work	Prince *et al.*[a]	Hitchcook *et al.*[b]		Defect / δ
C1s_{CO} excitation					
1	286.08	286.17 (+0.09)[c]	286.30 (+0.22)[c]	π^*_{CO}	0.69
	286.26	286.32 (+0.06)		+Δv (+0.18)	
	286.42	286.48 (+0.06)		+2Δv (+0.16)	
6	290.14	290.21 (+0.07)	290.12 (-0.02)	3$s\sigma$(CO)	1.12
7	290.96	291.02 (+0.06)	290.90 (-0.06)	3$p\sigma$(CO)	0.88
8	291.36	291.40 (+0.04)	291.20 (-0.16)	3$p\pi$(CO)	0.73
9	292.05	292.10 (+0.05)		4$s\sigma$(CO)	1.39
	(294.00)[c]	(294.00)[c]	(293.83)[d]	IP	
C1s_{CH3} excitation					
2	287.26	287.33 (+0.07)	287.2 (-0.06)	3$s\sigma$(CH_3)	1.17
	287.62	287.69 (+0.07)		+Δv (+0.36)	
3	288.30	288.36 (+0.06)	288.30 (0)	3$p\pi$(CH_3)	0.89
4	288.98	289.05 (+0.07)		3$p\sigma$(CH_3)	0.60
5	289.20	289.26 (+0.06)	289.5 (+0.30)	4s(CH_3)	1.48
	(291.35)[d]	(291.35)[d]	(291.19)[e]	IP	

[a]Prince K C *et al.* (Ref. [25])
[b]Hitchcock A P and Brion C E (Ref. [24])
[c]difference from the present work
[d]Jolly W L , Bomben K D and Eyerman C J (Ref. [26])
[e]Jolly W L and Schaaf T F (Ref. [27])

3.1.2. Partial-ion-yield (PIY) spectra of AAL molecules and clusters

As was the case for previous O1s excitation of AAL clusters [5,6], we recorded an intense series of protonated clusters M_nH^+ (M: CH_3CHO) in the TOF fragment-mass measurement of the present cluster beams (figure 2(a)). Pressure variations of the product distribution important for the AAL beam characterization have been recently described [5], where atomic and smaller fragments than the parent molecule such as C^+, O^+, and C_2^+ in the cluster beam were attributed to the free AAL origin whereas the protonated clusters M_nH^+ were originated from AAL clusters. Figure 2(b) shows the variations of fractional intensities of the monomer-origin (C^+/C_2^+) and cluster-origin (M_nH^+) products with beam stagnation pressure. While the pressure dependences similar to the previous beam characterization (Figure 2(b) in Ref. 5) were given, larger back-ground intensities were admitted in the present monomer-origin (C^+/C_2^+) products. This is only the difference in the cluster beam-characteristics found for the cluster photochemistry apparatus in the SPring-8 facility.

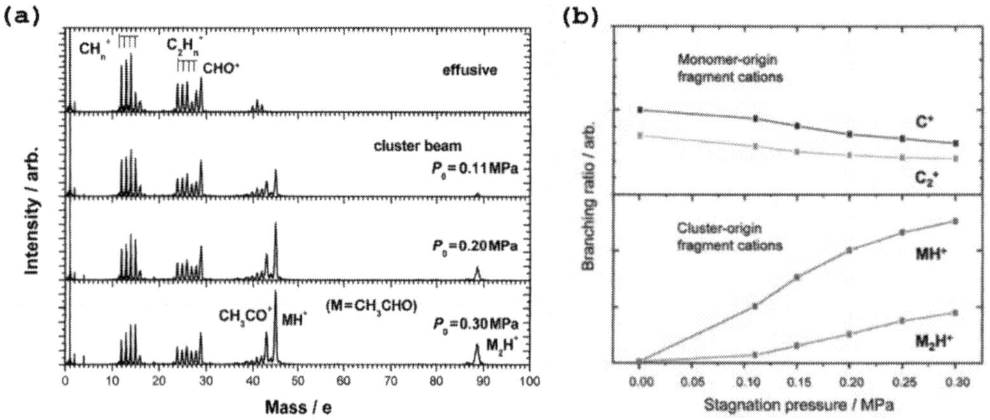

Figure 2. (a) Comparison of the TOF spectra among the effusive (top panel) and cluster (lower three panels) beams, with $C1s_{HCO} \rightarrow \pi^*_{CO}$ excitation at 286.1 eV. **(b)** Variations of the branching ratios for some monomer- and cluster-origin products with beam stagnation pressure.

By monitoring PIY intensities of the M_nH^+ products originating from the clusters, excitation spectra of AAL clusters could be recorded without any contribution of free molecule. In contrast, the PIYs of atomic and small fragments actually have the same band-features and peak-energies as those observed in the TIY of the effusive beam, it is also reasonable to conclude that they come from free AAL molecule. Figure 3 compares the PIY spectra in the C1*s* edge region for the monomer- and cluster-origin cations simultaneously recorded at the same energy points. In the present case where a fair amount of back-ground AAL can be admitted, we chose a moderate cluster-beam condition of P_0= 0.20 MPa in order to facilitate our signal accumulation over the entire C1*s* edge region (285–295 eV) with step-widths of ΔE=0.05–0.10 eV. The beam stagnation pressure is higher than that (P_0=0.10 MPa) used in the previous O1*s* excitation (PIY) measurement where the cluster beam was found to have the smallest cluster-size distribution with a prominent dimer-contribution [5,28].

The PIYs of the cluster-specific products, MH^+/M_2H^+ show distinct band-characteristics other than the first valence ($C1s_{HCO} \rightarrow \pi^*_{CO}$) transition; the $C1s_{HCO} \rightarrow$Rydberg transitions at ~290 eV and higher energies are found to lose their intensities whilst the $C1s_{CH3} \rightarrow$Rydberg transitions merely broaden their bands as the cluster-Rydberg bands generally show. Although the core-to-Rydberg ($C1s_{CH3} \rightarrow 3p\pi$: ~288.3 eV)/($C1s_{HCO} \rightarrow 3s\sigma$: ~290.1 eV) excitations in the present monomer system have relatively weak oscillator-strengths in origin, structural changes upon cluster formation could be thus extracted in the site-selective PIY measurement, under the present beam-stagnation condition. The present PIY spectra of small AAL clusters clearly show that there appears significant intensity reduction of the C1s_{HCO} core-to-Rydberg transitions, indicating that the site-dependent intermolecular interactions of AAL molecule are involved within the clusters.

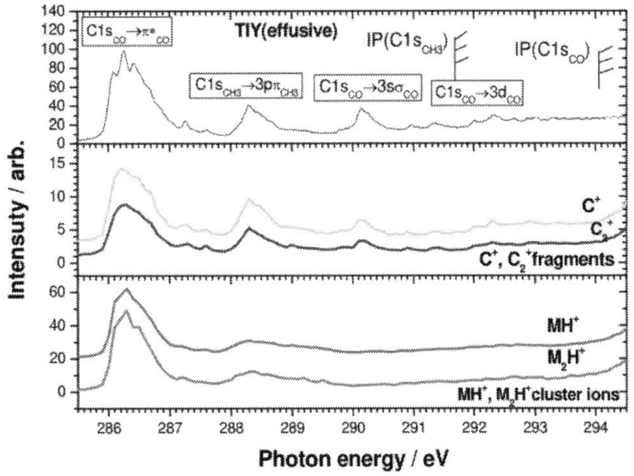

Figure 3. Comparison of the PIY spectra between monomer-origin (C^+, C_2^+) and cluster-origin (MH^+, M_2H^+) products in the carbon *K*-edge region, obtained in the cluster beam at P_0=0.20 MPa.

3.2. Computer modeling of XA spectra of AAL molecule and clusters by DFT calculations

3.2.1. CH···O interactions derived for the representative AAL dimers

In order to compare the experimental results with theoretical expectations and to give an interpretation for the intensity reduction of the "cluster" C1s_{HCO}→Rydberg bands from the "monomer" bands, model XA spectra were generated for the representative (AAL)$_n$ clusters using the DFT calculation. In the first stage, several dimer complexes with stable minima were examined to understand the outcomes of the interaction of C1(HCO)/C2(CH_3) core-atoms that appear in the calculated XAS. Examples of the dimer structures are shown in figure 4(a); i.e. **dimer 1**: the most stable non-planar (spatial) structure, and **dimer 2/dimer 3**: the planar structures with configurations of different types of CH···O interaction, optimized at the MP2/cc-pVTZ level of approximation. Some other configurations for AAL dimer have also been reported using ab initio MO and DFT calculations [29]. Among these dimers, non-planar structures were found to be more stable than the planar ones by ~0.5–1.0 kcal/mol depending on the level of calculation. We have generated several model XA spectra for the respective dimer configurations, and the representative cases are given in figure 4. It should be noted that the stabilization energy of the most stable **dimer 1** is calculated to be $\Delta E_b = -2.70$ kcal/mol and those for planar **dimer 2/dimer 3** are $\Delta E_b = -2.04/-1.87$ kcal/mol [30], respectively, including the BSSE and ZPVE corrections.

The overview XA spectra in the C *K*-edge region calculated for the representative dimers are shown in figure 4(b), where the comparison is also made with that of free molecule without any intermolecular interaction (top panel). Vertical bars represent the oscillator strengths of individual transitions from the constituent C atoms in the system. Although the low oscillator strengths of the core-to-higher Rydberg transitions are also reproduced in the calculated spectra, we focus our attention on the lowest-lying Rydberg transitions so that reliable comparison with the experimental transitions can be made. Since the C1(C2) atoms in non-planar **dimer 1** and planar **dimer 2** are not equivalent to the C3(C4) atoms under their chemical surroundings, their excitation energies have different magnitudes depending on the different influence of intermolecular (electrostatic etc.) multipoles and local interactions such as the H-bonding. The C1(C2) atom in planar **dimer 3**, in turn, is equivalent to the C3(C4) atom within the C_{2h} symmetry, so that their excitation energies are actually degenerate.

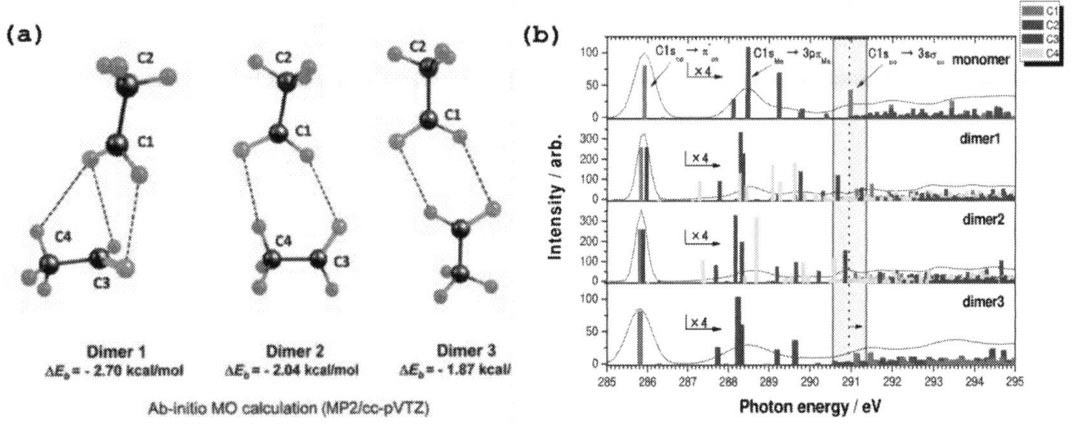

Figure 4. (a) Representative AAL dimer structures employed for the XAS modeling calculation in the carbon *K*-edge region. These structures were optimized using the MP2/cc-pVTZ level of ab initio MO calculation. **(b)** Comparison of the calculated XAS among the AAL molecule (top panel) and representative $(AAL)_2$ dimers (lower three panels); Vertical (colored) bars denote the band positions and transition intensities of the respective C1–C4 carbon core-atoms.

Following the argument in **section 3.1.2**, close inspection of the core-to-lower Rydberg transitions among the representative dimers reveals the following characteristics under the intermolecular CH···O H-bonding interaction. Firstly, relative intensity of the $C1s_{CH3}\rightarrow 3p\pi$ transition is mostly comparable to the monomer one, whether any methyl hydrogen of the C2(C4) site interacts with the carbonyl oxygen atom of the counterpart molecule or methyl hydrogen atoms of the site are free from the intermolecular CH···O interaction. Secondly, in contrast, oscillator strength of the $C1s_{HCO}\rightarrow 3s\sigma$ transition in the aldehyde C1(C3) site decreases significantly in the complex when its aldehyde hydrogen directly associates with the counterpart oxygen atom. Summarizing the above, the intensity of the C1*s* core-to-Rydberg transition has such site-specific character that the distinct reduction of the band intensity takes place at the aldehyde carbon site, derived from the definite CH···O interaction [30]. In the dimer complex where both aldehyde sides of the molecules sit facing antiparallel each other to give a planar ring structure with doubly bridged CH···O bonds (such as the case of planar **dimer 3**), the C1*s* core-to-Rydberg bands actually disappear in the computer modeling XA spectra. In this respect, the XA spectrum given for the most stable (T-shaped) **dimer 1** shows a less favorable comparison with the C1*s* excitation spectra observed. This is not surprising since the present C1*s* core-excitation spectra were recorded under the moderate beam-stagnation condition where the major contribution of clusters larger than the dimer can be expected. As far as the present fundamental analysis (between the molecular configurations and spectral patterns in the representative dimers) is concerned, the experimental C1*s* excitation spectra of AAL clusters strongly favour the unit-structures such as cyclic **dimer 3** obtained in the computer modeling calculation.

In the OH···O interaction that affects on the O1*s* excitation of H-bonded FA clusters [1,7], the intensity reduction as well as the energy shift of core-to-σ^*(OH) transition have been described. The antibonding σ^*(OH) orbital of H-donor oxygen site is strongly perturbed upon H-bond formation and the state is destabilized due to the $n\rightarrow\sigma^*$ interaction. Charge transfer made upon complexation induces delocalization of the σ^* orbital to reduce the core-to-σ^* band-intensities. The low-lying C-Rydberg states such as the 3*s* and $3p\sigma$ have more or less antibonding σ^*(CH) character in the orbitals, so that the band-characteristics similar to the above transition can be realized in the C1s_{HCO}→Ryd bands. In figure 4(b), in fact, one may find the quenching and blue-shifting of the core-to-3*s* Rydberg band in each of the C1 sites in the representative dimers where the distinct CH···O interaction can be admitted.

It should be added that we have not examined the core-to-valence transition, C1s_{HCO}→π* and its vibrational progression in the present calculation since the simulation program is not incorporated with those other than the electronic transition. The energy shift of the band involves explicitly both vibrational and electronic transitions within the specified molecular complex with a definite size and configuration, so that the details on the structural change of the C1s_{HCO}→π* band upon cluster formation are separated here and will be given in a future publication.

3.2.2. Interpretation of CH···O interactions in small AAL clusters

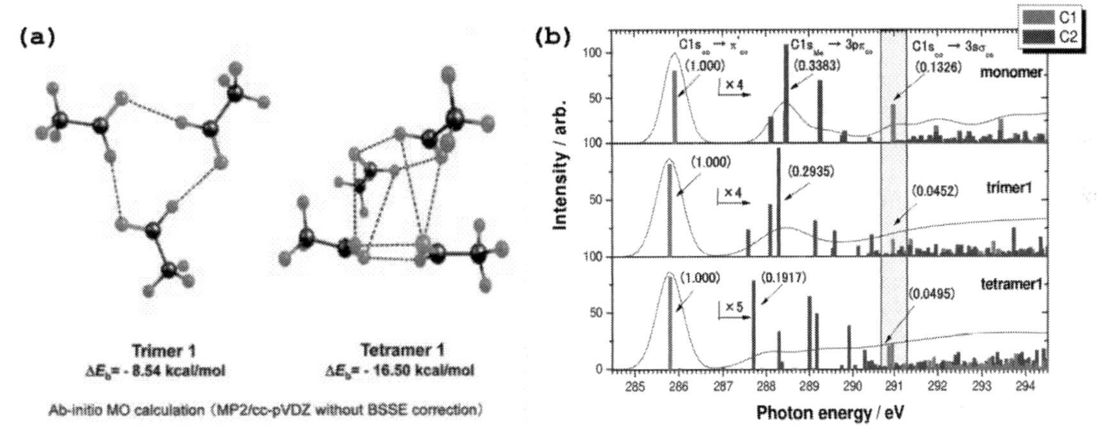

Figure 5. **(a)** Representative AAL trimer and tetramer structures employed for the XAS modeling in the carbon *K*-edge region. The structures were optimized using the MP2/cc-pVDZ level of approximation. **(b)** Comparison of the calculated XAS among the AAL molecule (top panel) and representative $(AAL)_3$ and $(AAL)_4$ clusters (lower panels); Vertical (colored) bars denote the band positions and transition intensities from the respective C1(HCO) and C2(CH_3) carbon sites of the constituent molecules.

To our knowledge, geometry optimization analyses of AAL clusters larger than the dimer are not available up to this time since AAL molecule has such a low symmetric structure and is expected to form a number of configurations as the $n(\geq 3)$-isomers. In order to rationalize the core-to-Rydberg band-disappearance observed in the small cluster regime of beam conditions, however, contributions of several clusters larger than AAL dimer are examined based on the representative unit-structures described in the previous analysis. Figure 5(a) shows the examples of the representative complexes (**trimer 1** and **tetramer 1**) exceedingly compatible with the above requirement, optimized at the MP2/pp-cVDZ level of approximation. Figure 5(b) compares the calculated XAS of these complexes with that of free molecule, where the intensities of the important transitions are scaled relative to that of the first C1s_{HCO}→π* band.

Trimer 1 extended from **dimer 3** in figure 4 has a planar cyclic structure of C_{3h} symmetry with an interaction energy of ~ –8.5 kcal/mol; three equivalent CH···O contacts (at each aldehyde site) are established in this minimum. Since the stabilization energy of **dimer 3** amounts to ΔE_b= –4.14 kcal/mol calculated with the same level of MP2/cc-pVDZ, the addition of a molecule to **dimer 3** to form **trimer 1** gains a larger energy-increment for a single CH···O interaction than that obtained for the formation of **dimer 3**. From a geometrical point of view, one could find a shortening of intermolecular CH···O distance in **trimer 1** with respect to the value obtained in **dimer 3**. In fact, the

CH···O distance (2.375 Å) in **trimer 1** is shortened by 0.073 Å from that (2.448 Å) derived in **dimer 3** [30], indicative of the increased strength of CH···O interaction. Such non-additive effect is due to the multipole interaction characteristic of the constituent molecule. In the case of cyclic dimer with doubly bridged H-bonds, it has been known that closer distance between the donor-hydrogen atoms (the acceptor-oxygens) of the counterpart-molecules destabilizes the system [31] via electrostatic repulsion to give weaker intermolecular interaction. **Dimer 3** is then the example that could not lead to the most stable structure. The cyclic structures of trimers, however, are often found to be the ones of the most favourable configurations (on the thermochemical ground) for the simple polar molecules such as acetonitrile [32], formic acid [1,7,33], formamide [31], etc. As an analogue of these complexes cyclic **trimer 1** is expected to be the most stable one of AAL trimer that may effectively contribute to the excitation spectra under the present cluster-beam condition. One can easily recognize that the calculated XAS of **trimer 1** (figure 5(b)) gives a satisfactory correspondence of the intensity-reduction features to the observed $C1s_{HCO}$→Ryd bands.

A representative of AAL tetramer was also obtained by geometry optimization starting from a planar cyclic structure, according to the analogy given for the lower member complexes as above. **Tetramer 1** shown in figure 5(a) gave the minimum of a stack-of-two dimers structure as a result. We tried the optimization several times by changing initial guess of the cyclic geometries to lead the same stacking structure. This strongly implies that the cyclic structure, if any, is less stable than the stacking structure in AAL tetramer. The cyclic dimer unit in the stacked complex has a configuration similar to that of **dimer 3** and the units are oriented in a twisted form; the unit is just a typical arrangement to meet the spectral requirements (figure 5(b)). Taking into account the stabilization energy (ΔE_b= −4.14 kcal/mol) of **dimer 3** and the complexation energy (ΔE_b= −16.50 kcal/mol) of **tetramer 1** calculated at the same level of approximation, significant magnitude of interaction is expected between the dimer pairs in the complex. Since the closest four CH···O distances between the stacking molecules are around 3.43 Å whilst those between the molecules in the cyclic dimers are ~2.45 Å, both H-donor and acceptor interactions between two monomers in the different pairs make tiny contributions to the total stabilization energy. As we alluded earlier, the charge distribution within AAL molecule is generally localized on the carbonyl C^+O^- site that actually defines the relatively large dipole moment, so that the antiparallel oriented dipole interaction between the molecules in the different pairs plays important roles in energetics. Another interaction to be mentioned is "through space" π-π interactions between the CO sites of the molecules in the different pairs. The dimer pairs in the complex are twisted such that the favourable contacts of the CO sites in the different pairs are doubly realized (figure 5(a)). The stabilization by stacking is thus interpretable as being both due to these multipole and through-space interactions between the opposite dimers. The significant stabilization given in **tetramer 1** leads to another strong candidate of the cluster to contribute to the site-specific C1s excitation spectra, observed under the present beam-stagnation condition.

The present interpretation of the quenching of the C1*s* core-to-Rydberg states for the aldehyde site is based primarily on the calculated XA spectra (figure 4) where the representative AAL dimer configurations are used as the unit-structures to model the cluster bonding and its geometric structures. The unit-structure of the planar cyclic dimer where both hydrogen and oxygen at the aldehyde site associate with those at the corresponding site leads to very weak oscillator strengths, due to the CH···O hydrogen-bonding interaction, for the excitation to the cluster-Rydberg states. In the present study, the cluster PIYs (figure 3) of the C1*s* core-to-Rydberg were recorded at a slightly higher stagnation pressure (P_0=0.20 MPa) than the previous O1*s* measurements (P_0=0.10 MPa), in order to facilitate the signal accumulation of the relatively weak core-to-Rydberg bands. Since cluster size distribution shifts to lager clusters with increasing stagnation pressure (i.e. increasing condensation parameter Γ^* [34]), predominant contribution of larger clusters than the dimer could be expected to the cluster PIYs observed. A less favorable agreement between the observed PIYs and calculated XA spectrum of the most stable **dimer 1** can be thus understood by the above argument. Here, we propose **trimer 1** and **tetramer 1** extended from the unit-structure of the planar cyclic dimer to account for the major sources of cluster that may contribute significantly to the present PIY spectra. These beam-stagnation conditions are both near the threshold of cluster formation, so that the present C1*s* excitation spectra reflect mostly the electronic properties of such small AAL clusters.

4. Conclusions

C1s core-electron excitation spectra of small acetaldehyde (AAL) clusters have been studied in the beam conditions. The band intensities of C1*s* core-to-Ryd transition of AAL clusters are found to change site-dependently upon cluster formation showing significant intensity reduction of $C1s_{HCO}\rightarrow$ Ryd bands relative to the monomer bands. Computer modeling XA spectra demonstrated that the band reduction is the outcome of the site- and geometry-specific CH···O interaction where the planar association between the HCO sites of different molecules takes place exclusively within the clusters. The intensity reduction of the bands is interpreted as being due to the antibonding σ^*_{CH} character in the Rydberg orbitals of HCO carbon site that may participate in the weak CH···O hydrogen bonds. As the representatives of small acetaldehyde clusters ($n\geq3$), two types of molecular complexes (a planar cyclic trimer and a tetramer with a stack of two dimers) with head(HCO site)-to-head association between AAL molecules were derived to rationalize the site-dependent core-to-Ryd band-features observed in the C1*s* excitation spectra. Significant contribution of the above mentioned complexes to the spectra could be justified under the present cluster-beam stagnation condition.

Acknowledgments

The present synchrotron radiation experiments were carried out at BL27SU in the SPring-8 facility with the approval of the SPring-8 Program Advisory committee (Proposal No. 2009A1077-NSb-np). One of the authors (K.T.) thanks the Research Center for Computational Science at Okazaki for the allocation of SGI Altix4700 computer time.

References

[1] Tabayashi K, Yamamoto K, Takahashi O, Tamenori Y, Harries J R, Gejo T, Iseda M, Tamura T, Honma K, Suzuki I H, Nagaoka S and Ibuki T 2006 *J. Chem. Phys.* **125** 194307
[2] Tabayashi K, Yamanaka T, Namatame H and Taniguchi M 2008 *Adv. Syn. Rad.* **1** 255
[3] Tamenori Y, Okada K, Takahashi O, Arakawa S, Tabayashi K, Hiraya A, Gejo T and Honma K 2008 *J. Chem. Phys.* **128** 124321
[4] Tamenori Y, Takahashi O, Yamashita K, Yamaguchi T, Okada K, Tabayashi K, Gejo T and Honma K 2009 *J. Chem. Phys.* **131** 174311
[5] Tabayashi K, Chohda M, Yamanaka T, Takahashi O and Yoshida H 2010 *Nucl. Instr. and Meth.* A**619** 388
[6] Tabayashi K, Chohda M, Yamanaka T, Tsutsumi Y, Takahashi O, Yoshida H and Taniguchi M 2010 *J. Phys. Conf. Ser.* **235** 012017
[7] Takahashi O, Yamanouchi S, Yamamoto K and Tabayashi K 2006 *Chem. Phys. Lett.* **419** 501
[8] Weast R C *ed.* 2000 *Handbook of Chemistry and Physics* (CRC Press, Boca Raton)
[9] Jeffrey G A 1997 *An Introduction to Hydrogen Bonding* (Oxford University Press, Oxford)
[10] Hobza P and Havlas Z 2000 *Chem. Rev.* **100** 4253
[11] Tamenori Y, Yamaguchi T, Okada K, Tabayashi K, Gejo T and Honma K 2005 *J. Electron Spectrosc. Relat. Phenom.* **144-147** 235
[12] Prince K C, Avaldi L, Coreno M, Camilloni R and de Simone M 1999 *J. Phys.* B: *At. Mol. Phys.* **32** 2551
[13] Frisch M J, Trucks G W, Schlegel H B *et al.* 2004 *GAUSSIAN 03*, Rev. C. 02, Gaussian, Inc., Wallingford, CT
[14] Sinha P, Boesch S E, Gu C, Wheeler R A and Wilson A K 2004 *J. Phys. Chem.* A **108** 9213
[15] Boys S F and Bernardi F 1970 *Mol. Phys.* **19** 533
[16] Hermann K, Pettersson L G M and Casida M E 2004 *STOBE-DEMON*, Version 2.0, Stobe Software

[17] Triguero L, Pettersson L G M and Ågren H 1998 *Phys. Rev.* B **58** 8097
[18] Triguero L, Plashkevych O, Pettersson L G M and Ågren H 1999 *J. Electron Spectrosc. Relat. Phenom.* **104** 195
[19] Takahashi O and Pettersson L G M 2004 *J. Chem. Phys.* **121** 10339
[20] Kutzelnigg W, Fleischer U and Schindler M 1990 *NMR-Basic Principles and Progress*, Springer-Verlag, Heidelberg
[21] Perdew J P and Wang Y 1986 *Phys. Rev.* B**33** 8800
[22] Perdew J P and Wang Y 1992 *Phys. Rev.* B**45** 13244
[23] Kolczewski C, Püttener R, Plashkexych O, Ågren H, Staemmler V, Martins M, Snell G, Schlachter A S, Sant'Anna M, Kaindl G and Pettersson L G M 2001 *J. Chem. Phys.* **115** 6426
[24] Hitchcock A P and Brion C E 1980 *J. Electron Spectrosc. Relat. Phenom.* **19** 231
[25] Prince K C, Richter R, de Simone M, Alagia M and Coreno M 2003 *J. Phys. Chem.* A **107** 1955
[26] Jolly W L, Bomben K D and Eyerman C J 1984 *At. Data Nucl. Tables* **31** 433
[27] Jolly W L and Schaaf T F 1976 *J. Am. Chem. Soc.* **98** 3178
[28] Tabayashi K, Maruyama T, Tanaka K, Namatame H and Taniguchi M 2007 *AIP Conference Proceedings* **879** 1788
[29] Hermida-Ramon J M and Rios M A 1998 *Chem. Phys. Lett.* **290** 431
[30] Tsutsumi Y 2010 *Graduate thesis,* Hiroshima University
[31] Yasunaga K 2000 *Graduate thesis*, Hiroshima University
[32] Cabaleiro-Lago E M, Hermida-Ramon J M, Pena-Gallego A, Martinez-Nunez E and Fernandez-Ramos A 2000 *J. Mol. Struct. (Theochem)* **498** 21
[33] Roy A K and Thakkar A 2004 *Chem. Phys. Lett.* **386** 162
[34] Hagena O F 1987 *Z. Phys.* D **4** 291

Journal of Physics: Conference Series **288** (2011) 012023 doi:10.1088/1742-6596/288/1/012023

Dissociation and recapture dynamics in H_2O following O 1*s* inner-shell excitation

T Gejo[1], M Oura[2] , M Kuniwake[1], K Honma[1], and J R Harries[3]

[1] *University of Hyogo, Kouto 3-2-1, Kamigori-cho, 678-1297, JAPAN*
[2] *RIKEN SPring-8 Center, Kouto 1-1-1, Sayo-cho, 679-5148, JAPAN*
[3] *JAEA, Kouto 1-1-1, Sayo-cho, 679-5148, JAPAN*

E-mail: gejo@sci.u-hyogo.ac.jp

Abstract. The emission of low–energy and Auger electrons from H_2O has been investigated following photo-excitation/ionization at photon excitation energies just below and just above the O 1s ionization threshold. It is found that neutral oxygen atoms in high Rydberg states are formed. Direct evidence of post-collision interaction (PCI) between the Auger electron and the photo-electron is observed. The initially-excited electron is captured into a Rydberg orbital of H_2O^+, and remains associated with the oxygen fragment even after the cleavage of both O-H bonds. This implies that the Rydberg states are stable, with the electron behaving as a spectator during the dissociation of H_2O^+.

1. Introduction

Study of the dynamics of atoms and molecules following photo-excitation in the vicinity of inner-shell ionization thresholds reveals many interesting properties [1-6]. When the excitation energy is within a few hundred meV below an ionization threshold energy, an electron can be excited to a high-n Rydberg state. For photon energies just above threshold, an electron can escape from the ion as a slow photoelectron. The core-hole ion, however, immediately undergoes Auger decay, emitting a fast Auger electron. This leads to a drastic increase in the strength of the Coulomb interaction between the slow out-going photoelectron and the residual ion. As a result, the photoelectron loses kinetic energy to the Auger electron. This phenomenon has been termed the "post collision interaction (PCI)" [1]. If the slow electron loses sufficient kinetic energy it is captured into a high-n ionic Rydberg state as a "recaptured" electron [2, 3].

Recently, the production of neutral hydrogen atoms following photo-excitation near the oxygen 1*s* ionization threshold of H_2O has been reported. Excited atomic hydrogen fragments have been found to be efficiently produced following spectator Auger decay [7-10]. This implies that the initially excited Rydberg electron can be trapped by a fragment H ion following ionization and dissociation. The simple view is that the ejected hydrogen ion captures the Rydberg electron, which remains in a highly excited state. Whereas these studies concentrated on capture by the fragment H ion, the question remained as to whether highly excited O or OH neutral fragments may be also generated through similar processes. If the highly excited O or OH fragment subsequently decays by autoionization, emitting the captured electron, the observation of this electron by electron spectroscopy should allow us to obtain information on the intermediate state of the fragment, and investigate the dynamics of the decay process when H_2O is excited in the vicinity of the ionization threshold.

Studying such processes as a function of excitation photon energy should provide highly detailed information. Of particular interest is the relation between PCI effects and the recapture processes. The photoelectron loses kinetic energy through the PCI effect when it departs from the ion. In the PCI process, there should be also a corresponding energy gain for the Auger electron. Therefore, by investigating the relationship between the energies of the recaptured photoelectrons and the Auger electrons, we should be able to obtain information on how the electrons interacted.

In this paper we first show that, when an inner-shell electron is ejected from the H_2O molecule near threshold, the slow photoelectron can be captured into a high-n Rydberg state and remain attached to the oxygen ion following the cleavage of both of the O-H bonds. Since excited hydrogen atoms are also generated in this energy region [7-10], this means that the fate of the electron depends on the state and position of the nuclei in the molecule. Preferential population of *d* orbital angular momentum Rydberg states is observed for the oxygen atoms.

2. Apparatus and experimental method

The experiment was performed at the b-branch of the soft X-ray undulator beamline BL17SU at SPring-8. The beamline is equipped with a high-resolution varied-line-spacing-plane-grating monochromator [11], which can provide a highly stable monochromatic soft x-ray beam with a resolving power of over 10,000 [12]. The polarization of the incident photon beam generated by the multi-polarization-mode undulator can be switched by changing the operational mode of the undulator [13-16], between linear, pseudo-vertical and helical modes. In the present study we have chosen the helical mode to achieve the highest possible photon flux. Energy calibration of the incident photon beam was carried out by recording the total ion yield spectrum from H_2O in the vicinity of the O 1s ionization threshold and comparing with the results of [17]. The photon flux was estimated to be of the order of 10^{11} photons/sec. The absolute accuracy of the incident photon energy is estimated to be about ±0.1eV.

To record electron kinetic energy spectra using circularly polarized light, we have used a hemispherical electron energy analyzer (VG-Scienta2002) equipped with a gas-cell: both were manufactured by Gammadata-Scienta Ab. The lens axis of the analyzer is fixed in the horizontal direction. The electron analyzer was calibrated by recording the low energy electron spectrum corresponding to the recapture process in Ne in the vicinity of the 1s ionization threshold and comparing to the results of [5]. The resolution of electron analyzer was set to be about 20 meV. The typical pressure in the main chamber housing the gas-cell was about 1.0×10^{-3} Pa during the measurements.

3 Results and discussion

Figure 1 shows the electron kinetic energy spectrum for H_2O recorded at a photon energy of 539.6 eV, 200 meV below the O 1s ionization threshold which is at 539.82 eV [17]. A Rydberg series can be seen converging to a kinetic energy of just over 3 eV. These peaks are due to the autoionization decay of neutral excited states in the oxygen atom which lie above the ionization threshold. The peaks can be assigned to an n*d* Rydberg series with n = 3, 4, 5, 6... These electrons are ejected when the core ion changes its electronic state from $^2D(2s^22p^3)$ to $^4S(2s^22p^3)$. The transition energy between the $^2D(2s^22p^3)$ and $^4S(2s^22p^3)$ states is 3.325 eV [18]. A further Rydberg series is also seen for kinetic energies of around 1.6 eV, although the intensity is weak. These peaks can be assigned as a Rydberg series arising from transitions from the core ionized $^2P(2s^22p^3)$ state to the $^2D(2s^22p^3)$ state. The transition energy between the $^2P(2s^22p^3)$ and $^2D(2s^22p^3)$ states is 1.692 eV [18]. No peaks at energies corresponding to the decay of OH fragments are observed. Table 1 summarizes the observed peaks, their assignments, and a comparison of the observed energies to the optical data from [18].

These results suggest that the initially-excited Rydberg electron remains attached to the oxygen fragment even after the cleavage of both O-H bonds. Even though the binding energy of the Rydberg

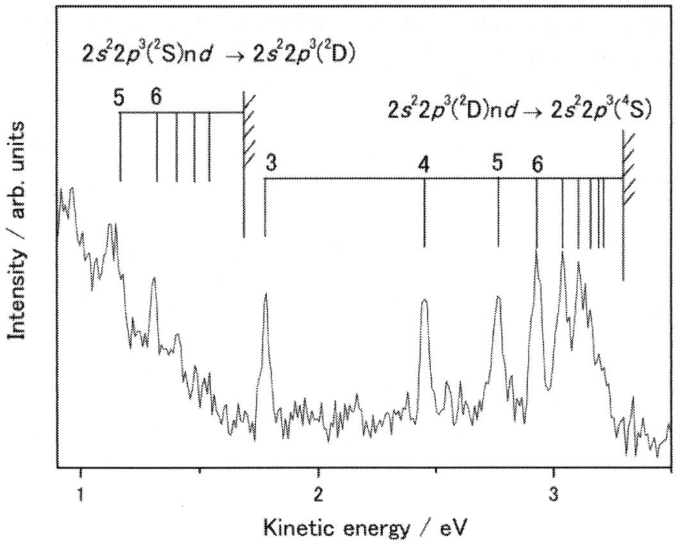

Figure 1. Electron spectra of H_2O when excited at a photon energy of 539.6 eV.

Table 1: Energy levels (eV) of autoionization states of oxygen atoms

Kinetic energy (eV)	state	optical data (eV)
1.12	$2s^22p^3(^2P)6s \rightarrow 2s^22p^3(^2D)$	1.102
1.15	$2s^22p^3(^2P)5d \rightarrow 2s^22p^3(^2D)$	1.143
1.31	$2s^22p^3(^2P)6d \rightarrow 2s^22p^3(^2D)$	1.311
1.41	$2s^22p^3(^2P)7d \rightarrow 2s^22p^3(^2D)$	1.413
1.48	$2s^22p^3(^2P)8d \rightarrow 2s^22p^3(^2D)$	1.478
1.54	$2s^22p^3(^2P)9d \rightarrow 2s^22p^3(^2D)$	1.524
	$2s^22p^3(^2P) \rightarrow 2s^22p^3(^2D)$	1.692
1.78	$2s^22p^3(^2D)3d \rightarrow 2s^22p^3(^4S)$	1.778
2.16	?	
2.39	$2s^22p^3(^2D)5s \rightarrow 2s^22p^3(^4S)$	2.380
2.46	$2s^22p^3(^2D)4d \rightarrow 2s^22p^3(^4S)$	2.467
2.55	?	
2.77	$2s^22p^3(^2D)5d \rightarrow 2s^22p^3(^4S)$	2.776
2.93	$2s^22p^3(^2D)6d \rightarrow 2s^22p^3(^4S)$	2.945
3.04	$2s^22p^3(^2D)7d \rightarrow 2s^22p^3(^4S)$	3.047
3.11	$2s^22p^3(^2D)8d \rightarrow 2s^22p^3(^4S)$	3.112
3.16	$2s^22p^3(^2D)9d \rightarrow 2s^22p^3(^4S)$	3.157
3.20	$2s^22p^3(^2D)10d \rightarrow 2s^22p^3(^4S)$	3.189
3.23	$2s^22p^3(^2D)11d \rightarrow 2s^22p^3(^4S)$	
	$2s^22p^3(^2D) \rightarrow 2s^22p^3(^4S)$	3.325

* The optical data energies have been averaged over the different spin multiplicities, weighted by the degeneracies.

electron is only around 1 eV, this implies that the highly-excited Rydberg states are very stable, behaving like "spectators" throughout the dissociation process. By studying the fluorescence decay of neutral hydrogen atoms formed when it is the H^+ fragment which captures the electron, Melero García *et al.* [9] observed that the orbital character of the initially-excited electron tends to keeps its orbital character (n, l value) when it is captured. Recently we have observed that metastable (with lifetimes longer than a few microseconds) hydrogen fragments can also be efficiently produced in this excitation energy region [10]. These two observations suggest that there are at least two different recapture decay paths: one in which an ejected hydrogen ion captures the Rydberg electron into an equivalent highly excited state, and one in which the electron remains in a Rydberg state and attached to the excited oxygen ion.

To study further the formation mechanisms of the excited oxygen atoms, we have recorded electron spectra for various photon energies between 539.1 eV (0.7 eV below threshold) and 540.4 eV (0.6 eV above threshold). Figure 2 summarizes these spectra. The spectra clearly show peaks appearing at higher kinetic energies as the photon energy is increased, and that increasing the excitation energy by just a few hundred meV changes the distribution drastically. This suggests that the fate of the initially excited electron is very sensitive to the PCI effect. An increase in excitation energy of only 400 meV (from 539.4 to 539.8 eV) change the peak of the Rydberg state distributions drastically, although the peaks are unresolved for n larger than around 8.

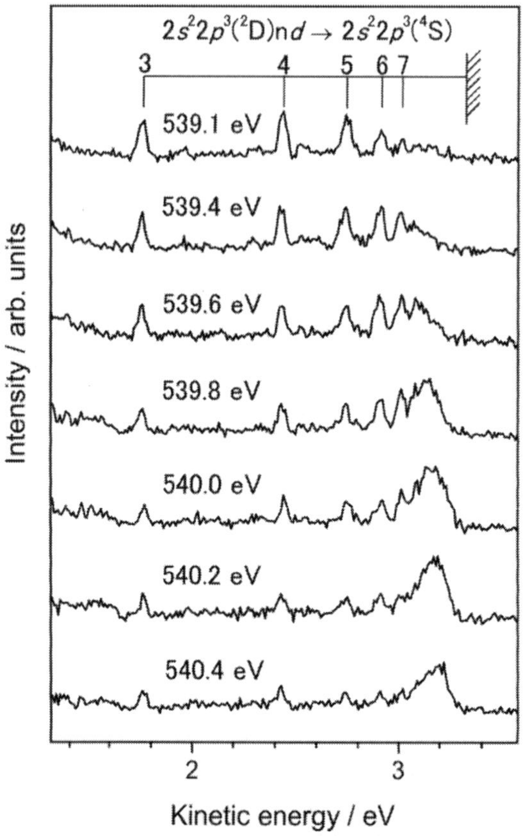

Figure 2. Electron spectra of H_2O for excitation energies between 539.1 and 540.4 eV.

The spectra shown in figure 1 and 2 are surprising in that it shows mainly transitions from n*d* Rydberg states. This is in contrast to similar studies in other systems [4,6]. For recapture following 1*s*-ε*p* excitation in the Ne atom, de Fanis *et al.* only observed a strong n*p* Rydberg series [4]. No n*d* states were observed. Since the Ne 1s electron can only be excited to an n*p* Rydberg state with a dipole transition, this implies that the photoelectrons are recaptured to Rydberg states with same orbital angular momentum. In molecules, core-excited N_2 shows also the formation of nitrogen atom fragments in high-n Rydberg states [6]. The assignment shows that both n*p* and n*d* Rydberg states are populated by the recapture of slow photoelectrons. In contrast, we observe mainly n*d* Rydberg states, with the n*p* Rydberg bands showing only low intensity. For example, the peak expected due to the $2s^22p^3(^2D)4p \rightarrow 2s^22p^3(^4S)$ transition at 1.98 eV is not observed in the spectrum of figure 1. One possible explanation for the domination of n*d* states is that the highly excited and recaptured photoelectron already has *d*-type character before the dissociation takes place, since in H_2O the two hydrogen atoms have small positive charges and should attract those electrons. Other possibility is that the Rydberg electrons change character from n*p* to n*d* due to the cleavage of the two O-H bonds. Since we do not have enough information to conclude that such processes occur, calculations of the dissociation dynamics will be necessary to understand our results.

Figure 3 shows the Auger electron spectra for several excitation energies between 539.6 and 550.0 eV. At excitation energies around 540 eV, a shift is seen. Since this shift is only observed in the vicinity of the 1s ionization threshold, it can be attributed to the PCI effect; the Auger electron gains kinetic energy due to the change in the Coulomb potential as it "overtakes" the slow photoelectron. This energy shift has been estimated by the shift at half-maximum, indicated by the black broken line in figure 3. The shift increases as the threshold energy is approached, reaching 0.58 eV at 539.6 eV (0.2 eV below the threshold). This implies that the average energy gain of the Auger electron due to the PCI effect is 0.58 eV at threshold.

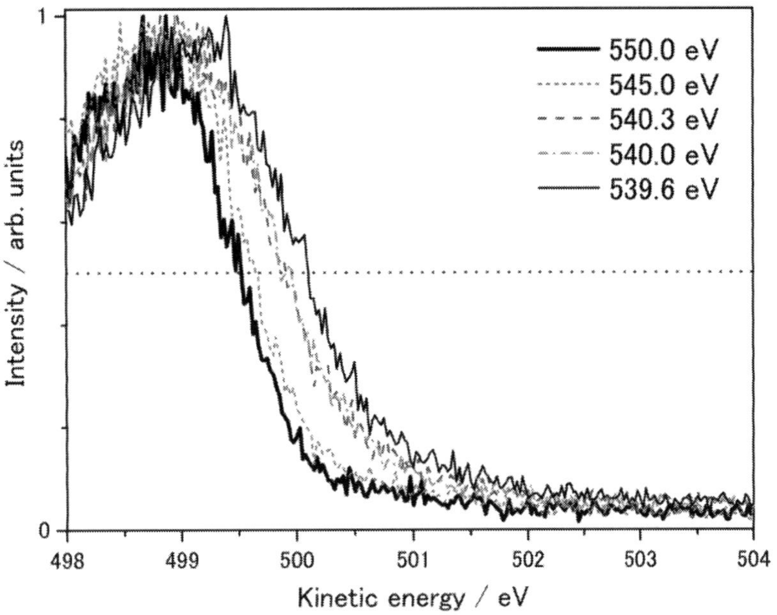

Figure 3. Auger electron spectra of H_2O when excited at photon energies of 539.6, 540.0, 540.3, 545.0 and 550.0 eV. The intensities are normalized at maximum. The black broken line shows where we have estimated the observed energy shifts.

Considering only the PCI effect the initially-excited electron should lose the same 0.58 eV kinetic energy. Assuming that the total energy is conserved, this should result in capture into an n = 7 Rydberg state. The H_2O^{++} ion core dissociates into two or three fragments which undergo separation mainly due to Coulomb repulsion. This further changes the interaction potential, but the Rydberg electron can be expected to stay in the same electronic state adiabatically. The peak of the Rydberg distribution seen in figure 1 is at n = 7, supporting this dynamical picture. A similar calculation for an incident photon energy of 540.0 eV (0.2 eV above the threshold) gives a final Rydberg state n = 8, since the average energy gain of the Auger electron due to the PCI effect is 0.43 eV at this energy. Deconvolution fitting analysis suggests that the Rydberg state distribution in the spectrum of figure 2 has a peak at around n = 8, again supporting our interpretation.

4. Summary

We have observed photoelectron recapture following O 1s ionization in H_2O by detecting autoionization and Auger electrons. In this paper we first investigate the dynamics of the inner-shell excited electron by detecting the slow electrons emitted when neutral excited oxygen fragments decay by autoionization. It is made clear that the dynamics of the electron depends on the initially excited state. An unsolved issue is what determines the final state of the Rydberg electron following cleavage of the two O-H bonds, and what determines whether it attaches to an H^+ fragment or to the O^+ fragment. This suggests that dynamical changes to the molecule structure upon the excitation of Rydberg states is very important.

5. Acknowledgements

The experiment was carried out at BL17SU in SPring-8, with the approval of RIKEN SPring-8 Center (Proposal No. 20090017). The work was supported by the Matsuo foundation and Hyogo prefecture scientific support.

References

[1] Eberhardt W, Bernstorff S, Jochims H W, Whitfield S B and Crasemann B 1988 *Phys. Rev. A* **38** 3808
[2] Aksela H, Kivilompolo M, Nõmmiste E and Aksela A 1977 *Phys. Rev. Lett.* **79** 4970
[3] Samson James A R, Stolte W C, He Z X, Cutler J N and D. Hansen 1977 *Phys. Rev. A* **56** R2530.
[4] Fanis A De, Prümper G, Hergenhahn U, Oura M, Kitajima M, Tanaka T, Tanaka H, Fritzsche S, Kabachnik N M and Ueda K 2004 *Phys. Rev. A* **70** 040702
[5] Fanis A De, Prümper G, Hergenhahn U, Kukk E, Tanaka T, Kitajima M, Tanaka H, Fritzsche S, Kabachnik N M. and Ueda K 2005 *J. Phys. B: At. Mol. Opt.* **38** 2229
[6] Shigemasa E, Kaneyasu T, Tamenori Y and Hikosaka Y 2007 *J. Elec. Spec. and Rel. Phen.* **156-158** 289-293
[7] Kivimäki A, Corero M, Richter R, Ruiz J Álvarez, García E Melero, Simone M de, Feyer V, Vall-llosera G 2006 *J. Phys. B: At. Mol. Opt.* **39** 1101
[8] Kivimäki A, Simone M de, Coreno M, Feyer V, García E Melero, Ruiz J Álvarez, Richter R. and Prince K C, 2007 *Phys. Rev. A* **75**, 014503
[9] García E M, Kivimäki A, Pettersson L G M, Álvarez J, Coreno M, Simone M de, Richter R and Prince K C 2006 *Phys. Rev. Lett.* **96** 063003
[10] Harries J R, Gejo T, Sullivan J P and Azuma Y in preparation.
[11] Ohashi H *et al.* 2007 *AIP Conf. Proc.* **879** 523
[12] Senba Y, Ohashi H, Kishimoto H, Miura T, Goto S, Shin S, Shintake T and Ishikawa T 2007 *AIP Conf. Proc.* **879** 718
[13] Tanaka T, Shirasawa K and Kitamura H, 2002 *Rev. Sci. Instrum.* **73** 1724
[14] Shirasawa K, Tanaka T, Seike T, Hiraya A and Kitamura H 2004 *AIP Conf. Proc.* **705** 203
[15] Shirasawa K, Hiraya A, Tanaka T and Kitamura H 2004 *Phys. Rev. ST Accel. Beams* **7** 020702

[16] Oura M, Nakamura T, Takeuchi T, Senba Y, Ohashi H, Shirasawa K, Tanaka T, Takeuchi M, Furukawa Y, Hirono T, Ohata T, Kitamura H and Shin S, 2007 *J. Synchrotron Radiat.* **14** 483
[17] Okada K, Ueda K, Tokushima T, Senba Y, Yoshida H, Shimizu Y, Simon M, Chiba H, Okumura H, Tamenori Y, Ohashi H, Saito N, Nagaoka S, Suzuki I H, Ishiguro E, Koyano I, Ibuki T and Hiraya A 2000 *Chem. Phys. Lett.* **326** 314
[18] http://physics.nist.gov/PhysRefData/ASD/levels_form.html.

Theoretical spectroscopy of O 1s and N 1s excited states of N_2O

M. Ehara[1,2,3,4*], T. Horikawa[3], R. Fukuda[1,2,4], H. Nakatsuji[4,5], T. Tanaka[6], M. Hoshino[6], H. Tanaka[6], K. Ueda[7]

[1] Institute for Molecular Science, 38 Nishigonaka, Myodaiji, Okazaki 444-8585, Japan
[2] Research Center for Computational Science, Okazaki 444-8585, Japan
[3] Graduate University for Advanced Studies, Okazaki 444-8585, Japan
[4] JST, CREST, Sanboncho-5, Chiyoda-ku, Tokyo 102-0075, Japan
[5] Quantum Chemistry Research Institute, Goryo Oohara 1-36, Nishikyo-ku, Kyoto 615-8245, Japan
[6] Department of Physics, Sophia University,Tokyo 102-8554, Japan
[7] Institute of Multidisciplinary Research for Advanced Materials, Tohoku University, Sendai 980-8577, Japan

E-mail: ehara@ims.ac.jp

Abstract. The O 1s and N 1s excited states of N_2O (nitrous oxide, N_t–N_c–O) have been investigated by the symmetry-adapted cluster–configuration interaction (SAC–CI) method. Our approach in this series of works using high-resolution angle-resolved ion-yield (ARIY) spectroscopy and the SAC–CI method is reviewed for the O 1s excited states of N_2O. The vibrational structure observed by ARIY spectroscopy was interpreted by two-dimensional *ab initio* potential energy surfaces (2D PESs). The valence–Rydberg coupling was analyzed by the electronic part of the second moment, $\langle r^2 \rangle$. The thermal effect in the core-electron excitation spectrum was examined by the PES calculations in the bending coordinate. The 2D PESs of the N_t and N_c excited states have also been calculated by the SAC–CI method and are discussed in detail.

1. Introduction

Molecules show the characteristic geometry changes by the core-electron excitations/ionizations, which originates from the significant orbital relaxation in the core-electron processes [1]. This characteristic geometry changes can be observed in the vibrational spectra by photoexcitation spectroscopy. Recently, much effort has been devoted to probing the geometry changes and dynamics by experiment [2]. To interpret these spectra and excited-state dynamics, however, accurate theoretical information on the molecular structure in the core-excited and ionized states is indispensable.

Recent state-of-the-art theories has enabled us to obtain precise knowledge of the core-electron excited or ionized states. Theoretical information is valuable in interpreting the wide variety of the high-resolution spectra. Theory predicts the fine details of the excitation and ionization processes. Therefore the interplay between experiment and theory has become important in modern molecular spectroscopy so that the predictive ability of a theory is an essential issue.

The vibrational spectra and geometry relaxation of the O 1s and N 1s excited states of N_2O (nitrous oxide) are of interest because the vibrational spectra in these states show different features. N_2O has two nonequivalent nitrogen atoms, the terminal N_t and the center N_c, and therefore, the chemical environment in these core-excited states is different. Adachi et al. [3] presented the first angle-resolved energetic-ion yield (ARIY) spectra of the N 1s and O 1s excited states of N_2O, and investigated them with *ab initio* SCF calculations. They paid special attention to the O $1s^{-1}$ π^*, N_c $1s^{-1}$ π^*, and N_t $1s^{-1}$ π^* excitations and showed that the Renner–Teller coupling via the (0, ν_2', 0) mode breaks the degeneracy of these Π states. Tanaka et al. [4] found that the Renner–Teller effect in these Π states is pronounced in a hot target molecule. Adachi further showed that no Renner–Teller effect is present in any of the Rydberg excited states. Prince et al. [5] reported total yield X-ray absorption fine structure spectra without symmetry resolution, but at high energy resolution near both the N 1s and O 1s edges of N_2O several new Rydberg states were found in their study. Thus, these spectra have been well investigated; however, for the detailed assignments and interpretation of the spectra, reliable theoretical calculations including electron correlation are still necessary. In particular, information on the potential energy surfaces (PESs) is relevant to understanding the vibrational spectra and geometry changes in the O 1s and N 1s excited states of N_2O. The geometry changes in these core-excited states can be qualitatively understood by the electrostatic force theory [6,7].

Theoretical spectroscopy on the core-electron states has been achieved by various types of electronic structure theories. Peyerimhoff and co-workers performed pioneering work using the multireference single and double configuration interaction (MRSDCI) method [8,9]. Cederbaum and co-workers have greatly contributed to the inner-shell electronic processes as well as the valence ionizations using the Green's function method called the algebraic diagrammatic construction (ADC) approach [10,11]. They introduced essential and useful chemical concepts for those electronic processes, which have accelerated the development of this field. Theories based on cluster expansion have also been applied to the core-electron processes such as the symmetry-adapted cluster–configuration interaction (SAC–CI) method originally developed by Nakatsuji [12,13] and the multireference coupled cluster method proposed by Jana and Mukherjee [14]. Recently, time-dependent density functional theories (TDDFT) with new functionals designed for the core-electron processes have also been developed [15].

The SAC–CI method [12,13] has been successfully applied to a wide range of chemistry and physics. The method has been established as a useful tool for investigating molecular excited states through numerous applications to molecular spectroscopy, biological chemistry, and surface chemistry [16,17]. The SAC–CI general-*R* method [18,19] is designed to investigate the complex multielectron processes such as shake-up satellite states and has been successfully applied to those processes [17]. Because the method is applicable to general electronic states, it provides a powerful tool for the inner-shell electronic processes. The SAC–CI method with the equivalent core model (ECM) is also useful for studying the vibrational structure and geometry relaxation in the core-electron states. There, we can use the SD-*R* method in the direct SAC–CI framework [20]. We have investigated various types of core-electron processes with the SAC–CI method [21-29]; for example, core-electron binding energies, inner-shell shake-up satellite spectra, vibrational spectra and geometry relaxation of satellites, g–u splitting, valence–Rydberg coupling, and its vibration-induced suppression.

In this article, we review our recent experimental and theoretical work on the O 1s excited states of N_2O [28] and present a SAC–CI study on the N 1s excited states; namely, the two-dimensional potential energy surfaces (2D PESs) of the N_t 1s and N_c 1s excited states with regard to π^*, $np\pi$ (n = 3–5), $ns\sigma$ (n = 3–5), and $np\sigma$ (n = 3, 4) transitions.

2. Theory

2.1. SAC/SAC–CI theory

The 2D PESs of the ground, O 1s excited, and N 1s excited states were calculated along the direction of the normal coordinates q_1 and q_3 corresponding to the quasisymmetric (ν_1') and quasi-antisymmetric (ν_3') stretching vibrational motions, respectively, in the ranges R_{NN} = 1.00–1.30 Å and R_{NO} = 1.00–1.55 Å. The basis sets were the correlation-consistent polarized valence triple zeta (cc-pVTZ) basis sets proposed by Dunning without f function, namely [4s3p2d] [30] plus Rydberg functions [5s5p] [31] placed on the central N atom for describing n = 3, 4, 5 (s, p, and d functions for O 1s excited states, and s and p functions for N 1s excited states).

The ground and core-excited states of N_2O were calculated by the SAC and SAC–CI methods, respectively. The ground-state geometry was calculated to be R_{NN} = 1.122 and R_{NO} = 1.184 Å, in good agreement with the experimental values [32] of 1.127 and 1.185 Å, respectively. To calculate the 2D PESs of the core-excited states, the ECM was adopted. For the O 1s, N_t 1s, and N_c 1s excited states, the excited states of NNF, ONO, and NOO molecules, respectively, were calculated by the SAC–CI method. SAC–CI SD-*R* calculations in which single (S) and double (D) excitations were adopted for the *R*-operators were performed and the excited states of the neutral radicals ONO and NOO were calculated by electron attachment to ONO^+ and NOO^+, respectively. We employed the algorithm calculating σ-vectors directly that includes all of the S_2R_1 and S_2R_2 nonlinear terms [20]. All of the *S*- and *R*-operators were included without perturbation selection in the SAC–CI calculations. The equilibrium geometries of each state were obtained by the analytical energy gradients of the SAC–CI method. The optimized structures were confirmed to be local minima by calculating the PESs in the bending coordinate.

The SAC–CI calculations were performed with the development version of the Gaussian09 suite of programs [33].

2.2. 2D vibrational analysis

The vibrational analysis was performed using the SAC–CI 2D PESs of the O 1s and N 1s excited states of N_2O. To calculate the Franck–Condon (FC) factors of the vibrational spectrum, the vibrational wave function was obtained by the grid method, in which the Lanczos algorithm was adopted for the diagonalization. The 2D PESs of the N_2O molecule were described in the bonding coordinates where the bond distances are r_1 and r_2, the bond angle being fixed. In this coordinate, the kinetic part of the Hamiltonian of the vibrational motion is given by:

$$T = \frac{p_1^2}{2\mu_{N_cN_t}} + \frac{p_2^2}{2\mu_{N_cO}} + \frac{p_1 p_2}{m_{N_c}},\ p_k = -i\frac{\partial}{\partial r_k},\ k = 1,2,$$

and $\mu_{N_cN_t}$ is the reduced mass of N_t and N_c atoms. The coordinates r_1 and r_2 are represented by the Hermite discrete variable representation [34]. The 2D PESs were fitted by the analytical functions of the 2D Morse expansion:

$$V(r_1,r_2) = \sum_{i,j=0}^{n} B_{ij}(1-e^{-a_1(r_1-r_{e_1})})^i(1-e^{-a_2(r_2-r_{e_2})})^j,$$

where r_{e_1} and r_{e_2} are equilibrium distances and are determined by the analytical energy gradients of the SAC/SAC–CI method. This function is suitable for describing the PESs of the bound state [35]. The vibrational spectra were calculated in the framework of the FC approximation. These calculations were conducted using the MCTDH program system developed by the Heidelberg group [36].

3. Applications to core-electronic processes

3.1. O 1s excited states of N_2O

ARIY spectra were measured in the O K edge of N_2O [28]. Figure 1 shows the ARIY spectra measured at angles of 0° and 90° relative to the polarization vector along with the total ion yield measurement. The measurement at 0°, ARIY(I(0°)), mainly observes the Σ states, while the measurement at 90°, ARIY(I(90°)), observes the Π states. In the ARIY(I(0°)), the Σ and Π states interfere with each other; therefore, the Σ spectrum cannot be resolved completely. In the O $1s^{-1}$ 4sσ state, irregular excitation to the higher vibrational levels was observed, while the vibrational spectrum of the 3pπ state was a normal Rydberg spectrum.

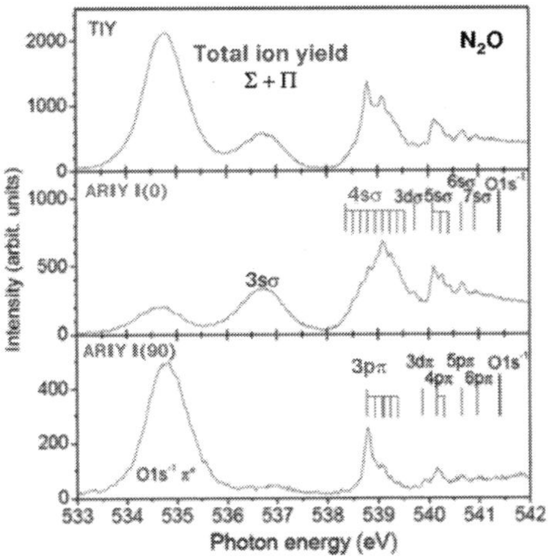

Figure 1. ARIY spectra of O 1s excitation of N_2O measured at 0° and 90° relative to the polarization vector [28].

To clarify this irregular vibrational progression, the PESs of the O 1s core-excited states were examined [28]. Figure 2 show the PESs and the electronic part of the second moments $<r^2>$ of the O 1s excited states of Σ and Δ symmetry along the R_{NO} distance. The second moment $<r^2>$ of the Rydberg states increases monotonically with R_{NO}, while that of the O $1s^{-1}$ 4sσ state decreases drastically at large R_{NO}. This shows that strong valence–Rydberg coupling occurs in the O $1s^{-1}$ 4sσ state having a σ* component. Because of this strong valence–Rydberg coupling, the O $1s^{-1}$ 4sσ state has a potential energy minimum at large R_{NO}. This shows that the irregular Rydberg behavior in the O $1s^{-1}$ 4sσ state is caused by the strong valence–Rydberg coupling.

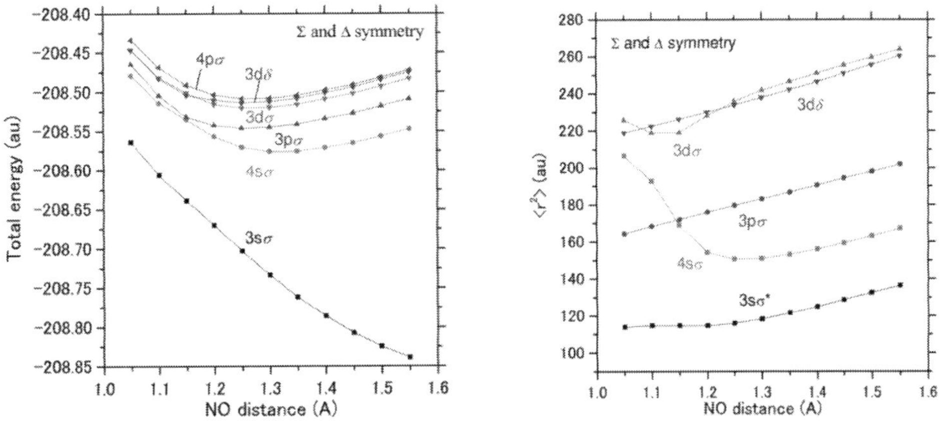

Figure 2. Potential energy curves (left) and second moments $<r^2>$ (right) of the O 1s excited states of N_2O [28].

The FC analysis based on the SAC–CI 2D PESs was performed for the O $1s^{-1}$ $4s\sigma$ and $3p\pi$ states [28]. In Figure 3, theoretical spectra are compared with the ARIY $I(0°)$ and $I(90°)$ spectra for the O $1s^{-1}$ $4s\sigma$ and $3p\pi$ states, respectively. Theory reproduced the observed vibrational excitations accurately. The higher vibrational levels are excited in the O $1s^{-1}$ $4s\sigma$ state, while the vibrational spectrum of the $3p\pi$ state shows the standard Rydberg-type vibrational progression.

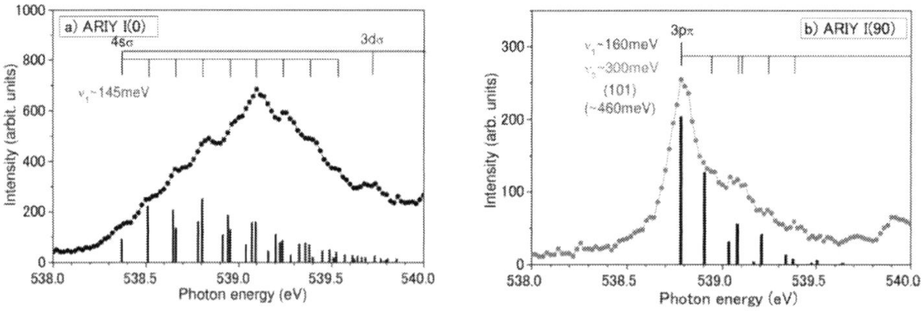

Figure 3. Vibrational spectra of a) O $1s^{-1}$ $4s\sigma$ (ARIY, $I(0°)$) and b) O $1s^{-1}$ $3p\pi$ (ARIY, $I(90°)$) excited states of N_2O. The SAC–CI spectra are shown by the vertical solid lines [28].

3.2. Thermal effects in core-electron excitation spectra

Absorption spectra of vibrationally excited "hot" molecules were observed for the O 1s excitation of N_2O [29]. The absorption spectra were measured at 300/700 K and the vibrationally excited absorption spectrum was extracted assuming the Boltzmann distribution (Figure 4) [29]. Based on the Boltzmann distribution, the bending mode is dominantly excited in the ground state. Comparing the spectra of (0, 0, 0) and (0, ν_2, 0), the intensity of the $ns\sigma$ Rydberg series is significantly suppressed and the peaks show a blue shift. The absorption of the $A'(\pi)$ state, on the other hand, is enhanced for the excitation from the vibrationally excited states, (0, ν_2, 0), and shows a red shift. These observations suggest that the decrease in the bond angle causes a decrease in the mixing of the valence character that enhances the transition probability to the Rydberg states.

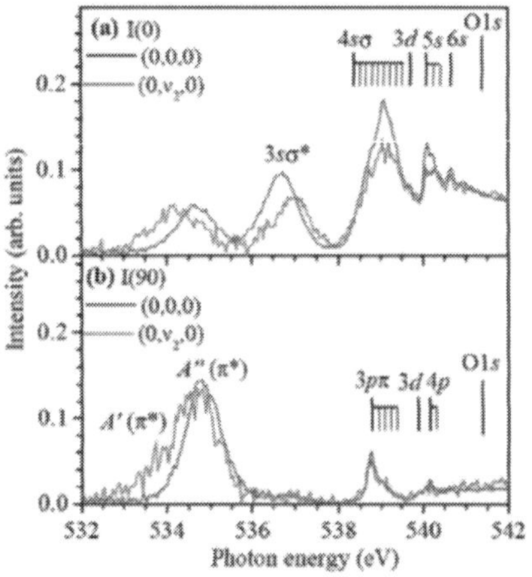

Figure 4. ARIY spectra in the N_2O O 1s excitation region: blue and red lines show the spectra from the vibrational ground and excited states, respectively [29].

To understand these phenomena, we performed SAC–CI calculations of the energies and the second moments $\langle r^2 \rangle$ of the O 1s excited states varying the bond angle [29]. Figure 5 shows a cut of the calculated PESs of the O 1s excited states of A' symmetry. The $1A'$ state is correlated with the π^*

state and stabilizes along the bending coordinate. All of the other states are stable in the linear structure. A characteristic curve crossing occurs between the σ and π Rydberg states along the bending coordinate. These PESs explain the red shift of the π* state and the blue shift of the 3sσ and 4sσ states. To analyze the valence–Rydberg coupling, we examined the electronic part of the second moment $<r^2>$, which is anticorrelated with the amount of valence character (Figure 5). The second moment of the 3sσ, 4sσ, and 5sσ states becomes larger as the molecule becomes more bent. This indicates that the mixing of the valence character in these states becomes less as the bond angle decreases. Consequently, the absorption intensity of the *n*sσ Rydberg states weakens. These results explain the interpretation of the intensity changes observed for excitation from vibrationally excited molecules.

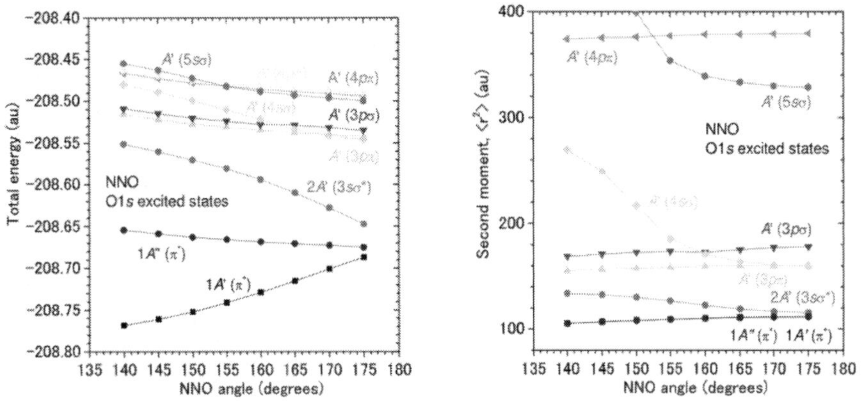

Figure 5. Potential energy curves (left) and second moments $<r^2>$ (right) of the low-lying O 1s excited states of N_2O along the bending coordinate [29].

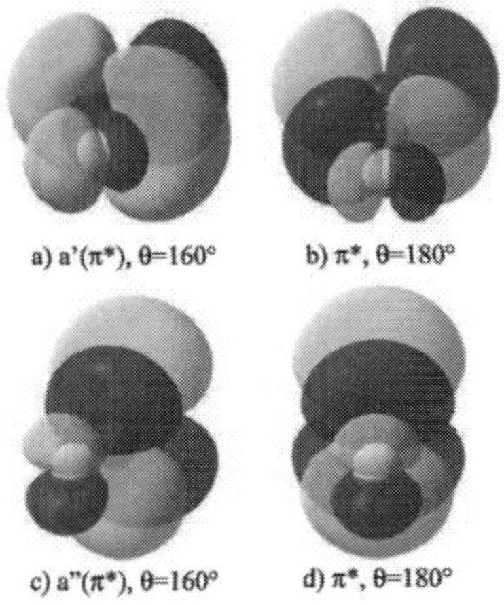

Figure 6. Calculated in-plane *a*'(π*) and out-of-plane *a*"(π*) MOs at bond angles of θ=160° and 180°[29].

The *A*'(π*) state stabilizes along the bending coordinate whereas the 3s*A*' state destabilizes more than the *A*"(π*) state. This anticorrelation indicates that the *A*'(π*) and 3s*A*' states are strongly coupled. This coupling opens a flow of valence character from the 3s*A*', 4s*A*', and 5s*A*' Rydberg states to the *A*'(π*) state. Analyzing the MOs that contribute to the excitations, we conclude that the counterpart of the decrease in the mixing of the valence character in the 3s, 4s, and 5s states is an increase in the sσ-type character of the *a*'(π*) orbital. The π* MO is displayed in Figure 6. Figures 6(c) and 6(d) show that the out-of-plane *a*"(π*) MO has a pure pπ-type orbital independent of bond angle. For the in-plane

$a'(\pi^*)$ MO, on the other hand, the sσ-type orbital component contributes to the MO, as can be seen in Figures 6(a) and 6(b).

Based on the SAC–CI calculations of the second moment $\langle r^2 \rangle$, the suppression is interpreted as being the result of a decrease in the mixing of the valence character in the *n*sσ Rydberg states with decreasing bond angle.

3.3. N_t 1s excited states of N_2O

The ARIY spectra of the N 1s excited states of N_2O were also observed in the energy range 399–413 eV. The ARIY spectra and its theoretical interpretations are presented in a separate note [37]. In this article, we discuss the results of the 2D PESs of the N_t 1s and N_c 1s excited states regarding the π*, *n*pπ (*n* = 3–5), *n*sσ (*n* = 3–5), and *n*pσ (*n* = 3, 4) transitions. We also compare those PESs with the PESs of the O 1s excited states.

The 2D PESs of the N_t 1s excited states were investigated by the SAC–CI method within the ECM approximation. Figures 7 and 8 show the 2D PESs of these states with Π- and Σ-symmetry, respectively, for R_{NN} = 1.00, 1.10, 1.20, and 1.30 Å. The N_t $1s^{-1}$ π* excited state shows a bound state in the N–O distance as in the O $1s^{-1}$ π* excited state. Other *n*pπ Rydberg excited states have almost the same shape for the PESs converging to the N_t $1s^{-1}$ ionized state. The PESs of the N_t $1s^{-1}$ *n*sσ and *n*pσ states show a characteristic shape. In the range R_{NN} = 1.00–1.20 Å, the N_t $1s^{-1}$ 3sσ state has a flat region around the FC region in the PES. For larger internuclear distances $R_{NN} \geq 1.30$ Å, the PES of this state becomes bound. The energy minimum in the N_t $1s^{-1}$ 3sσ state exists for a very large R_{NN} distance. The PESs of the N_t $1s^{-1}$ 4sσ and 3pσ states have an avoided crossing in the region of R_{NN} =

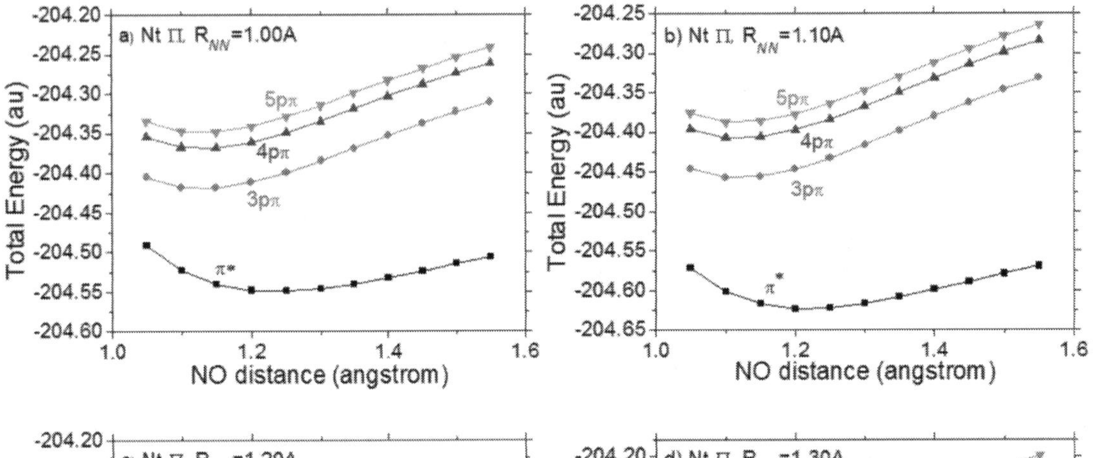

Figure 7. The 2D PESs of the low-lying N_t 1s excited states (π* and *n*pπ (*n* = 3–5) transitions) of N_2O in the R_{NO} coordinate for R_{NN} = 1.00, 1.10 [37], 1.20, and 1.30 Å.

1.00–1.20 Å. The valence–Rydberg coupling of the N_t $1s^{-1}$ 4sσ state, however, is not so large in the FC region, and therefore the 3pσ state is more stable than the 4sσ state. This situation of the N_t $1s^{-1}$ 4sσ state is different from that in the O $1s^{-1}$ 4sσ state shown in Figure 2. For the large internuclear distance of $R_{NN} \geq 1.30$ Å, valence–Rydberg coupling of the N_t $1s^{-1}$ 4sσ state becomes large and its PES is much lower in energy than that of the N_t $1s^{-1}$ 3pσ state. The PESs of the N_t $1s^{-1}$ 5sσ and 4pσ states also interact with each other in the same manner.

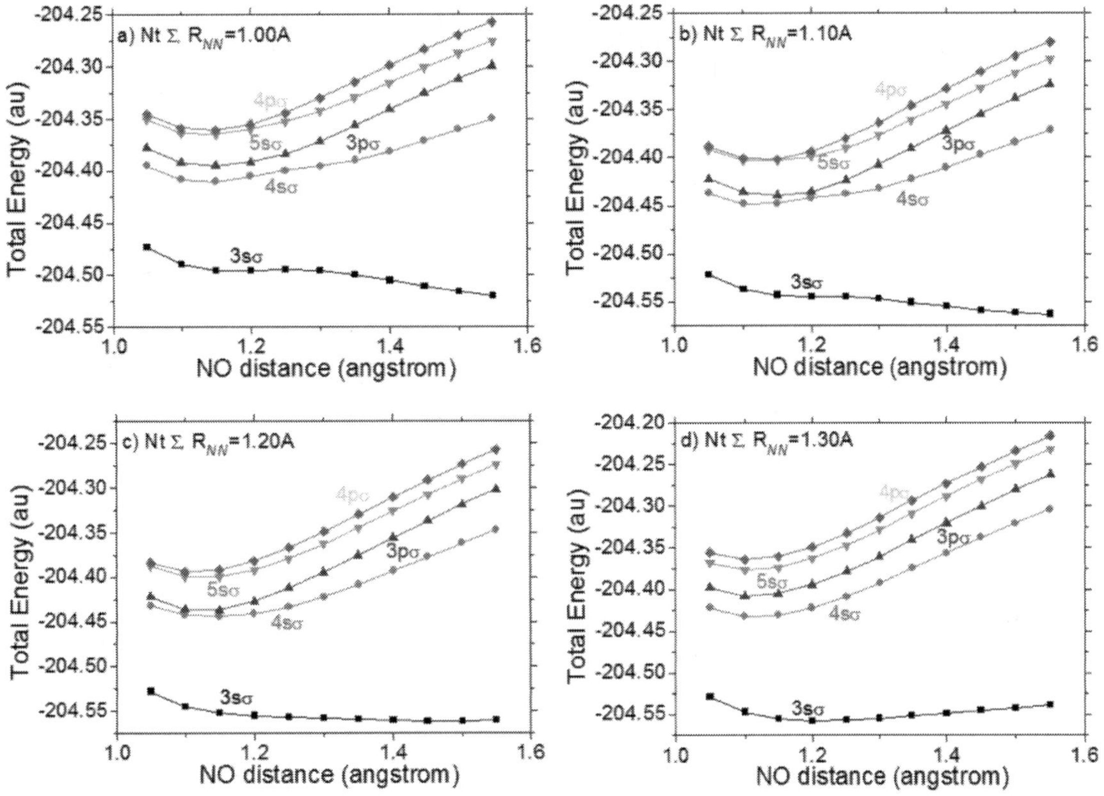

Figure 8. The 2D PESs of the low-lying N_t 1s excited states (*n*sσ (*n* = 3–5) and *n*pσ (*n* = 3, 4) transitions) of N_2O in the R_{NO} coordinate for R_{NN} = 1.00, 1.10 [37], 1.20, and 1.30 Å.

3.4. N_c 1s excited states of N_2O

The 2D PESs of the N_c 1s excited states were also calculated by the SAC–CI method in the same manner. The PESs are shown in Figures 9 and 10 for Π- and Σ-symmetry, respectively, for R_{NN} = 1.00, 1.10, 1.20, and 1.30 Å.

The PES of the N_c $1s^{-1}$ π^* excited state is repulsive in the N–O coordinate, which is different from those of the O $1s^{-1}$ π^* [28] and N_t $1s^{-1}$ π^* excited states. Other *n*pπ Rydberg excited states are bound; however, their PESs are flat in the R_{NO} coordinate compared with the O 1s and N_t 1s Rydberg states. These PESs converge monotonically to the N_c $1s^{-1}$ ionized state. The PESs of the N_c $1s^{-1}$ *n*sσ and *n*pσ states have repulsive character compared with those of the O $1s^{-1}$ and N_t $1s^{-1}$ Rydberg excited states for the *n*sσ and *n*pσ transitions. The N_c $1s^{-1}$ 3sσ state is intrinsically repulsive, while other states are bound, with their equilibrium R_{NO} distances being relatively large. In the range R_{NN} = 1.00–1.20 Å, the PES of the N_c $1s^{-1}$ 4sσ state has an avoided crossing with the N_c $1s^{-1}$ 3pσ state. In the N_c $1s^{-1}$ 4sσ state, valence–Rydberg coupling is strong even in the FC region and the repulsive nature is remarkable. The valence–Rydberg mixing of other states is not so strong from the nature of their PESs. Therefore, the equilibrium bond length of the N–O bond becomes large. Other higher N_c $1s^{-1}$ $ns\sigma$ and $np\sigma$ states are flat. The PESs of the N_c $1s^{-1}$ $5s\sigma$ and $4p\sigma$ states also interact at small R_{NN} and R_{NO} distances.

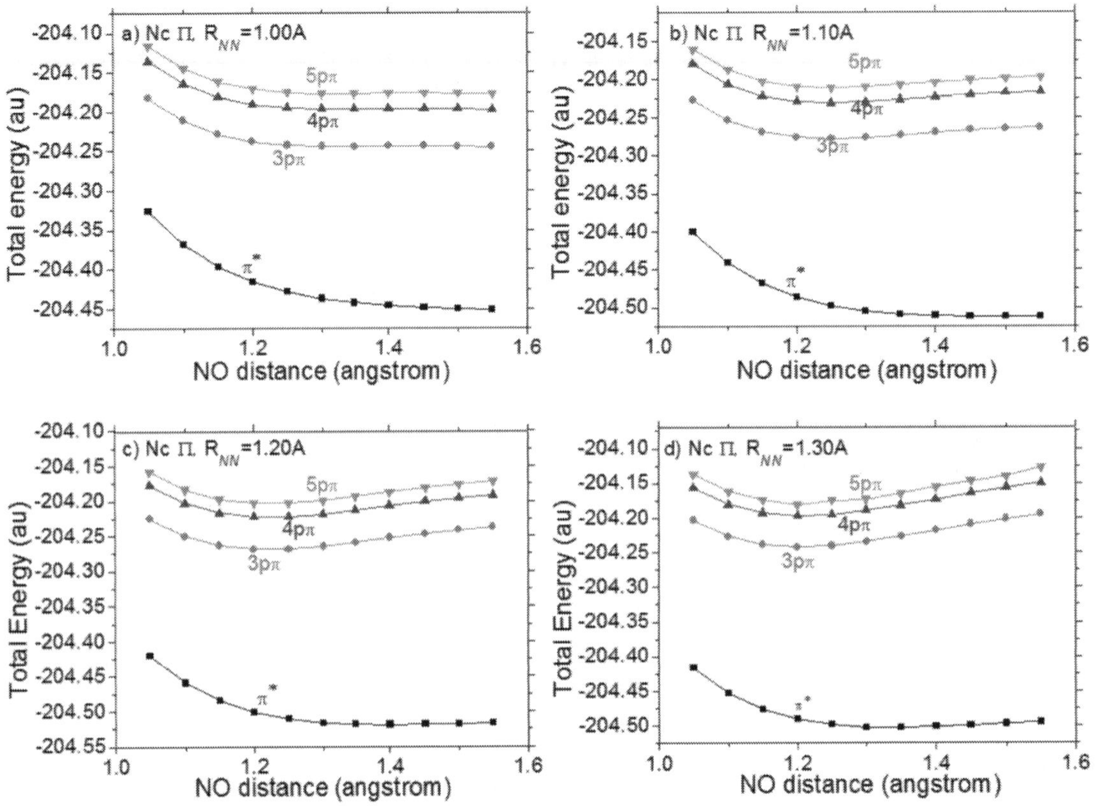

Figure 9. The 2D PESs of the low-lying N_c 1s excited states (π^* and $np\pi$ (n = 3–5) transitions) of N_2O in the R_{NO} coordinate for R_{NN} = 1.00, 1.10 [37], 1.20, and 1.30 Å.

4. Summary

In this article, we present joint work on the O 1s and N 1s excited states of N_2O using high-resolution ARIY spectroscopy and the SAC–CI method.

The high-resolution ARIY was observed in the region of the O 1s edge of N_2O. The vibrational excitations were found to be specific to the individual Rydberg states in the Σ-symmetry spectrum. The FC analysis based on the SAC–CI 2D PESs well reproduces the observed state-specificity of the vibrational excitations. The irregular vibrational excitations along the same $ns\sigma$ Rydberg series were attributed to valence–Rydberg coupling.

A significant suppression of the intensity in the O 1s → $ns\sigma$ Rydberg series in N_2O for excitation from the bending vibrationally excited ground states was observed. Based on the second moment analysis, this suppression was attributed to a decrease in the mixing of the valence character in the $ns\sigma$ Rydberg states with decreasing bond angle.

The 2D PESs of the low-lying N_t 1s and N_c 1s excited states of N_2O have also been investigated by the SAC–CI method. The bound nature of the PESs of the N_t 1s excited states is strong compared with those of the O 1s excited states, while the PESs of the N_c 1s excited states have a repulsive character.

The valence–Rydberg coupling in the N_c 1s excited states is strong, in particular in the N_c $1s^{-1}$ 3sσ and 4sσ states. Detailed assignments of the ARIY spectra are presented in a separate note [37].

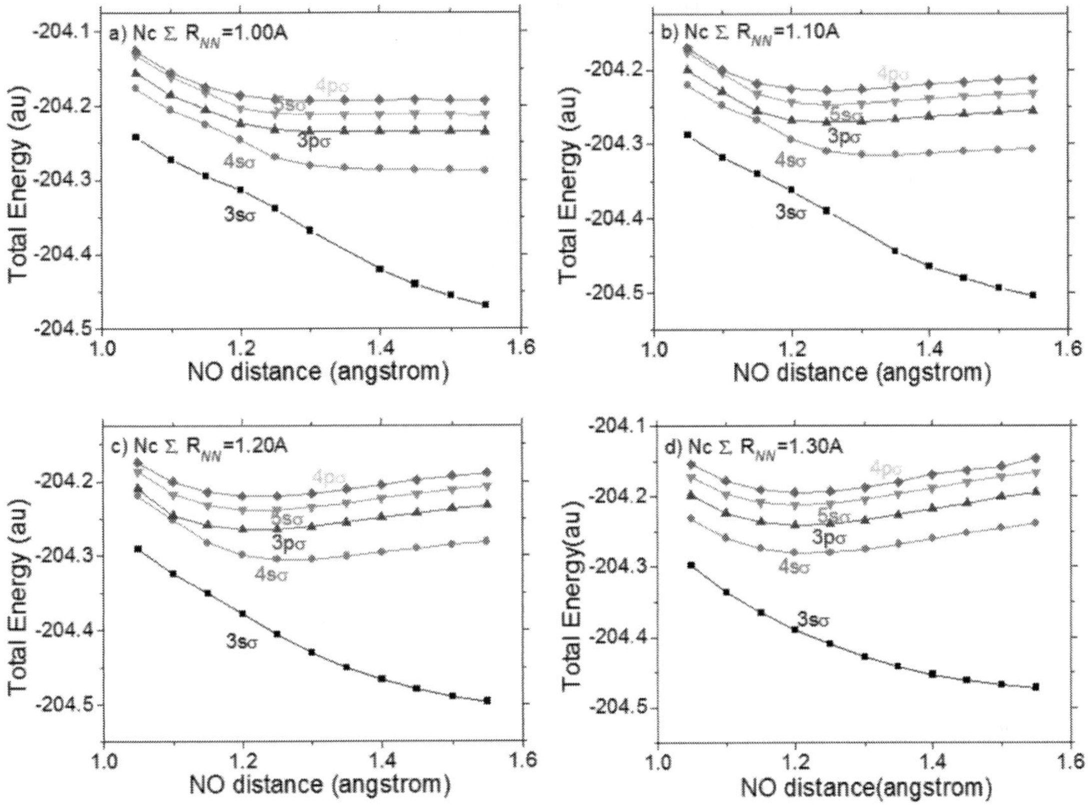

Figure 10. The 2D PESs of the low-lying N_c 1s excited states (*n*sσ (*n* = 3–5) and *n*pσ (*n* = 3, 4) transitions) of N_2O in the R_{NO} coordinate for R_{NN} = 1.00, 1.10 [37], 1.20, and 1.30 Å.

Acknowledgments
The experiment was carried out with the approval of the SPring-8 program review committee. This study was supported by JST-CREST, a Grants-in-Aid for Scientific Research from the Japanese Society for the Promotion of Science (JSPS) and the Next Generation Supercomputing Project. The computations were partly performed at Research Center for Computational Science, Okazaki, Japan.

References
[1] Siegbahn K, Nordling C, Johansson G, Hdman J, Heden P F, Hamrin K, Gelius U, Bergmark T, Werme L O, Manne R and Baer Y 1969 *ESCA Applied to Free Molecules* (North-Holland, Amsterdam)
[2] Ueda K 2003 *J. Phys. B: At. Mol. Opt. Phys.* **36** R1
[3] Adachi J, Kosugi N, Shigemasa E and Yagishita A 1995 *J. Chem. Phys.* **102** 7369
[4] Tanaka T et al. 2006 *Chem. Phys. Lett.* **428** 34
[5] Prince K C, Avaldi L, Coreno M, Camilloni R and Simone M d 1999 *J. Phys. B:At. Mol. Opt. Phys.* **32** 2551
[6] Nakatsuji H 1973 *J. Am. Chem. Soc.* **95** 345

[7] Nakatsuji H and Koga T 1981 *The Force Concept in Chemistry* (Van Nostrand Reinhold Company, New York)
[8] Peyerimhoff S D 1975 *Adv. Quantum Chem.* **9** 69
[9] Buenker R J and Peyerimhoff S D 1968 *Theor. Chim. Acta* **12** 183
[10] Cederbaum L S and Domcke W 1977 *J. Chem. Phys.* **66** 5084
[11] Cederbaum L S, Domcke W, Schirmer J and von Niessen W 1986 *Adv. Chem. Phys.* **65** 115
[12] Nakatsuji H 1978 *Chem. Phys. Lett.* **59** 362
[13] Nakatsuji H 1979 *Chem. Phys. Lett.* **67** 329
[14] Jana D and Mukherjee D 2005 *J. Chem. Phys.* **122** 234101
[15] Nakata A, Tsuneda T and Hirao K 2009 *J. Comp. Chem.* **30** 2583
[16] Nakatsuji H 1992 *Acta Chim. Hungarica, Models in Chemistry* **129** 719
[17] Ehara M, Hasegawa J and Nakatsuji H 2005 *SAC-CI Method Applied to Molecular Spectroscopy, in Theory and Applications of Computational Chemistry: The First 40 Years, A Volume of Technical and Historical Perspectives* (Elsevier) p. 1099
[18] Nakatsuji H 1991 *Chem. Phys. Lett.* **177** 331
[19] Ehara M and Nakatsuji H 1998 *Chem. Phys. Lett.* **282** 347
[20] Fukuda R and Nakatsuji H 2008 *J. Chem. Phys.* **128** 094105
[21] Sankari R, Ehara M, Nakatsuji H, Senba Y, Hosokawa K, Yoshida H, Fanis A D, Tamenori Y, Aksela S and Ueda K 2003 *Chem. Phys. Lett.* **380** 647
[22] Kuramoto K, Ehara M and Nakatsuji H 2005 *J. Chem. Phys.* **122** 014304
[23] Ueda K, Hoshino M, Tanaka T, Kitajima M, Tanaka H, Fanis A D, Tamenori Y, Ehara M, Oyagi F, Kuramoto K and Nakatsuji H 2005 *Phys. Rev. Lett.* **94** 243004
[24] Ehara M, Nakatsuji H, Matsumoto M, Hatamoto T, Liu X-J, Lischke T, Pruemper G, Tanaka T, Makochekanwa C, Hoshino M, Tanaka H, Harries J R, Tamenori Y and Ueda K 2006 *J. Chem. Phys.* **124** 124311
[25] Ehara M, Kuramoto K, Nakatsuji H, Hoshino M, Tanaka T, Kitajima M, Tanaka H, Tamenori Y, Fanis A D and Ueda K 2006 *J. Chem. Phys.* **125** 114304
[26] Ehara M, Tamaki R, Nakatsuji H, Lucchese R R, Soderstrom J, Tanaka T, Hoshino M, Kitajima M, Tanaka H, Fanis A D and Ueda K 2007 *Chem. Phys. Lett.* **438** 14
[27] Ehara M and Nakatsuji H 2008 *Coll. Czech Chem. Commun.* **73** 771
[28] Tanaka T, Ueda K, Feifel R, Karlsson L, Tanaka H, Hoshino M, Kitajima M, Ehara M, Fukuda R, Tamaki R and Nakatsuji H 2007 *Chem. Phys. Lett.* **435** 182
[29] Tanaka T, Hoshino M, Kato H, Ehara M, Yamada N, Fukuda R, Nakatsuji H, Tamenori Y, Harries J R, Prumper G, Tanaka H and Ueda K 2008 *Phys. Rev. A* **77** 012709
[30] Dunning Jr. T H 1989 *J. Chem. Phys.* **90** 1007
[31] Kaufmann K, Nager C and Jungen M 1985 *Chem. Phys.* **95** 385
[32] Teffo J L and Chedin A 1989 *J. Mol. Spectrosc.* **135** 389
[33] Frisch M J et al 2010 Gaussian Development Version, Revision H. 06, Gaussian Inc., Wallingford CT.
[34] Lill J V, Parker G A and Light J C 1982 *Chem. Phys. Lett.* **89** 483
[35] Beck M H and Meyer H-D 2001 *J. Chem. Phys.* **114** 2036
[36] Worth G A, Beck M H, Jackle A and Meyer H-D 2003 *The MCTDH Package, Version 8.3, University Heidelberg,* Heidelberg, Germany
[37] Ehara M, Horikawa T, Fukuda R, Nakatsuji H, Tanaka T, Hoshino M, Tanaka H and Ueda K *submitted for publication*

Int. Conf. on Many Particle Spectroscopy of Atoms, Molecules, Clusters and Surfaces IOP Publishing
Journal of Physics: Conference Series **288** (2011) 012025 doi:10.1088/1742-6596/288/1/012025

Interference and non-Franck-Condon effects in ionization of H_2 molecules by photon impact

O. A. Fojón[1], J. Fernández[2], A. Palacios[2] and F. Martín[2]
[1]Instituto de Física Rosario (CONICET-UNR) 2000 Rosario, Argentina
[2]Departamento de Química C-9, Universidad Autónoma de Madrid, 28049 Madrid, Spain

E-mail: fojon@ifir-conicet.gov.ar

Abstract. We analyze theoretically interference effects in the spectra of electrons emitted in the H_2 photoionization by high energy linearly polarized photons. Molecular bound and continuum states are accurately described by means of B-spline basis allowing the inclusion of the nuclear degrees of freedom. One interesting feature is observed: the usual Franck-Condon behavior is not followed when the H_2 internuclear axis is parallel to the polarization direction. Moreover, this is related to the fact that for this molecular orientation and under certain conditions, the electron cannot be emitted in the direction of the radiation field. On the contrary, for H_2 molecules perpendicular to the polarization direction, the angular distribution of electrons is analogous to the one observed in the two slits Young's experiment.

1. Introduction

Nowadays, the study of interference effects in the electronic emission from diatomic molecules is a very active subject of research (see for instance [1]). In the sixties, Cohen and Fano explained anomalies in the photoelectron spectra of diatomic molecules [2]. They argued that the coherent emission from both molecular centers produced interferences when the de Broglie wavelength of the ejected electrons is of the order of the internuclear distance of the molecule in analogy with the Young's two slit experiment. They elaborated a simple model based on the following assumptions: i) ionization is basically a *one*-electron process, ii) the lowest molecular orbital ψ_g is well described in the Linear combination of atomic orbitals (LCAO) approximation ($\psi_g \sim 1s_A + 1s_B$, where $1s_A$ and $1s_B$ are atomic orbitals centred in nucleus A and B, respectively) and iii) the ionized electron is well described by a single-centre plane wave. In this way, Cohen and Fano showed that the total photoionization cross section is given by the analytical form

$$\sigma = \sigma_H \left[1 + \sin(k_e R)/(k_e R)\right]/(1 + S) \tag{1}$$

where σ_H is the photoionization cross section of a hydrogen atom with effective charge Z_{eff}, R is the equilibrium internuclear distance, S is the overlap between $1s_A$ and $1s_B$, and k_e is the electron wave vector. This formula shows the typical oscillatory behaviour associated with interferences.

However, it was not until recently that this kind of interference was experimentally detected in the differential cross sections of ionization of diatomic molecules by heavy ion impact [3]. The detection of these effects is not an easy task as the cross section as a function of the energy decreases quickly. In order to circumvent this inconvenient in those experiments, the ratio between molecular cross sections and adequate atomic cross sections (two isolated atoms for instance, for which no interferences are produced) was considered. The coherent emission was then traced by analyzing oscillations in the mentioned ratios. In the same way, interferences effects were observed subsequently in the case of electronic impact [4, 5]. Moreover, the case of photoionization of hydrogen molecules was revisited in order to provide a realistic description of the photoionization spectra including the influence of the nuclear motion. In this work we review some of the remarkable results obtained for the case of impact of linearly polarized photons on hydrogen molecules. The Cohen and Fano predictions were studied concluding that they are valid at high emission energies. At low energies, deviations are observed mainly due to the electronic correlation and the nature of the molecular potential [6]. Interestingly, it was also observed that the usual Franck-Condon behavior is not followed when the H_2 internuclear axis is parallel to the polarization direction for linear polarized photons [7]. The origin of this anomaly may be related to interference and confinement effects [8]. Such effects are more easily evidenced in the electron angular distributions associated with specific vibrational states of the residual H_2^+ ion. For H_2 molecules oriented perpendicular to the polarization direction, the distributions are close to the ones obtained in the Young's double-slit experiment. In the case of molecules parallel to the polarization direction and under certain conditions, the electron cannot be emitted in the classical direction given by the radiation field.

Atomic untis are used unless otherwise explicitly stated.

2. Theoretical methods

In this section, the theoretical methods employed in the previous works are described [7].

Photoionization cross sections have been evaluated in the framework of the dipole and the adiabatic Born-Oppenheimer (BO) approximations. The fully differential photoionization cross section, i.e., differential in both the energy and direction of the ejected electron and the energy and orientation of the residual H_2^+ molecular ion, is given by Dill's formula [9]:

$$\frac{d\sigma_\alpha^{\mu_0}(\omega)}{d\Omega_n d\Omega_e d\varepsilon} = \frac{4\pi^2\omega}{c} \sum_{\mu_a,\mu_b} \sum_{\ell_a,m_a} \sum_{\ell_b,m_b} i^{(\ell_a-\ell_b)} e^{i(\hat{\sigma}_{\ell_b}(\varepsilon)-\hat{\sigma}_{\ell_a}(\varepsilon))} (-1)^{m_b+\mu_a-\mu_0} T^*_{\alpha\ell_a m_a\mu_a}(\varepsilon) T_{\alpha\ell_b m_b\mu_b}(\varepsilon)$$

$$\times \sum_{L_e} \left[\frac{(2\ell_a+1)(2\ell_b+1)}{(2L_e+1)}\right]^{\frac{1}{2}} C(\ell_a,\ell_b,L_e;-m_a,m_b,M_e) C(\ell_a,\ell_b,L_e;0,0,0) Y_{L_e}^{M_e\,*}(\theta_e,\phi_e)$$

$$\times \sum_{L_\gamma} \left[\frac{1}{(2L_\gamma+1)}\right]^{\frac{1}{2}} C(1,1,L_\gamma;-\mu_a,\mu_b,M_\gamma) C(1,1,L_\gamma;-\mu_0,\mu_0,0) Y_{L_\gamma}^{M_\gamma}(\theta_n,\phi_n), \quad (2)$$

with $\mu_{a,b} = 0, \pm 1$, $M_\gamma = -\mu_a + \mu_b$, $M_e = -m_a + m_b$, $\ell_a + \ell_b \geq L_e \geq |\ell_a - \ell_b|$ and $2 \geq L_\gamma \geq 0$. In this equation, $\mu_0 = 0$ for linearly polarized light and $\mu_0 = \pm 1$ for circularly polarized light, α denotes the electronic state of the residual molecular ion, $\hbar\omega$ is the photon energy, ε is the photoelectron energy, $\Omega_e = (\theta_e, \phi_e)$ is the photoelectron emission direction in the molecular frame (θ_e and ϕ_e are the polar angles), $\Omega_n = (\theta_n, \phi_n)$ is the polarization direction with respect to the molecular axis z, c is the speed of light, $C(j_1, j_2, j; m_1, m_2, m)$ denotes a Clebsch-Gordan coefficient, Y_L^M is a spherical harmonic, $\hat{\sigma}_\ell(\varepsilon)$ is the Coulomb phase shift and $T_{\alpha\ell m\mu}(\varepsilon)$ is the transition dipole matrix element given by:

$$T_{\alpha\ell m\mu}(\varepsilon) = \int dR \langle \Psi_{g\nu}(\mathbf{r}, R) | \mathbf{e}_\mu \cdot \mathbf{D} | \Psi^+_{\alpha v_\alpha \ell m\varepsilon}(\mathbf{r}, R) \rangle, \quad (3)$$

where $\Psi_{g\nu}$ is the ground molecular state of energy $W_{g\nu}$, $\Psi^+_{\alpha v_\alpha \ell m\varepsilon}$ is the final molecular state of energy $W_{v_\alpha} + \varepsilon$ representing a molecular ion in the v_α vibronic state (either dissociative or non dissociative) and an ionized electron of energy ε and angular momentum ℓm, $\mathbf{r}$ represents the electronic coordinates, R is the internuclear distance, $\mathbf{e}_\mu$ is the photon polarization vector, and $\mathbf{D}$ is either $\sum_i \mathbf{r}_i$ (length gauge) or $(\hbar\omega)^{-1}\sum_i \nabla_i$ (velocity gauge). Energy conservation implies that $W_{g\nu} + \hbar\omega = W_{v_\alpha} + \varepsilon$. Neglecting rotational effects, the wave functions $\Psi_{g\nu}$ and $\Psi^+_{\alpha v_\alpha \ell m\varepsilon}$ are evaluated in the adiabatic (Born-Oppenheimer, BO) approximation

$$\Psi_{nv_n}(\mathbf{r}, R) = R^{-1}\chi_{v_n}(R)\psi_n(\mathbf{r}, R), \tag{4}$$

where ψ_n and χ_{v_n} are the usual electronic and nuclear BO wave functions [10, 11]. For each value of R, the electronic continuum states must satisfy the usual outgoing boundary conditions of electron-molecule scattering.

Integrating equation (2) over one or several differential magnitudes leads to partially differential or total cross sections, each one representing a specific experimental situation [12]. In particular, integrating equation 2 over the solid angle Ω_e, leads to the cross section differential in the nuclear solid angle and in the energy of the ejected electron irrespective of the electron emission direction [12]:

$$\frac{d\sigma^{\mu_0}_\alpha(\omega)}{d\Omega_n d\varepsilon} = \frac{1}{4\pi}\frac{d\sigma^{\mu_0}_\alpha(\omega)}{d\varepsilon}[1 + \beta^{\mu_0}_{\alpha,n}(\varepsilon)P_2(\cos\theta_n)], \tag{5}$$

where P_2 is the Legendre polynomial of order 2, $d\sigma^{\mu_0}_\alpha(\omega)/d\varepsilon$ is the cross section differential in the energy of the ejected electron,

$$\frac{d\sigma^{\mu_0}_\alpha(\omega)}{d\varepsilon} = \frac{4\pi^2\omega}{3c}\sum_{\ell m\mu}|T_{\alpha\ell m\mu}(\varepsilon)|^2, \tag{6}$$

and $\beta^{\mu_0}_{\alpha,n}(\varepsilon)$ is the nuclear asymmetry parameter

$$\beta^{\mu_0}_{\alpha,n}(\varepsilon) = \frac{3\mu_0^2 - 2}{2} \frac{\sum_{\ell m\mu}|T_{\alpha\ell m\mu - M_i}(\varepsilon)|^2[3(m - M_i)^2 - 2]}{\sum_{\ell m\mu}|T_{\alpha\ell m\mu - M_i}(\varepsilon)|^2}, \tag{7}$$

where M_i is the projection of the initial-state angular momentum. As we pointed out before, in this paper, we restrict our study to linearly polarized light, i.e., $\mu_0 = 0$, and to H_2 molecules initially in the ground state $X^1\Sigma_g^+$, i.e., $M_i = 0$. The above equations can also be used to obtain the cross sections differential in the energy of the residual molecular ion (non dissociative case) or the ejected proton (dissociative case) by using the energy conservation relation

$$W_{g\nu} + \hbar\omega = W_{v_\alpha} + \varepsilon \tag{8}$$

that relates ε to W_{v_α}. Since the ground state of H_2 has $^1\Sigma_g^+$ symmetry, the dipole selection rule implies that only continuum states of $^1\Sigma_u^+$ and $^1\Pi_u$ symmetries can be populated. Hence, the cross section given in Eq. (6) can be written as the sum of $^1\Sigma_u^+$ and $^1\Pi_u$ cross sections

$$\frac{d\sigma_\alpha(\omega)}{d\varepsilon} = \frac{d\sigma^\Sigma_\alpha(\omega)}{d\varepsilon} + 2\frac{d\sigma^\Pi_\alpha(\omega)}{d\varepsilon}. \tag{9}$$

In this paper, we will analyze in detail two particular molecular orientations: parallel ($\theta_n = 0$) and perpendicular ($\theta_n = \pi/2$) to the polarization vector. Hence, from Eq. (5), it can be easily seen that

$$\left.\frac{d\sigma_\alpha(\omega)}{d\varepsilon}\right|_{\theta_n=0} = \frac{\pi\omega}{c}\sum_\ell |T_{\alpha\ell 00}(\varepsilon)|^2 \equiv \frac{1}{4\pi}\frac{d\sigma_\alpha^\Sigma(\omega)}{d\varepsilon} \tag{10}$$

and

$$\left.\frac{d\sigma_\alpha(\omega)}{d\varepsilon}\right|_{\theta_n=\frac{\pi}{2}} = \frac{\pi\omega}{c}\sum_\ell |T_{\alpha\ell 11}(\varepsilon)|^2 \equiv \frac{1}{4\pi}\frac{d\sigma_\alpha^\Pi(\omega)}{d\varepsilon}. \tag{11}$$

Summation (integration) in Eqs. (9), (10) and (11) over the energy of the ejected electron when the ionized molecule if left in a non dissociative (dissociative) state leads to the integrated cross section

$$\sigma_\alpha(\omega) \equiv \sum\!\!\!\!\!\!\int d\varepsilon \frac{d\sigma_\alpha^{\mu_0}(\omega)}{d\varepsilon} = \sigma_\alpha^\Sigma(\omega) + 2\sigma_\alpha^\Pi(\omega) \tag{12}$$

and the corresponding ones for molecules oriented parallel and perpendicular to the polarization vector, respectively,

$$\sigma_\alpha(\omega)|_{\theta_n=0} = \frac{1}{4\pi}\sigma_\alpha^\Sigma(\omega) \tag{13}$$

and

$$\sigma_\alpha(\omega)|_{\theta_n=\frac{\pi}{2}} = \frac{1}{4\pi}\sigma_\alpha^\Pi(\omega). \tag{14}$$

The total photoionization cross section is obtained by summing over all open channels α the integrated cross sections given in Eq. (12):

$$\sigma(\omega) = \sum_\alpha \sigma_\alpha(\omega). \tag{15}$$

Starting again from equation (2), one can integrate over the solid angle Ω_n to obtain the cross section differential in the energy and solid angle of the ejected electron irrespective of the molecular orientation [13, 14]:

$$\frac{d\sigma_\alpha^{\mu_0}(\omega)}{d\Omega_e d\varepsilon} = \frac{1}{4\pi}\frac{d\sigma_\alpha^{\mu_0}(\omega)}{d\varepsilon}[1 + \beta_{\alpha,e}^{\mu_0}(\varepsilon)P_2(\cos\theta_e)], \tag{16}$$

where $d\sigma_\alpha^{\mu_0}(\omega)/d\varepsilon$ is the cross section differential in the energy of the ejected electron given in Eq. (6) and $\beta_{\alpha,e}^{\mu_0}(\varepsilon)$ is the electron asymmetry parameter

$$\begin{aligned}\beta_{\alpha,e}^{\mu_0}(\varepsilon) &= \frac{1}{5}\frac{C(1,1,2;-\mu_0,\mu_0,0)}{\sum_{\ell m\mu}|T_{\alpha\ell m\mu}(\varepsilon)|^2}\sum_{\mu_a,\mu_b}\sum_{\ell_a,m_a}\sum_{\ell_b,m_b} i^{(\ell_a-\ell_b)}e^{i(\hat\sigma_{\ell_a}(\varepsilon)-\hat\sigma_{\ell_b}(\varepsilon))}(-1)^{m_b+\mu_a}\\ &\times\ T^*_{\alpha\ell_a m_a\mu_a}(\varepsilon)T_{\alpha\ell_b m_b\mu_b}(\varepsilon)\\ &\times\ [(2\ell_a+1)(2\ell_b+1)]^{\frac{1}{2}}C(\ell_a,\ell_b,2;-m_a,m_b,M_e)C(\ell_a,\ell_b,2;0,0,0)C(1,1,2;-\mu_a,\mu_b,M_\gamma)\\ &\times\ \delta_{m_a+\mu_a,m_b+\mu_b}.\end{aligned} \tag{17}$$

The evaluation of the wave functions involved in the $T_{\alpha\ell m\mu}$ matrix elements requires the use of computational techniques. It is worthy to be mentioned that such techniques have been employed with succes to analyze many ionization reactions of H_2 targets. Among them, we cite resonant dissociative photoionization [11, 15, 16] and ion impact ionization [17]. Moreover, they allowed the obtention of the first numerical solution of the complete photo-induced break-up of H_2 [18]. More specifically, the vibrational (bound and dissociative) wave functions have been expanded in a basis of 280 B-splines of order $k = 8$ contained in a box of 12 a.u.. The electronic wave functions have been evaluated as described in detail in Refs. [10, 11]. Briefly, the ground state Ψ_g results from a configuration interaction (CI) calculation in which the H_2 hamiltonian has been diagonalized in a basis of 321 configurations built from products of one-electron H_2^+ orbitals and pseudo-orbitals. The calculated energy at the equilibrium internuclear distance is -1.8865023 a.u., to be compared with the exact non relativistic value -1.88876138 a.u. [19]. All these orbitals have been represented through a one-center expansion that includes spherical harmonics up to $\ell = 16$. The corresponding radial parts have been expanded in a basis of 310 B-splines of order $k = 8$ in a box of radial length of 60 a.u.. The final electronic continuum state $\Psi^+_{\alpha v_\alpha \ell m \varepsilon}$ results from a close coupling calculation that includes all partial waves with $\ell \leq 7$ associated with the four lowest ionization thresholds of H_2: $X^2\Sigma_g^+(1s\sigma_g)$, $^2\Sigma_u^+(2p\sigma_u)$, $^2\Pi_u(2p\pi_u)$, and $^2\Sigma_g^+(2s\sigma_g)$. For every value of R, these continuum states satisfy the usual boundary conditions corresponding to (i) one electron in a bound electronic state of H_2^+ and (ii) the other electron in a single outgoing spherical wave with a well defined value of the angular momentum ℓ plus a combination of incoming spherical waves for all accessible electronic states of H_2^+ and all possible values of the angular momentum of the ejected electron compatible with the molecular symmetry [10] (At the photon energy considered in this work, for a given energy, there is always a continuum state for each electronic state α of the residual H_2^+ ion and angular momentum ℓ of the ionized electron).

In this way, we represent in a realistic way the molecular nature including electron correlation and the two-center character of the molecular potential in all bound and continuum wavefunctions. Moreover, interferences among the various ionization thresholds and angular momenta of the ejected electron was taken into account through $\Psi^+_{\alpha v_\alpha \ell m \varepsilon}$. The convergence of the close coupling expansion was verified by performing calculations (for a reduced number of photon energies) in which six additional ionization thresholds were considered [6]. Those tests do not exhibit relevant variations in the dominant channel, i.e., the $^2\Sigma_g^+(1s\sigma_g)$ one.

3. Results

3.1. *Validity of the Cohen and Fano model*

To analyze the validity of equation (1), we calculated cross sections ratios dividing H_2 results by twice the ones of atomic H [6]. The result is shown in Fig. 1 as a function of photon energy.

Strong oscillations are clearly visible in both 4- and 10-channel calculations. We include also in Fig. 1 a fitting to the Cohen-Fano formula (1), leaving $a_0 = \sigma_H/(1+S)$ and $a_1 = R$ as free parameters. The fit produces $a_0 = 1.46$ and $a_1 = 1.57$. At high photon energies, the fitting backs the Cohen-Fano model. Extensions of this model have been proposed to interpret similar oscillatory behaviours observed in ionization of D_2 by fast heavy ions [3] and electrons [4]. As in the Cohen-Fano model [2], these theories are based on a one-electron description of ionization and an LCAO representation of the initial molecular state. This has led to results in reasonable agreement with experiments for large k_e. However, none of these theories have been able to reproduce the fall of the experimental $\sigma/2\sigma_H$ ratio when k_e decreases (see Fig. 1). The present results do show such a fall, in excellent agreement with the measurements of Samson and Haddad [23] (see Fig. 1). This is the first theoretical evidence of these observations.

The failure to reproduce the observed behaviour at relatively low photon energies using the fitting function given in equation (1) indicates that at least one of the basic assumptions of the

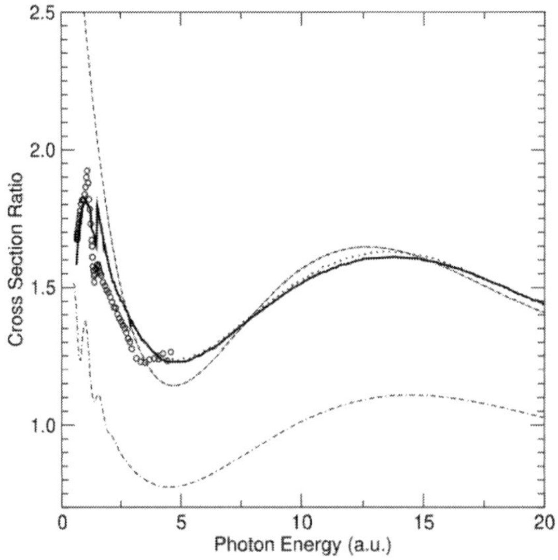

Figure 1. The $\sigma_{H_2}/2\sigma_H$ photoionization ratio. 4-channel (full line) and 10-channel (dotted line) results.Circles, experiments by Samson and Haddad taken from [20] and divided by $2\sigma_H$. Dashed line, fit of $\sigma_{H_2}/2\sigma_H$ to the Cohen-Fano formula (1) (see text). Results for a fictitious one electron molecule (dashed and dotted line); for the sake of clarity, the curve has been multiplied by 0.25 (see text for details).

Cohen-Fano model does not hold. To clarify this fact, we show in Fig. 1 B-spline results for a *fictitious* one-electron diatomic molecule with nuclear charges $Z = 0.5$ and equilibrium distance $R = 1.4$ au. This system differs from the real H_2 molecule in the absence of correlation and screening whereas all the other aspects are treated as in H_2. At high electron energies, this model calculation reproduces qualitatively those of H_2 but at lower energies it fails to reproduce the fall of the $\sigma/2\sigma_H$ ratio. Therefore, the origin of this fall must be electron correlation and/or screening. Indeed, at low k_e, the ejected electron is more sensitive to details of the potential near the nuclei. In particular, in a real H_2 molecule, a slow electron feels a charge larger than 0.5 due to incomplete screening of the inner electron, which must lead to a decrease of the ionization cross section. Electron correlation is expected to play a significant role only in the vicinity of the doubly excited states of H_2, where the nuclear motion cannot be neglected. Fig. 1 shows some structure near the ionization threshold on the curve of the fictitious one-electron molecule.

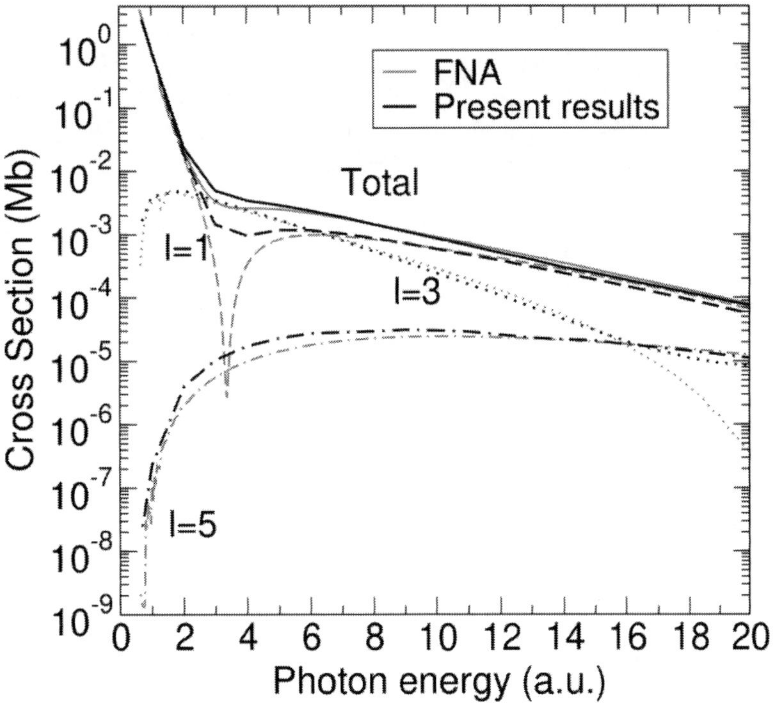

Figure 2. (Color online) Partial wave contributions to the H_2 photoionization cross section associated with the ${}^2\Sigma_g^+(1s\sigma_g)$ ionization threshold. Results for ${}^1\Sigma_u^+$ symmetry. Results from the FNA are shown by red curves.

This is probably provoked by the two-centre nature of the electronic continuum that is absent in the Cohen-Fano model but must be visible for electrons with small kinetic energy.

3.2. Confinement effects

In Fig. 2, we show results for the different partial waves associated with the lowest ionization channel ${}^2\Sigma_u^+(1s\sigma_g)$ for the ${}^1\Sigma_u^+$. It can be seen that results including the nuclear motion do differ significantly from the FNA ones [6]. Discrepancies are more evident at photon energies of $\sim$ 3.5 a.u. and $>$ 20 a.u. where the $\ell = 1$ and $\ell = 3$ partial waves contributions exhibit a minimum, respectively. Analogous minima were observed for FNA results for the case of H_2^+ [21]. For photon energies around 3.5 a.u., it can be seen from 2 that the minimum observed for the $\ell = 1$ wave in the H_2 FNA results is more pronounced than in calculations including the nuclear motion. On the contrary, its position is only slighlty changed by the nuclear motion. As shown in our previous work [8], such minima are expected to appear when the momentum of the ejected electron satisfies the condition:

$$k_e R_e \sim \ell\pi \quad (\ell = 2n + 1) \tag{18}$$

where k_e and R_e are the ejected electron momentum and the equilibrium internuclear distance,respectively. The condition requires that the product $k_e R_e$ be equal to an odd number times π. Incidentally, the l angular momenta associated to the partial waves used in the description of the final continuum states must be also odd due to the symmetry requirements imposed by the selection rules. So, we use these odd l values to express the mentioned condition.

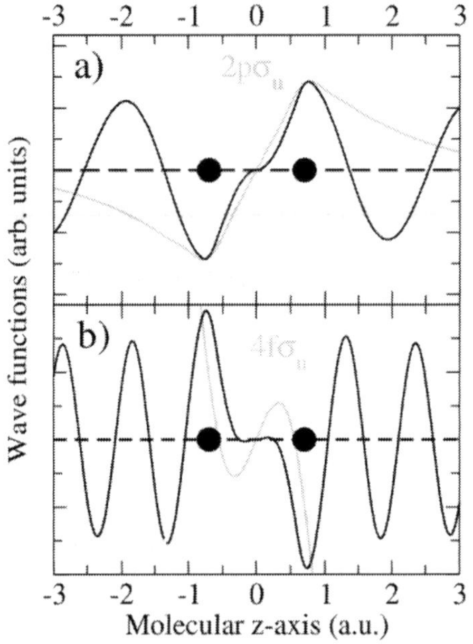

Figure 3. (Color online) Wave functions of the ejected electron (black line) compared with the lowest H_2^+ molecular orbitals (green line) along the internuclear z-axis for the k and ℓ values satisfying approximately the confinement condition. The wave functions of the ejected electron, obtained at the equilibrium distance, are the real K-matrix standing waves that correspond to the calculated complex S-matrix scattering wave functions. Panel a, $\varepsilon = 2.6$ a.u. and $\ell = 1$; panel b, $\varepsilon = 19$ a.u. and $\ell = 3$.

According to this simple formula, the minimum in the $\ell = 1$ partial wave should appear at a photon energy of ~ 3.1 a.u. and that of the $\ell = 3$ one at ~ 23 a.u.. These values are in reasonable agreement with the actual ones observed in Fig. 2.

The electron energies, $\varepsilon = k_e^2/2$, obtained by using Eq. 18 are the same as the ones of an electron confined in a one-dimensional infinite-square well potential of width R_e. This coincidence inspired us the word 'confinement' to refer to this interference effect. In fact, for the particular k_e values given by Eq. 18, a system of standing waves is created in the inner region $|r| < R_e$ leading to destructive interferences outside this region and producing consequently no propagating waves. At high energies, the minima location is also the one found in the transmission function for a one-dimensional potential made of two delta functions separated by a distance R_e. The $k_e R_e = \ell\pi$ condition obtained from these two models where $l = 2n + 1$ indicates an odd number (of course, in all these analogies, l is not any angular momentum, it is just an odd number) can be easily understood considering that in three dimensions (a real situation), the electrons tend to move in the directions given by the electric field of the incident radiation. As shown by Eq. (13), the ${}^1\Sigma_u^+$ cross section represents photoionization of H_2 molecules oriented parallel to the polarization axis, then electrons are mostly forced to move along the internuclear axis. Owing to the fact that the ejected electron has a large kinetic energy, it will only be affected by the potential in the vicinity of the two nuclei, which is similar to what happens in a one-dimensional infinite-square well or in the scattering by two delta functions. Hence the larger k_e the more accurate the formula $k_e R_e \sim \ell\pi$. This is indeed what happens in H_2 photoionization even for the lowest value of k_e that arise from this formula ($\ell = 1$) and also, as shown in [6] and [24], in H_2^+ photoionization.

Journal of Physics: Conference Series **288** (2011) 012025 doi:10.1088/1742-6596/288/1/012025

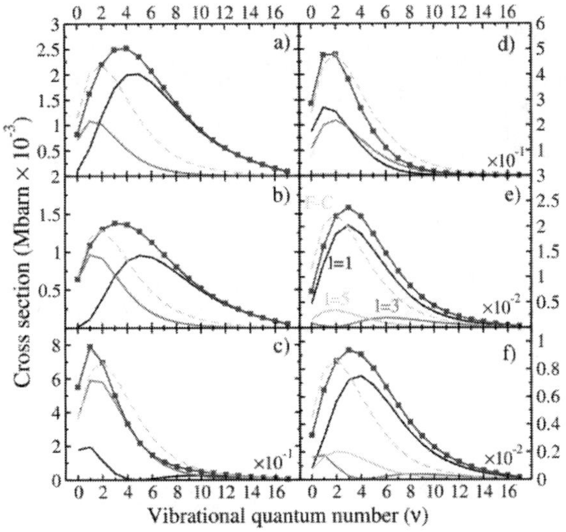

Figure 4. (Color online) Non dissociative photoionization cross section of H_2 of $^1\Sigma_u^+$ symmetry, as a function of the vibrational quantum number v of the residual H_2^+ ion at photon energies 2, 2.5, 3.5, 6.0, 16.5 and 20 a.u., panels a, b, c, d, e and f, respectively. Blue line with squares, total cross section; black line, $\ell = 1$ partial wave contribution; red line, $\ell = 3$; green line, $\ell = 5$. Dashed cyan line, Frank-Condon distribution (FC).

In Fig. 3, it can be seen that the reduced electron continuum wave function at $R_e = 1.4$ a.u (the equilibrium internuclear distance of H_2) associated with the $\ell = 1$ partial wave is very similar in fashion to the $2p\sigma_u$ orbital of H_2^+ (with $R_e = 2$ a.u.) in the region between both molecular centers. In turn, the reduced continuum wave function associated with the $\ell = 3$ partial wave has the same nodal structure as the $4p\sigma_u$ orbital in between the nuclei. This similarity between reduced electron continuum waves and the H_2^+ molecular orbitals gives additional evidence for the image of partial-wave confinement. In Ref. [24], the minima in the $^1\Sigma_u^+$ cross section have been interpreted as Cooper-like minima similar to those found in atomic photoionization. The continuum wave functions of H_2^+ were unfortunately compared with the $1s\sigma_g$ orbital, which does not have the correct nodal symmetry imposed by the dipole selection rule. Moreover, this interpretation does not bring a simple prediction for the minima as the one given by our Eq. 18.

3.3. Non Franck-Condon effects

In what follows, we consider only non dissociative photoionization that is the most important contribution to the total cross section.

At low photon energies (e.g., 20 eV), the vibrational distribution of residual H_2^+ follows approximately a Franck-Condon distribution, which predicts that the $v = 2, 3$ vibrational levels should be the most populated ones.

In Fig. 4 we show cross sections differential in the vibrational energy of the residual H_2^+ ion (viz. the reverse of the energy of the ejected electron) for the non dissociative case and for both parallel ($^1\Sigma_u^+$) molecular orientations.

It can be seen that the population of the final vibrational states does not follow a Franck-Condon behavior, even for a photon energy of 2 a.u.. At this energy, the $\ell = 1$ partial wave is dominant for almost all v, but, surprisingly, the $\ell = 3$ partial wave dominates for the lowest v's. As the photon energy increases reaching the region in which the $\ell = 1$ partial wave has a

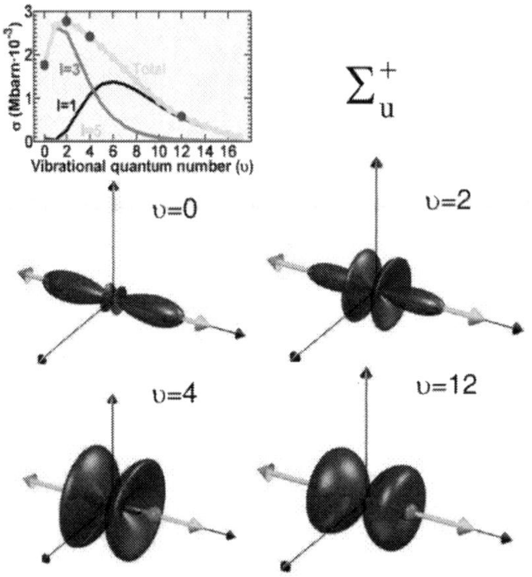

Figure 5. (Color online) Electron angular distributions for non-dissociative photoionization ($H_2 + \hbar\omega \rightarrow H_2^+(v) + e^-$) of H_2 oriented parallel (Σ_u^+ symmetry) to the polarization direction at a photon energy of 2.5 a.u.. Angular distributions corresponding to leaving the residual H_2^+ ion in different vibrational states are shown by three dimensional plots in blue. For a better visualization, these distributions have been renormalized so that their maximum value is always 1. The polarization direction is indicated by the double arrow (green). The two nuclei are indicated by two small spheres (red). The angle-integrated cross sections, including the partial wave decomposition, as a function of the final quantum vibrational number v are shown in the two-dimensional plots.

minimum (see Fig. 2), the $\ell = 3$ partial wave becomes dominant at higher and higher v until, at energies above 6 a.u., the $\ell = 1$ partial wave becomes again the dominant one. A more careful inspection of Fig. 4 shows that the $\ell = 1$ partial wave has a minimum around $v = 5$ at a photon energy of 3.5 a.u.. This is a photon energy that lies in the minimum of the cross section (see Fig. 2). Another minimum is clearly visible for the $\ell = 5$ partial wave at photon energies of 16.5 and 20 a.u but now located around $\nu = 4$. From this analysis one may conclude that this deviation from the FC predictions is related to the confinement effects discussed above.

3.4. Interference effects in the angular distribution of electrons
For the perpendicular orientation, one can see that the electronic emission is mainly produced, as expected, in the direction of the polarization vector. As the photon energy increases, additional lobes appear producing interference patterns analogous to the ones of the Young's double-slit experiment given by $R \sin\theta_e = n\lambda_e$ [8].

In Fig. 5, we show the electron angular distributions for non-dissociative photoionization of H_2 for molecules parallel (Σ_u^+ symmetry) to the polarization direction at a photon energy of 2.5 a.u.. The corresponding angular distributions are only shown for the case of H_2^+ ions left in the $v = 0$, 2, 4, and 12 vibrational levels. The first three values of v correspond to vibrational levels that are close to or within the FC region, while the $v = 12$ vibrational level is clearly outside this region. The distribution has an almost perfect f shape ($\ell = 3$) when H_2^+ remains in a low vibrational level. As the vibrational quantum number of the remaining H_2^+ increases, a complicated angular pattern emerges as the result of the interference between the $\ell = 1$ and $\ell = 3$ partial waves. For $v = 4$, interferences lead to almost no emission in the direction of the polarization vector, whereas, for $v = 12$, the distribution is almost p-like. In Ref. [8], this sudden change of the angular distribution with v has been attributed to confinement because, around this photon energy, the $\ell = 1$ partial component of the $^1\Sigma_u^+$ integrated cross section exhibits a pronounced minimum (see Fig. 2).

In the case of H_2^+ photoionization, the angular distributions approximately follow the formula [25] $(\mathbf{e}_\mu \cdot \mathbf{k}_e)^2 \cos^2(\mathbf{k}_e \cdot \mathbf{R}/2)$. If $\mathbf{e}_\mu$ and $\mathbf{k}_e$ are parallel to the molecular axis, this formula leads to zero when $k_e R = \pi, 3\pi, ...$, i.e., no electron emission along the molecular axis in agreement with the image of confinement.

4. Conclusions

Photoionization of H_2 ionization by photons of a few hundred eV has been studied. By using B-spline functions to evaluate bound and continuum states and within the framework of the Born-Oppenheimer approximation, we obtain a realistic description of the molecular sates including its nuclear degrees of freedom.

The Cohen and Fano predictions were analyzed showing that they are valid at sufficiently high photon energies. Failures are related to electronic correlation and/or screening [6].

In the case of molecules oriented parallel to the polarization direction, the partial-wave cross sections exhibit pronounced minima at electron energies determined closely by $k_e R_e = \ell\pi$, where ℓ is an odd number and R_e is the equilibrium internuclear distance [7, 8].

The fully differential angular distribution presents in a clear way the effects of confinement for molecules aligned parallel to the polarization direction. Young's double slit interferences are clearly visible in the perpendicular case when the electron wavelength is comparable to the internuclear distance [7, 8] .

From the analysis of the vibrational distribution of the residual H_2^+ ion for the dominant non-dissociative channel, it is found that molecules parallel to the polarization direction transitions do not follow the typical Franck Condon (FC) distribution. Deviations from the FC distribution are more pronounced in the regions where confinement occurs. On the contrary, no deviations are observed for molecules perpendicular to the polarization direction [7].

Acknowledgments

Work partially supported by the Spanish Ministerio de Ciencia e Innovación (contract FIS2007-60064), the European Science Foundation (COST action CM0702), and the Spanish Subdirección General de Cooperación Internacional. OAF acknowledges the kind hospitality in the group of 'Computations in Atomic and Molecular Physics of Unbound Systems' of the Universidad Autónoma de Madrid (Spain), as well as financial support from the Secretaría de Estado de Educación y Universidades (Spain) through a grant for foreign researchers. Calculations were performed at the Barcelona Supercomputer Center Mare Nostrum and the Centro de Computación Científica UAM (Spain). OAF acknowledges partial financial support from CONICET (PIP 1026) and ANPCyT (PICT 1912).

References

[1] D. Akoury *et al*, *Science* **318** (2007) 949.
[2] Cohen H D and Fano U 1966 *Phys. Rev.* **150** 30
[3] Stolterfoht *et al* 2001 *Phys. Rev. Lett.* **87** 023201
[4] O. Kamalou, J.-Y. Chesnel, D. Martina, F. Frémont, J. Hanssen, C. R. Stia, O. A. Fojón, R. D. Rivarola 2005 *Phys. Rev. A* **71** 010702(R).
[5] S. Chatterjee, S. Kasthurirangan, A. H. Kelkar, C. R. Stia, O. A. Fojón, R. D. Rivarola, and L. C. Tribedi 2009 *J Phys. B: At. Mol. Opt. Phys.* **42** 065201

[6] O. A. Fojón, J. Fernández, A. Palacios, R. D. Rivarola, F. Martín 2004 *J. Phys. B: At. Mol. Opt. Phys.* **37** 3035.
[7] J. Fernández, O. A. Fojón and F.Martín 2009 *Phys. Rev. A* **79** 023420
[8] J. Fernández, O. A. Fojón, A. Palacios, and F. Martín 2007 *Phys. Rev. Lett.* **98** 043005
[9] Dill D 1976 *J. Chem. Phys.* **65** 1130
[10] Martín F 1999 *J. Phys. B: At. Mol. Opt. Phys.* **32** R197
[11] Bachau H, Cormier E, Decleva P, Hansen J E and Martín F 2001 *Rep. Prog. Phys.* **64** 1815
[12] S. Wallace and D. Dill 1978 *Phys. Rev. B* **17** 1692 .
[13] J. C. Tully, R. S. Berry, and B. J. Dalton 1968 *Phys. Rev.* **176** 95
[14] I. Cacelli, V. Carravetta, A. Rizzo, and R. Moccia 1991 *Phys. Rep.* **205** 283
[15] I. Sánchez and F. Martín 1999 *Phys. Rev. Lett.* **82** 3775
[16] F. Martín et al. 2007 *Science* **315** 629
[17] Laurent, Fernández, Legendre, Tarisien, Adoui, Cassimi, Fléchard, Frémont, Gervais, Giglio et al. F 2006 *Phys. Rev. Lett.* **96** 173201
[18] Vanroose, Martín, Rescigno, and McCurdy 2005 *Science* **310** 1787
[19] Kolos W, Szalewicz K and Monkhorst H J 1986 *J. Chem. Phys.* **84** 3278
[20] Chung Y M, Lee E M, Masuoka T and Samson J A R 1993 *J. Chem. Phys.* **99** 885
[21] O. A. Fojón, A. Palacios, J. Fernández, R. D. Rivarola, and F.Martín 2006 *Phys. Lett. A* **350** 371
[22] F. Martín 1999 *J. Phys. B: At. Mol. Opt. Phys.* **32** R197
[23] Samson J A R and Haddad G N 1994 *J. Opt. Soc. Am. B* **11** 277
[24] R. Della Picca, P. D. Fainstein, M. L. Martiarena, and A. Dubois 2008 *Phys. Rev. A* **77** 022702
[25] Walter M and Briggs J 1999 *J. Phys. B: At. Mol. Opt. Phys.* **32** 2487

Journal of Physics: Conference Series **288** (2011) 012026 doi:10.1088/1742-6596/288/1/012026

Coherence and partial localization in electron emission from asymmetric diatomic molecules

C A Tachino[1], M E Galassi[1], F Martín[2] and R D Rivarola[1*]

[1] Instituto de Física Rosario, Consejo Nacional de Investigaciones Científicas y Técnicas, Universidad Nacional de Rosario, Av. Pellegrini 250, 2000 Rosario, Argentina

[2] Departamento de Química, C-9, Universidad Autónoma de Madrid, 28049 Madrid, Spain

E-mail: *rivarola@fceia.unr.edu.ar

Abstract. The presence of interference patterns in the spectra of emitted electrons from HeH^+ molecular ions due to the impact of bare ions is analysed. It is shown that these oscillations, which are explained in terms of the coherent emission of electrons from the vicinities of the target nuclei, are preserved even after averaging over all molecular orientations. The influence of partial localization of electrons around the alpha particle center is studied. The relationship between electron ionization by ion impact and photoionization is investigated.

1. Introduction

Interference effects due to coherent emission of electrones from the neighbourhood of the nuclei of diatomic targets have been studied during the last 50 years. The first theoretical predictions concerning this phenomenon were made by Cohen and Fano in the decade of 1960 for photoionization of H_2 molecular ions [1]. After this pioneering work, many theoretical studies were carried out to investigate the presence of interference in ionization of single diatomic molecules by impact of photons and electrons [2–5]. Nevertheless, the existence of this effect was confirmed fourty years later, when cross sections for single electron ionization from H_2 targets by 60 MeV/u Kr^{34+} ion impact were measured [6]. Since then, numerous theoretical [7–20] and experimental [21–35] investigations were carried out for impact of photons, electrons and ions on diatomic molecules. These investigations were mainly oriented to the study of coherent electron emission from homonuclear targets such as H_2, N_2 and O_2, and to a lesser extent, for heteronuclear molecules.

The aim of the present work is to present the results obtained for single ionization of the heteronuclear molecular ion HeH^+ by impact of bare ions. It has been shown that, in spite of the highly asymmetry of the electron distribution of this target, fingerprints of coherent emission of electrons from the vicinities of the molecular centres were clearly seen in the angular distributions [36] and that interferences are preserved even after averaging the corresponding cross sections over all possible molecular orientations [37]. Details concerning the theoretical model have already been given previously; thus, only a general description of the method will be presented in the following sections. The straight line version of the impact parameter method will be employed.

Atomic units are used throughout unless otherwise stated.

Journal of Physics: Conference Series **288** (2011) 012026 doi:10.1088/1742-6596/288/1/012026

2. Theory

Single ionization of HeH^+ molecular ions by impact of bare ions at intermediate and high collision energies is investigated. The incident particle moves with a velocity $\vec{v}$ with respect to a reference frame whose origin is fixed at the centre of mass of the target. At the high impact energies here considered, the vibrational and rotational times of the molecule are much larger than the collision one, so it can be assumed that the nuclei of the target remain frozen in their initial positions during the reaction. Also, we consider that the residual electron remains in its initial state, in such a form that the two-electron problem can be reduced to the analysis of a one-electron system. It is possible to show that this hypothesis is equivalent to the one of an independent electron model where the passive electron may occupy any state of HeH^{2+}. With these assumptions in mind, the five-fold differential cross section (FDCS) as a function of the electron energy ε_k and the orientation of the ionized electron Ω_k and of the molecule Ω_b can be obtained as:

$$\sigma^{(5)}(\varepsilon_k, \Omega_k, \Omega_b) = 2k \int d\vec{\rho}\, |a^+_{i,f}(\vec{\rho})|^2 \tag{1}$$

with $a^+_{i,f}$ the post-form of the transition amplitude for the active electron. In equation (1) $k = \sqrt{2\varepsilon_k}$ is the modulus of the electron momentum $\vec{k}$, with ε_k the electron emission energy, and $\vec{\rho}$ is the impact parameter vector. The double differential cross sections (DDCS) as a function of the energy and solid angle of the emitted electron can be then obtained by averaging equation (1) over all possible molecular orientations:

$$\sigma^{(2)}(\varepsilon_k, \Omega_k) = \frac{2k}{4\pi} \int d\vec{\rho}\, d\Omega_b\, |a^+_{i,f}(\vec{\rho})|^2 \tag{2}$$

Into the straight line version of the impact parameter approximation, the expression for a^+_{if} is:

$$a^+_{if}(\vec{\rho}) = -i \int_{-\infty}^{+\infty} dt \left\langle \chi^-_f \middle| W^\dagger_f \middle| \chi^+_i \right\rangle \tag{3}$$

where $W^\dagger_f$ is the perturbation operator corresponding to the final channel, and χ^+_i (χ^-_f) are the initial (final) wavefuntion with correct asymptotic outgoing (incoming) boundary conditions. For the continuum distorted wave-eikonal initial state (CDW-EIS) approximation, these wavefuntions are chosen as:

$$\chi^{+,EIS}_i(\vec{x}, t) = \varphi_i(\vec{x}) \exp(-i\,\varepsilon_i\, t)\, \mathcal{L}^{+,EIS}_i(\vec{s}) \tag{4}$$

$$\chi^{-,CDW}_f(\vec{x}, t) = \varphi_f(\vec{x}) \exp(-i\,\varepsilon_k\, t)\, \mathcal{L}^{-,CDW}_f(\vec{s}) \tag{5}$$

In the preceding expressions, φ_i and φ_f are bound and continuum states of the active electron in the entrance and exit channels, respectively, ε_i is the initial orbital energy, and $\vec{x}$ ($\vec{s}$) is the electron coordinate with respect to the centre of mass of the target (projectile nucleus). Also in equations (4) and (5), $\mathcal{L}^{+,EIS}_i$ and $\mathcal{L}^{-,CDW}_f$ represents projectile eikonal and continuum distortion functions, respectively.

In order to describe the initial molecular orbital of the HeH^+ ion, a basis set of Slater-type orbitals (STO) ϕ^{STO}_h centred on each molecular nuclei is employed:

$$\varphi_i(\vec{x}) = \omega_1 \phi^{STO}_1(\vec{x}_1) + \omega_2 \phi^{STO}_2(\vec{x}_2) \tag{6}$$

where the subscripts 1 and 2 denote the He^{2+} and H^+ nuclei, respectively, and $\vec{x}_h$ represents the electron coordinate with respect to the nucleus h (h = 1, 2). The values of both the STO exponents and the coefficients ω_h of equation (6) were optimized by employing the quantum chemistry program Gaussian 98 [38]. These quantities were calculated by using a STO-6G basis set within the Hartree-Fock (HF) approximation, where each Slater-type orbital is described as a linear combination of 6 Gaussian-type orbitals (GTO) (see references [39,40] for more details). The equilibrium internuclear distance b and the orbital energy ε_i were calculated by using a $6-311G^*$ basis set, also within the HF formalism.

The final continuum wavefunction in the exit channel is described by employing the CDW-EIS model and the two-effective centre (TEC) approximation [7,10], so that the expression for φ_f is chosen as:

$$\varphi_f(\vec{x}) = (2\pi)^{-3/2} \exp[i\vec{k}\cdot(\vec{x}_h + \vec{b}_h)] N^*(\xi)\, {}_1F_1(-i\xi; 1; -ik\,x_h - i\vec{k}\cdot\vec{x}_h) \tag{7}$$

when the component ϕ_h^{STO} of the initial orbital wavefunction (see equation (6)) is considered. In equation (7), $\vec{b}_h$ represents the position of the h-nucleus with respect to the centre of mass of the molecular nuclei, and $\xi = k^{-1}\sqrt{-2\varepsilon_i}$.

Considering the scattering matrix element $\mathcal{R}^+_{if}(\vec{\eta})$, which is the Fourier transform of the transition amplitude $a^+_{if}(\vec{\rho})$, and the well-known Parseval-Plancherel identity, the expressions for the differential cross sections of equations (1) and (2) can be rewritten as it follows:

$$\begin{aligned} \sigma^{(5)}(\varepsilon_k, \Omega_k, \Omega_b) &= 2k \int d\vec{\eta}\, |\mathcal{R}^+_{i,f}(\vec{\eta})|^2 & (8)\\ \sigma^{(2)}(\varepsilon_k, \Omega_k) &= \frac{2k}{4\pi} \int d\vec{\eta}\, d\Omega_b\, |\mathcal{R}^+_{i,f}(\vec{\eta})|^2 & (9) \end{aligned}$$

respectively, where $\vec{\eta}$ is the component of the total momentum transfer $\vec{K}$ perpendicular to the direction of the incident particle.

In a previous work [36], it was shown that, in the TEC approximation, the scattering matrix element $\mathcal{R}^+_{i,f}(\vec{\eta})$ can be expressed as a linear combination of effective one-centre scattering matrix elements $\mathcal{R}_h^{+,eff}$ associated with the complete basis of STO centred on each molecular nuclei. According to that, and within the CDW-EIS model, the following expression for the square modulus of $\mathcal{R}^+_{if}(\vec{\eta})$ is obtained:

$$\begin{aligned} |\mathcal{R}^{+,CDW-EIS}_{i,f}(\vec{\eta})|^2 &= \omega_1^2\, |\mathcal{R}_1^{+,eff}(\vec{\eta})|^2 + \omega_2^2\, |\mathcal{R}_2^{+,eff}(\vec{\eta})|^2 \\ &+2\,\omega_1\,\omega_2\, \cos[(\vec{k}+\vec{K})\cdot\vec{b}][\Re e\, \mathcal{R}_1^{+,eff}(\vec{\eta})\, \Re e\, \mathcal{R}_2^{+,eff}(\vec{\eta}) + \Im m\, \mathcal{R}_1^{+,eff}(\vec{\eta})\, \Im m\, \mathcal{R}_2^{+,eff}(\vec{\eta})] \\ &+2\,\omega_1\,\omega_2\, \sin[(\vec{k}+\vec{K})\cdot\vec{b}][\Re e\, \mathcal{R}_1^{+,eff}(\vec{\eta})\, \Im m\, \mathcal{R}_2^{+,eff}(\vec{\eta}) - \Im m\, \mathcal{R}_1^{+,eff}(\vec{\eta})\, \Re e\, \mathcal{R}_2^{+,eff}(\vec{\eta})] \end{aligned} \tag{10}$$

It can be seen that if in equation (10) we set $\omega_1 = \omega_2$ and $\mathcal{R}_1^{+,eff}(\vec{\eta}) = \mathcal{R}_2^{+,eff}(\vec{\eta})$, the expression corresponding to the homonuclear case is recovered [7].

3. Results and discussions

Our main interest is focused on interference effects in the single ionization reaction of the HeH^+ molecular target by impact of protons and bare carbon ions. We will analyse firstly the angular distribution of target electrons, considering the case of a coplanar geometry in which the molecule, the emitted electron and the projectile are all in the same plane. In order to reveal the oscillatory structures associated with the coherent emission of electrons from both molecular centres, we calculate the ratios:

$$R_{HeH^+}(\varepsilon_k, \Omega_k) = \frac{\sigma^{(5)}}{2k\omega_1^2 \int d\vec{\eta} |\mathcal{R}_1^{+,eff}(\vec{\eta})|^2} \tag{11}$$

In the preceding expression, the five-fold differential cross sections $\sigma^{(5)}$ of equation (8) is divided by a FDCS obtained by integrating the first addend of equation (10). This choice is supported by the fact that the electron density around the helium nucleus largely dominates the molecular electron density and it is not very different from that of the atomic helium orbital [36].

In figures 1 and 2 we present the ratios obtained for impact of C^{6+} at collision energies of 1, 13.7 and 100 MeV/u, when the molecule is aligned parallel and perpendicular to the direction of the projectile, respectively. These results are compared with the angular distributions produced by two emitters of spherical waves of amplitudes ω_1 and ω_2 separated a distance b:

$$I(\theta) \sim \omega_1^2 + \omega_2^2 + 2\omega_1\omega_2 \cos(\vec{k} \cdot \vec{b}) \tag{12}$$

where the coefficients ω_h ($h = 1, 2$) are defined in equation (6). This last formula describes correctly the ionization of HeH^+ molecular ions by circularly polarized light at high photon energies, with the incidence direction of the light perpendicular to the plane formed by the molecular axis and the electron ejection direction. A similar expression was employed by Akoury *et al.* [34] to describe the photoionization of H_2 molecules.

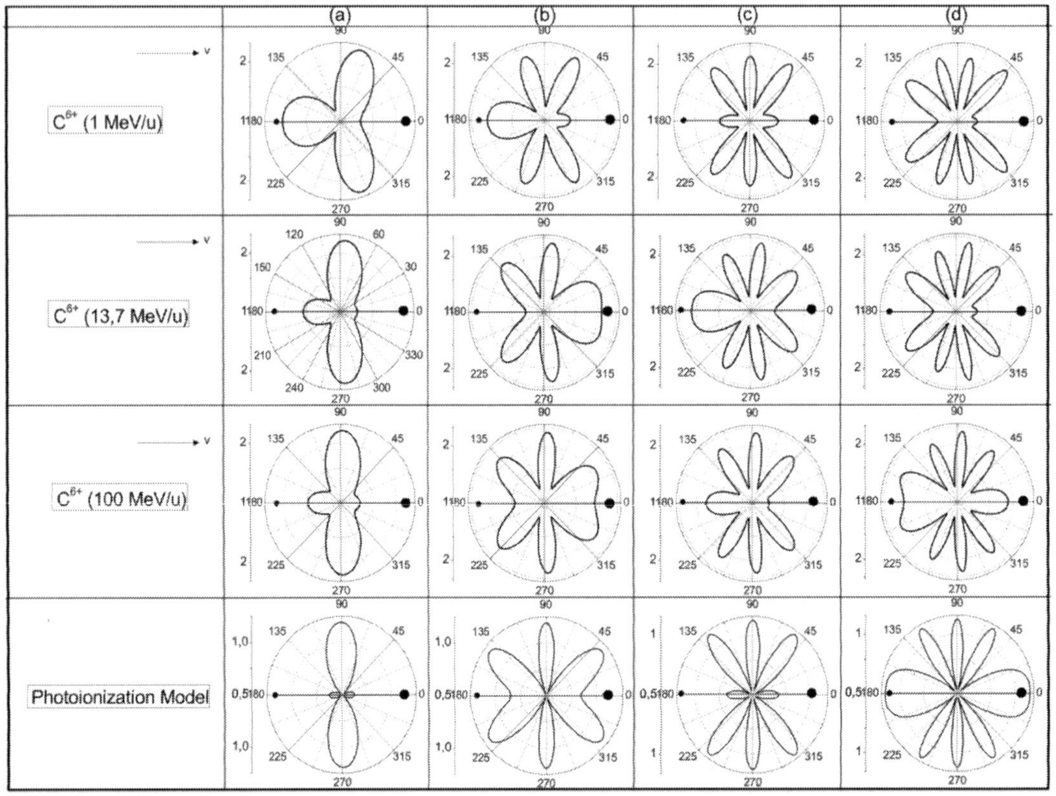

Figure 1. Five-fold differential cross section ratios and photoionization model calculations (equation (12)) for the HeH^+ aligned parallel to the incident beam, for different values of the electron energy: (a) 100 eV, (b) 400 eV, (c) 700 eV and (d) 1000 eV. The alpha particle is considered to be placed on the semi-axis coinciding with the emission angle $\theta = 0°$.

Figure 1 shows clearly the presence of interference patterns. For each impact velocity the effect is more noticeable as the energy of the ionized electron increases. Moreover, for parallel orientation, the differential cross section ratios tend to the angular distributions predicted by the photoionization model as the impact energy increases. This behaviour can be easily explained from equation (10) considering that the component of the momentum transfer $\vec{K}$ perpendicular to the impact velocity does not contribute to the oscillatory factors and the parallel one depends as v^{-1}. In consequence, for high energy impact, the arguments of the trigonometrical functions presents in equation (10) can be reduced to $\vec{k}\cdot\vec{b}$, and the resulting expression for $|\mathcal{R}_{i,f}^{+,CDW-EIS}(\vec{\eta})|^2$ is analogue to the one corresponding to the photoionization model of equation (12).

On the other hand, when the molecular ion is perpendicular to the direction of incidence, interference patterns are observed in the FDCS ratios. However, they do not show a clear tendence to the photoionization model results as the impact energies increases, as for parallel orientation. This behaviour can be explained again considering that for this particular molecular orientation, the transverse momentum transfer does not become negligible with respect to the electron momentum k at high energy impact.

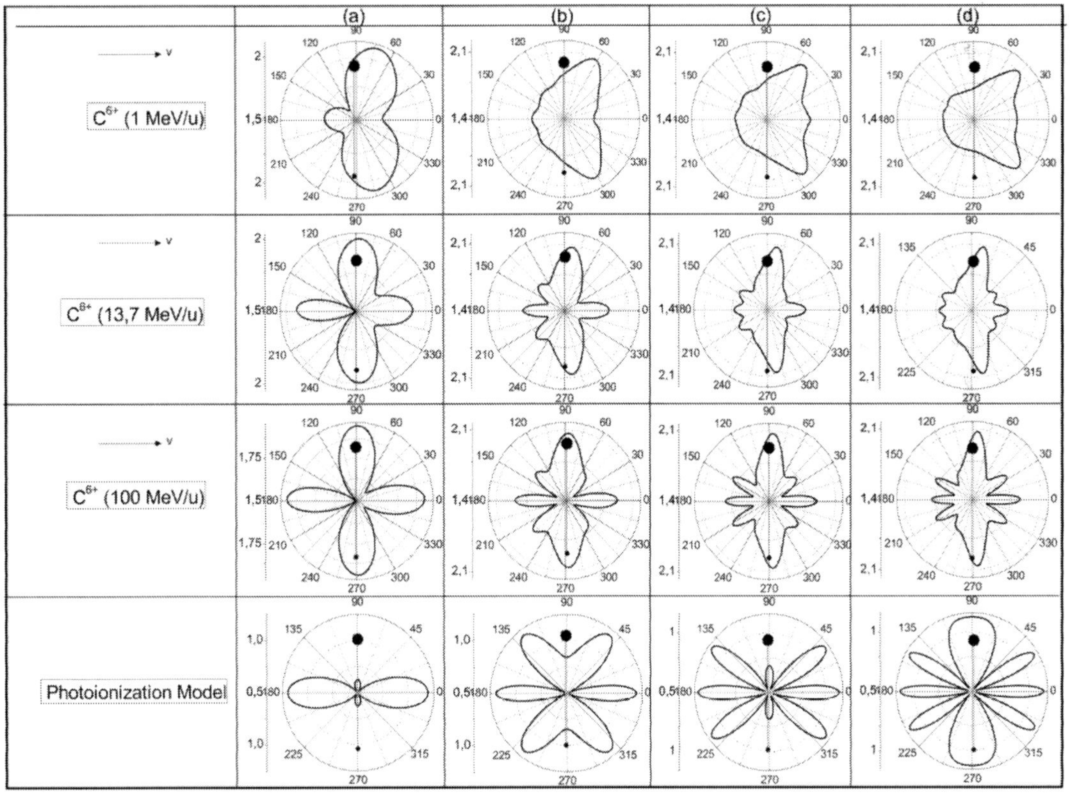

Figure 2. Five-fold differential cross section ratios and photoionization model calculations (equation (12)) for the HeH^+ aligned perpendicular to the incident beam, for different values of the electron energy: (a) 100 eV, (b) 400 eV, (c) 700 eV and (d) 1000 eV. The alpha particle is considered to be placed on the semi-axis coinciding with the emission angle $\theta = 90°$.

In order to test the validity of the explanation given in the preceding paragraph, we have plot in figure 3 the five-fold differential cross section ratios for perpendicular orientation but taking the total momentum trasfer $K = 0$ in the trigonometric functions of the third and fourth addends of equation (10). As it can be seen, there is a very good qualitative agreement between the new angular spectra and the photoionization results. The number of lobes is the same, and their positions are almost coincident. The small diferences between ion and photon impact can mainly be attributed to the different mechanisms leading to electron ionization of the target, being for ion impact the momentum transfer from the projectile the most important one.

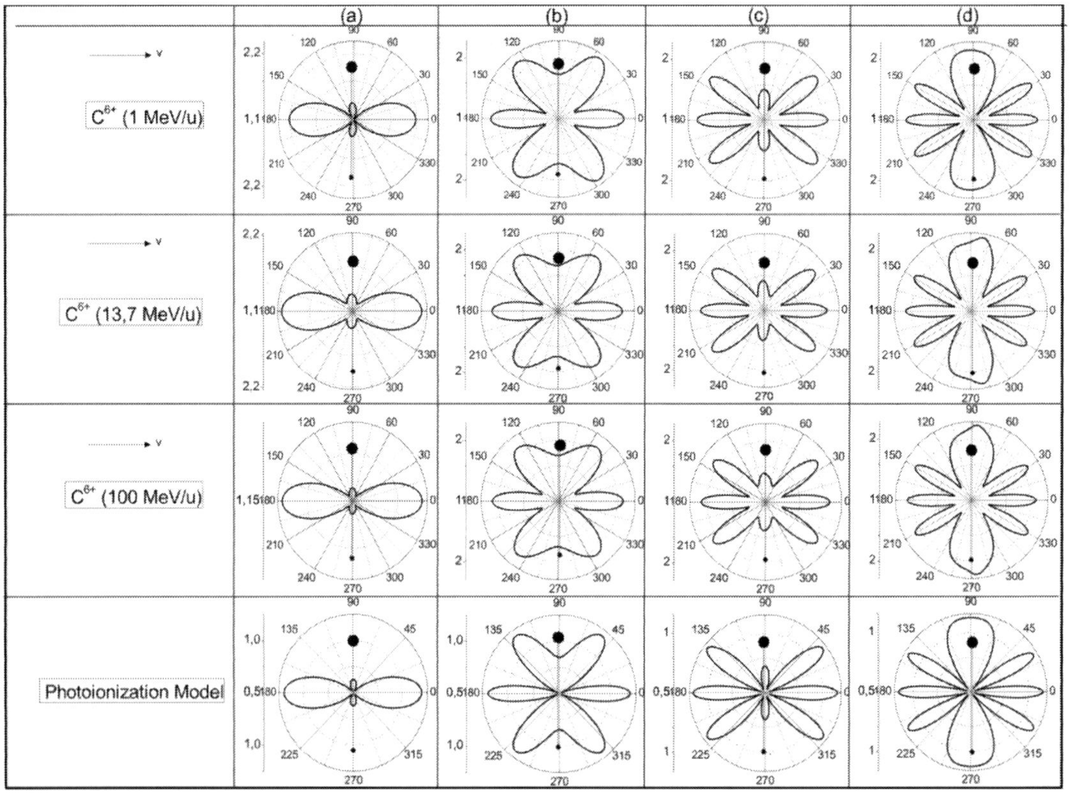

Figure 3. Same as figure 2, but considering $K = 0$ in the calculation of the trigonometric factors of equation (10).

In order to determine if interference patterns are preserved after averaging on all possible molecular orientations, we study the behaviour of double differential cross sections $\sigma^{(2)}(\varepsilon_k, \Omega_k)$. Employing equations (5) and (10), they can be expressed as:

$$\begin{aligned}
\sigma^{(2)}_{HeH^+}(\varepsilon_k, \Omega_k) &= 2\,k\,\omega_1^2 \int d\vec{\eta} \left|\mathcal{R}_1^{+,eff}(\vec{\eta})\right|^2 + 2\,k\,\omega_2^2 \int d\vec{\eta} \left|\mathcal{R}_2^{+,eff}(\vec{\eta})\right|^2 \\
&+ 2\,k\,\omega_1\,\omega_2 \int d\vec{\eta} \Big[\Re e\mathcal{R}_1^{+,eff}(\vec{\eta})\Re e\mathcal{R}_2^{+,eff}(\vec{\eta}) \\
&\quad + \Im m\mathcal{R}_1^{+,eff}(\vec{\eta})\Im m\mathcal{R}_2^{+,eff}(\vec{\eta})\Big] \frac{\sin[|\vec{k}+\vec{K}|b]}{|\vec{k}+\vec{K}|b} \\
&= S_{d_{He^{2+}}}(\varepsilon_k, \Omega_k) + S_{d_{H^+}}(\varepsilon_k, \Omega_k) + S_{i_{HeH^+}}(\varepsilon_k, \Omega_k) \qquad (13)
\end{aligned}$$

where $S_{d_{He^{2+}}}(\varepsilon_k, \Omega_k)$ and $S_{d_{H^+}}(\varepsilon_k, \Omega_k)$ represents the direct terms related to the ionization process of each one of the molecular nuclei, whereas $S_{i_{HeH^+}}(\varepsilon_k, \Omega_k)$ represents the interference term and contains the information arising from coherent electron emission [37]. For the particular case of the homonuclear H_2 molecule, the DDCS results in:

$$\begin{aligned}\sigma^{(2)}_{H_2}(\varepsilon_k,\Omega_k) &= 4\,k\,\omega^2\int d\vec{\eta}\left|\mathcal{R}^{+,eff}(\vec{\eta})\right|^2 + 4\,k\,\omega^2\int d\vec{\eta}\left|\mathcal{R}^{+,eff}(\vec{\eta})\right|^2\frac{\sin[|\vec{k}+\vec{K}|b]}{|\vec{k}+\vec{K}|b}\\ &= 2S_{d_{H^+}}(\varepsilon_k,\Omega_k) + S_{i_{H_2}}(\varepsilon_k,\Omega_k)\end{aligned} \tag{14}$$

where, analogously to the HeH^+ case, $S_{d_{H^+}}$ represents the direct term corresponding to the electron emission from the proximities of both H^+ nuclei and $S_{i_{H_2}}$ in the interference term.

To put in evidence the characteristic undulations associated to interference effects, we calculate the ratios:

$$\tilde{R}_{HeH^+}(\varepsilon_k,\Omega_k) = \frac{\sigma^{(2)}_{HeH^+}}{S_{d_{He^{2+}}}} = 1 + \frac{S_{d_{H^+}}}{S_{d_{He^{2+}}}} + \frac{S_{i_{HeH^+}}}{S_{d_{He^{2+}}}} \tag{15}$$

for HeH^+ molecular ions, and:

$$\tilde{R}_{H_2}(\varepsilon_k,\Omega_k) = \frac{\sigma^{(2)}_{H_2}}{2S_{d_{H^+}}} = 1 + \frac{S_{i_{H_2}}}{2S_{d_{H^+}}} \tag{16}$$

for hydrogen molecules.

It is evident from equation (15) that only the term $S_{i_{HeH^+}}/S_{d_{He^{2+}}}$, which will be henceforth denoted as $R_{i_{HeH^+}}$, can give contributions to coherent emission oscillations in the ratios $\tilde{R}_{HeH^+}$. Analogously, in equation (16), the important term is the quotient $S_{i_{H_2}}/2S_{d_{H^+}}$ which will be henceforth denoted as $R_{i_{H_2}}$. In figure 4 we plot the ratios $R_{i_{HeH^+}}$ and $R_{i_{H_2}}$, for proton impact at 13.7 and 100 MeV/u, at electron emission angles $\theta_k = 30°$ and $\theta_k = 150°$. It can be seen that smooth undulations appear in the $R_{i_{HeH^+}}$ spectra. If we consider a certain value of the electron emission angle θ_k and a fixed projectile energy, we can observe that the frequency of the oscillations increases as ε_k increases. Moreover, for each impact velocity, the ratio frequency is larger for the backward emission angles than for the forward ones as it has previously been observed for H_2 targets.

The corresponding ratios for the H_2 target have an oscillatory behaviour analogous to that of the HeH^+ case. It can be observed in figure 4 that the amplitudes of $R_{i_{HeH^+}}$ are smaller than those of $R_{i_{H_2}}$, especially at low electron emission energies. This difference can be attributed to the partial localization of the emitted electron around the alpha particle. The frecuency of the oscillations in these ratios are similar for the homonuclear and for the heteronuclear molecules; in particular, the positions of the respective maxima and minima are almost coincident.

Thus, in spite of the highly asymmetry of the electronic distribution of the HeH^+ molecular ion, interference patterns due to coherent emission of electrons from the proximities of the target nuclei can be observed in the doubly differential cross sections, even after averaging over all molecular orientations. These oscillations are less pronounced than the ones corresponding to the homonuclear case

Analogously to the FDCS case, we will compare the DDCS ratios calculated for the HeH^+ ion with photoionization results obtained following the work of Walter and Briggs [3]. In this model for linearly polarized photons, whose details have been already given in [37], the expression for the doubly differential cross sections is:

$$\sigma^{(2)}_{PH} = \frac{4\pi^2\alpha}{\omega_{PH}}|\mathcal{T}_{if}|^2 \tag{17}$$

where α is the fine-structure constant, ω_{PH} is the angular frequency of the incident photon and $\mathcal{T}_{if}$ is the transition matrix element given by:

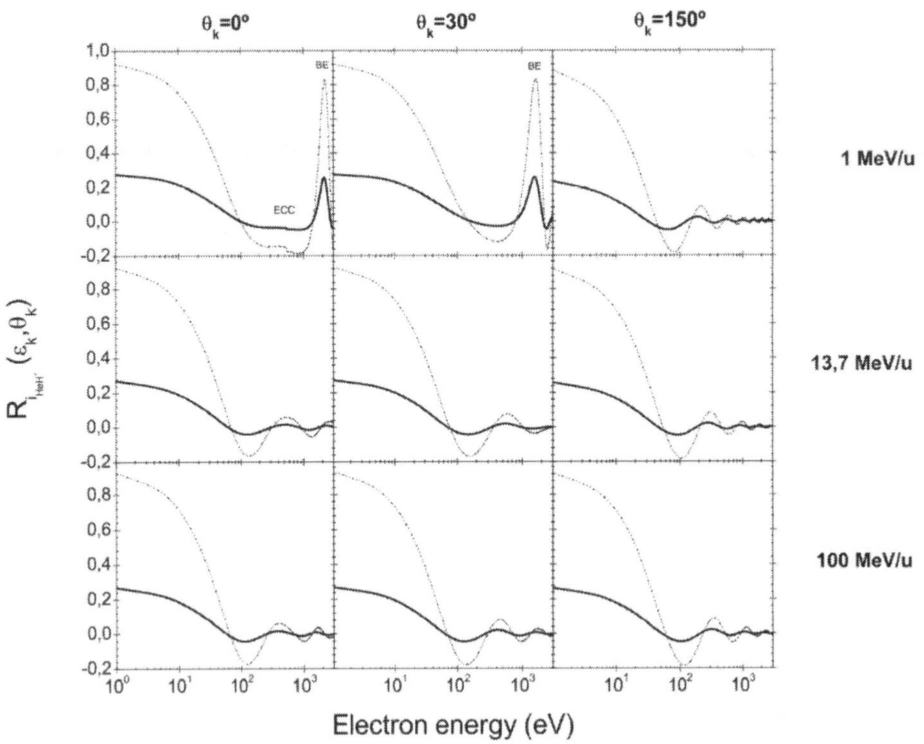

Figure 4. Comparison between the cross section ratios Ri_{HeH^+} (solid line) and Ri_{H_2} (dash dot dot line) for different impact energies and different values of θ_k. ECC and BE denote the electron capture to the continuum and binary encounter regions, respectively.

$$\mathcal{T}_{if} = \langle \varphi_f | \hat{\varepsilon} \cdot \nabla | \varphi_i \rangle \tag{18}$$

with $\hat{\varepsilon}$ the polarization vector perpendicular to the z axis, φ_i the initial bound state (equation (6)), and φ_f the final continuum wavefunction (equation (7)) described within the TEC approximation and the CDW-EIS model. The ratios obtained for this photoionization model are proportional to the factor $sin(|\vec{K} + \vec{k}|b)/(|\vec{K} + \vec{k}|b)$, which is also present in the integrand of the expression to calculate $\sigma^{(2)}$ and thus influencing the cross section ratios of equation (15) for the ion impact case.

A comparison of the ratios obtained for 100 MeV/u energy protons impacting on HeH^+ with those calculated with the photoionization model is given in figure 5. It can be seen that there is a qualitative agreement between both set of data. Still some differences are visible, in particular a shift in the oscillations with electron energy. In order to understand the origin of this discrepancy, we plot in figure 6 the cross section ratios for proton impact under the same conditions considered in figure 5, but taking the total momentum transfer $K = 0$. The agreement with photoionization results is now very good, being the frequency of the oscillations and the positions of the maxima and minima almost the same. Thus, the differences between ion and photon impact can be attributed to the nature of the ionizing mechanism. Again, it can be concluded that the influence of the momentum transferred by the projectile to the target is very important in the ionization reaction by ion impact.

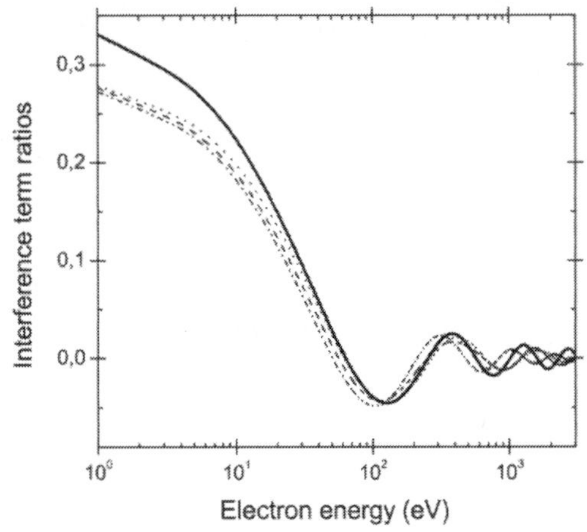

Figure 5. Comparison between the photoionization model (thick solid line) and the ratios $R_{i_{HeH^+}}$ for 100 MeV proton impact and $\theta_k = 0°$ (dash line), $\theta_k = 30°$ (dot line), and $\theta_k = 150°$ (dash dot dot line).

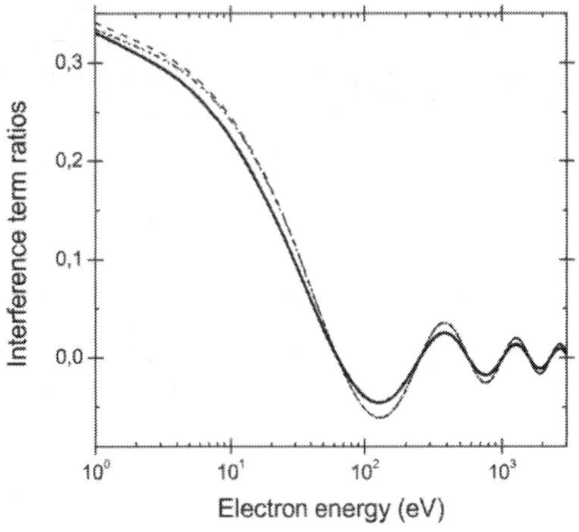

Figure 6. The same as figure 6 but considering $K = 0$ in the calculation of $R_{i_{HeH^+}}$.

4. Conclusions

Ionization of the heteronuclear HeH^+ molecular ion by bare ion impact was theoretically analysed. Five-fold differential cross sections were calculated for the target oriented parallel and perpendicular to the incident beam of particles in a complete coplanar geometry. We have shown that, in spite of the highly asymmetry of the target, there are fingerprints of coherent electron emission in the angular spectra of emitted electrons. Also, a comparison with photoionization with circularly polarized light was made, showing that the results corresponding to ion impact tend to the photoionization ones at high energy impact, when the molecule is aligned parallel to the incident beam. Nevertheless, this tendency is not observed in the angular spectra obtained for perpendicular orientation. This effect can be attributed to the momentum transferred by the ion beam to the target.

Also, doubly differential cross sections for impact of protons on HeH^+ were obtained. We shown that interference patterns are preserved even after averaging the angular spectra over all possible molecular orientations. The observed interference patterns for this heteronuclear target are less pronounced than for the case of the homonuclear H_2 target. This effect can be attributed to the partial localization of electrons in the initial state around the alpha particle of the HeH^+ molecular ion and can be associated with the Young interference experiment with circular holes of different diameters. We should remark that the final state of the residual target is not discriminated in our model. Comparison of these results with those corresponding to photoionization with linearly polarized light shows a qualitative agreement at very high proton energy. The small differences in DDCSs for the proton and photon cases have their origin in the transfer of momentum induced by the charged projectile, which is absent in the photoionization case.

Acknowledgments

This work was partially developed within the activities of the project PICT N° 1912 of the Agencia Nacional de Promoción Científica y Tecnológica and PIP 1026/10 of Consejo Nacional de Investigaciones Científicas y Técnicas of Argentina, FIS2007-60064, ACI2008-0777 and

CSD2007-00010 of MICINN (Spain), and the European COST Action CM0702. We thank the CCC-UAM for generous allocation of computer time. C A Tachino thanks to the Fundación Josefina Prats from Rosario (Argentina) for financial support.

References

[1] Cohen H D and Fano U 1966 *Phys. Rev.* **150** 30
[2] Kaplan I G and Markin A P 1969 *Sov. Phys. Dokl.* **14** 36
[3] Walter M and Briggs J 1999 *J. Phys. B: At. Mol. Opt. Phys.* **32** 2487
[4] Hanssen J, Joulakian B, Rivarola R D and Allan R J 1996 *Phys. Scr.* **53** 41
[5] Joulakian B, Hanssen J, Rivarola R D and Motassim A 1996 *Phys. Rev.* A **54** 1473
[6] Stolterfoht N *et al.* 2001 *Phys. Rev. Lett.* **87** 023201
[7] Galassi M E, Rivarola R D, Fainstein P D and Stolterfoht N 2002 *Phys. Rev.* A **66** 052705
[8] Nagy L, Kocbach L, Pora K and Hansen J 2002 *J. Phys. B: At. Mol. Opt. Phys.* **35** L453
[9] Laurent G, Fainstein P D, Galassi M E, Rivarola R D, Adoui L and Cassimi A 2002 *J. Phys. B: At. Mol. Opt. Phys.* **35** L495
[10] Galassi M E, Rivarola R D and Fainstein P D 2004 *Phys. Rev.* A **70** 032721
[11] Ciappina M F and Rivarola R D 2008 *J. Phys. B: At. Mol. Opt. Phys.* **41** 015203
[12] Alexander J S, Laforge A C, Hasan A, Machavariani Z S, Ciappina M F, Rivarola R D, Madison D H and Schulz M 2008 *Phys. Rev.* A **78** 060701
[13] Stia C R, Fojón O A, Weck P F, Hanssen J and Rivarola R D 2003 *J. Phys. B: At. Mol. Opt. Phys.* **36** L257
[14] Gao J, Madison D H and Peacher J L 2005 *Phys. Rev.* A **72**, 032721
[15] Fojón O A, Fernández J, Palacios A, Rivarola R D and Martín F 2004 *J. Phys. B: At. Mol. Opt. Phys.* **37** 3035
[16] Fojón A, Palacios A, Fernández J, Rivarola R D and Martín F 2006 *Phys. Lett.* A **350** 371
[17] Della Picca R, Fainstein P D, Martiarena M L and Dubois A 2006 *J. Phys. B: At. Mol. Opt. Phys.* **39** 473
[18] Yudin G L, Chelkowski S and Bandrauk A D 2006 *J. Phys. B: At. Mol. Opt. Phys.* **39** L17
[19] Fernández J, Fojón O A, Palacios A and Martín F 2007 *Phys. Rev. Lett.* **198** 043005
[20] Della Picca R, Fainstein P D, Martiarena M L, Sisourat N and DuBois A 2009 *Phys. Rev.* A **79** 032702
[21] Stolterfoht N *et al.* 2003 *Phys. Rev.* A **67** 030702 Tanis J A
[22] Hossain S, Alnaser A S, Landers A L, Pole D J, Knutson H, Robison A, Stamper B, Stolterfoht N and Tanis J A 2003 *Nucl. Instrum. Methods* B **205** 484
[23] Stolterfoht N, Sulik B, Skogvall B, Chesnel J Y, Frémont F, Hennecart D, Cassimi A, Adoui L, Hossain S and Tanis J A 2004 *Phys. Rev.* A **69** 012701
[24] Misra D, Kadhane U, Singh Y P, Tribedi L C, Fainstein P D and Richard P 2004 *Phys. Rev. Lett.* **92** 153201
[25] Hossain S, Landers A L, Stolterfoht N and Tanis J A 2005 *Phys. Rev.* A **72** 010701 (R)
[26] Misra D, Kelkar A, Kadhane U, Kumar A, Tribedi L and Fainstein P D 2006 *Phys. Rev.* A **74** 060701
[27] Misra D, Kelkar A, Kadhane U, Kumar A, Singh Y P, Tribedi L and Fainstein P D 2007 *Phys. Rev.* A **75** 052712
[28] Kamalou O, Chesnel J Y, Martina D, Hanssen J, Stia C R, Fojón O A, Rivarola R D and Frémont F 2005 *Phys. Rev.* A **71** 010702
[29] Milne-Brownlie D S, Foster M, Gao J, Lohmann B and Madison D H 2006 *Phys. Rev. Lett.* **96** 233201
[30] Staicu-Casagrande E M *et al.* 2008 *J. Phys. B: At. Mol. Opt. Phys.* **41** 025204 Fursa D V and Bray I
[31] Chatterjee S, Misra D, Kelkar A H, Tribedi L, Stia C R, Fojón O A and Rivarola R D 2008 *Phys. Rev.* A **78** 052701
[32] Chatterjee S, Kasthurirangan S, Kelkar A H, Stia C R, Fojón O A, Rivarola R D and Tribedi L C 2009 *J. Phys. B: At. Mol. Opt. Phys.* **42** 065201.
[33] Rolles D *et al.* 2005 *Nature* **437** 711
[34] Akoury D *et al.* 2007 *Science* **318** 949 C L, Osipov T, Adaniya H, Thompson J C, Prior M H, Belkacem A, Landers A L, Schmidt-Böcking H Dörner R
[35] Zimmermann B *et al.* 2008 *Nature Physics* **4** 649
[36] Tachino C A, Galassi M E, Martín F and Rivarola R D 2009 *J. Phys. B: At. Mol. Opt. Phys.* **42** 075203
[37] Tachino C A, Galassi M E, Martín F and Rivarola R D 2010 *J. Phys. B: At. Mol. Opt. Phys.* **43** 135203
[38] Frisch M J *et al.* 1999 *Gaussian 98, revision A11* (Gaussian Inc., Pittsburgh PA)
[39] O-ohata K, Taketa H and Huzinaga S 1966 *J. Phys. Soc. Japan* **21** 2306
[40] Hehre W J, Stewart R F and Pople J A 1969 *J. Chem. Phys.* **51** 2657

Multiple ionization and fragmentation of polyatomic molecules by fast heavy ions

A. Itoh, T. Mizuno* and T. Majima

Department of Nuclear Engineering, Kyoto University, Kyoto 606-8501, Japan

E-mail: itoh@nucleng.kyoto-u.ac.jp

Abstract. We investigated molecular fragmentation for CO, C_2H_2 and C_{60} molecules induced by impact of various fast heavy ions. Fragment ions produced in electron capture and loss collisions of projectile ions were measured in coincidence with final projectile charge states. As for C_{60}target, the number distribution of secondary electrons emitted in single collision event were also measured by means of a triple coincident method. The charge state distribution of transiently formed C_{60}^{r+} was successfully reproduced by our new model calculation on the basis of statistical energy deposition model developed for polyatomic molecules. Data acquisition using position sensitive detection system allows us to obtain 3D momentum imaging of fragment ions from CO and C_2H_2 molecules. Typical results relevant to molecular orientation effect and kinetic energy release (KER) are discussed.

1. Introduction

Molecular fragmentation by an external force exerted from a charged particle and a photon has been the subject of increasing attention to date [1, 2, 3]. In a collision with an energetic swift ion, a target molecule receives a considerable amount of electronic energy, leading to molecular excitation, fragmentation and multiple ionization as well. Furthermore, the amount of electronic energy deposition into a target is supposed to vary considerably depending on the impact parameter between collision partners and on the orientation of the target molecule. Hence, acquisition of various information about these quantities is important to achieve complete understanding of collision-induced molecular excitation and relaxation mechanisms.

On the other hand, recent and rapid development of multicoincidence and momentum imaging techniques allows one to deduce important parameters such as molecular orientation before collisions, kinetic energy of fragment ions, initial charge state of molecules prior to dissociation, and so forth. Investigation using this technique extends from diatomic to polyatomic molecules including C_{60} and biomolecules [4, 5]. To date, however, collision induced molecular fragmentation using MeV energy heavy ions is not fully understood yet.

In this paper, we investigate fragmentation as well as ionization of diatomic through C_{60} molecules induced by fast heavy ions with velocities higher than the Bohr velocity 2.2×10^8cm/s. In order to gain an insight into collision dynamics in molecular fragmentation, measurements were made under charge-changing conditions where the incident charge state of a projectile particle changes via electron capture/loss collisions with a target particle.

Journal of Physics: Conference Series **288** (2011) 012027 doi:10.1088/1742-6596/288/1/012027

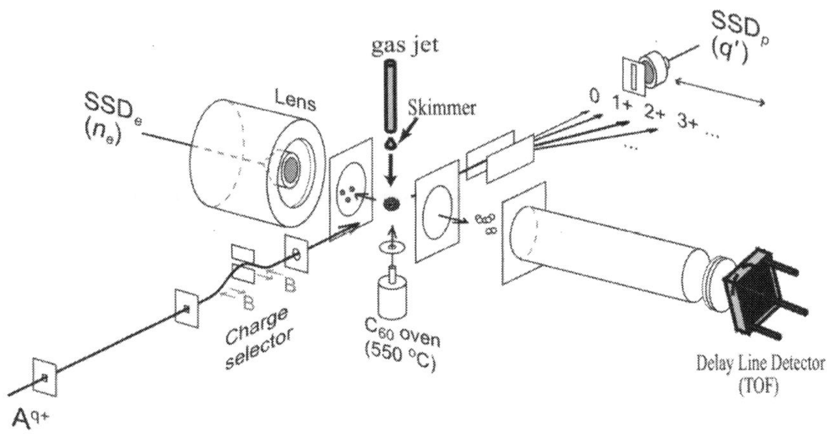

Figure 1. Schematic of experimental setup of the TOF-n_e coincidence measurements.

2. Experimental

Experiments were performed at the QSEC 1.7-MV tandem accelerator facility of Kyoto University. The experimental apparatus and method are described elsewhere [6, 7], and an essential outline is given here. Figure 1 shows a sketch of our experimental setup. A beam of A^{q+} extracted from the accelerator was carefully collimated to about 0.1 mm in diameter and was charge purified with a magnetic charge-selector before entering a collision chamber. Here A stands for a projectile species with a charge state q. After passing through a gaseous jet target the projectile particles were charge separated horizontally by an electrostatic deflector and detected by a movable solid state detector (SSD) which had a rectangular entrance slit of 0.5mm× 5 mm. The acceptance angle of SSD for scattered particles were about ±0.4 mrad (horizontal) and ±4 mrad (vertical), respectively.

Projectile ions studies in this work are 2.0MeV C^+, 6.0MeV O^{4+}, and 2.0-6.0 MeV $Si^{2+,3+}$ ions. Gaseous molecular jet targets are CO, C_2H_2 and C_{60}. As for a C_{60} target, we produced an effusive molecular beam by sublimation of high-purity powder in a quartz oven heated at 550°C as shown in Fig. 1. The diameter of the C_{60} beam in a collision region was estimated to be smaller than 4 mm. CO and C_2H_2 targets were produced from a nozzle with a skimmer.

Fragment ions were extracted perpendicular both to the incident beam and the molecular beam and were detected by a two-stage multichannel-plate detector. Mass-to-charge of product ions was measured with a time-of-flight (TOF) spectrometer operated under a Wiley-McLaren spatial-focusing condition. Data acquisition was made by using a fast-multichannel scaler (LN-6500,Labo.) for C_{60} experiments and a position-sensitive delay-line detector system (DLD-40, Roentdeck) [8] for other molecules. Both detection systems enable us to detect plural fragment ions produced in a single collision event.

Secondary electrons emitted from C_{60} were also measured with a SSD on which a positive voltage of 30 kV was applied. Hence, the pulse height of signals from this SSD is proportional to a total energy $30n_e$ keV when n_e electrons are detected simultaneously. Hence, a pulse height distribution provides a number distribution of secondary electrons. The position-sensitive DLD system provides initial momentum vectors of plural fragment ions and an information about angular correlation between them can be derived. Measurements were made in coincidence with

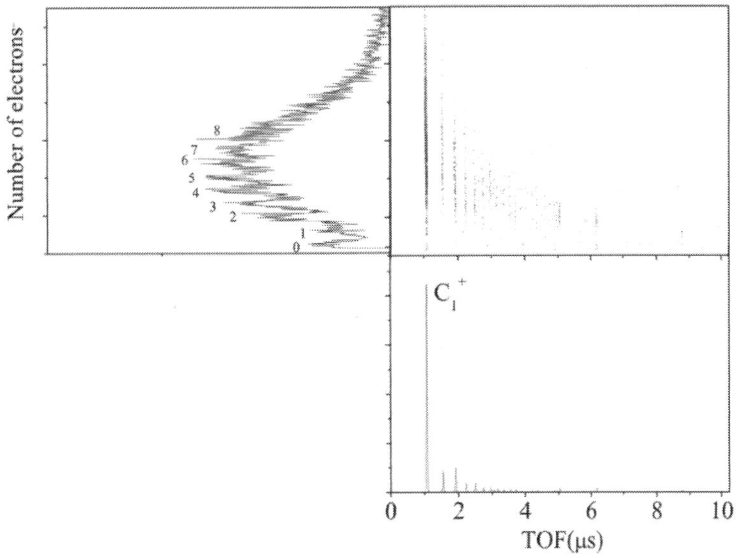

Figure 2. (Color online) Fragment ion spectrum and the number distribution of electrons in 1e-loss collisions of 6.0 MeV Si^{3+} with C_{60}.

charge-selected projectile particles at different scattering angles.

3. Results and discussion

3.1. C_{60} fragmentation

Figure 2 presents a typical example of correlation map between fragment ions (horizontal) and simultaneously emitted secondary electrons (vertical), measured for single electron loss collisions of 6.0 MeV Si^{3+} with C_{60}. It can be seen clearly that C_{60} is completely disintegrated into small-size fragment ions dominated by C^+ and the secondary electron spectrum exhibits a broad Gaussian-type distribution centered at the number of electrons of about 7.

From this map one can derive a series of n_e-distributions correlated with individual fragment ions and vice versa. It is emphasized that one can specify the transiently formed charge states of C_{60}^{r+} as we know the final charge state of projectiles and the total number of ionized electrons in addition to the charge state of fragment ions. We found that the multifragmentation of C_{60} begins at small degree of ionization of $r = 3$ [9]. It turned out that the key quantity charactering the multifragmentation of C_{60} is the internal excitation energy rather than the initial charge state r. The internal energy E_{int} is a part of total energy deposition into a C_{60} molecule E_d which varies significantly at different impact parameters b between collision partners. Figure 3 shows how the fragment ion distribution changes so dramatically when the impact parameter is different [6]. Here, two TOF spectra were obtained at scattering angles of 0 mrad and 0.8 mrad in two-electron loss collisions of 2MeV Si^{2+} with C_{60}. One can see obviously that only small fragment ions, resulting from complete molecular disintegration, are produced at 0.8 mrad, while at 0 mrad intact parent ions C_{60}^{r+} with $r = 1 - 4$ are also produced rather strongly. This completely different spectral feature indicates clearly that the value of E_d in close collisions is undoubtedly much larger than that in distant collisions. Also, we found that

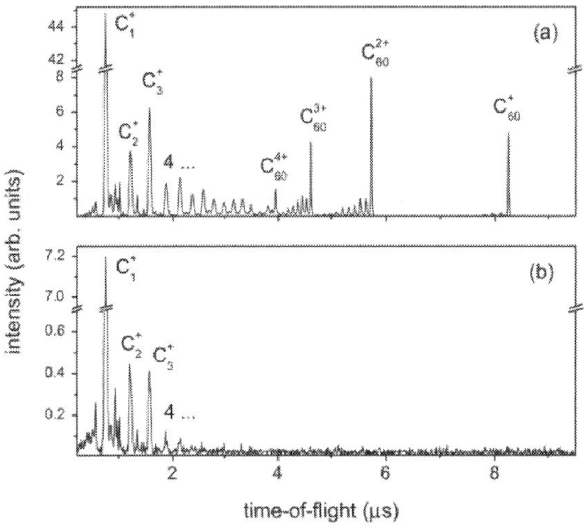

Figure 3. Comparison of TOF spectra for $2e$ loss collisions measured at two scattering angles; (a) $\theta = 0 \pm 0.4$mrad, and (b) $\theta = 0.8 \pm 0.4$mrad. 2.0 MeV Si^{2+} with C_{60}.

the mass distribution changes significantly in electron loss and capture collisions. Qualitatively to say, electron capture can take place in relatively distant collisions and electron loss may be limited to only close collisions because the electron loss is the ionization of a charged projectile ion by a neutral target molecule.

As the geometrical structure of C_{60} is known precisely, the yield of outgoing projectile particles scattered into various angles can be calculated numerically. The scattering yield of 2.0 MeV C^+ into θ = 0, 0.5 and 1.0 mrad are depicted in Fig. 4 as a function of the impact parameter b measured from the C_{60} center [10]. The number density of carbon atoms is depicted in the upper figure. We found that scattering into $\theta \neq 0$ is essentially limited to $b \leq 7$ corresponding to the radius of C_{60}. Together with the data of TOF spectra at different θ, we found that small fragment ions C_m^+ ($m = 1-3$) are produced in cage-penetration collisions and intact parent ions C_{60}^{r+} are produced in peripheral collisions with C_{60}.

As mentioned above, r-distribution associated with individual C_m^+ can be derived from 2D-map of TOF and n_e distribution as shown in Fig.2. Figure 5 shows the results measured for 0.5 MeV Si^+ and 6 MeV Si^{3+} impacts [11]. It should be pointed out that from r-distributions one can obtain information about the ionization energy E_{ion} which is the second part of the total energy deposition as $E_d = E_{int} + E_{ion}$. This was done by our model calculation [10] based on the statistical energy deposition (SED) model [12, 13]. With the aide of the r-hold ionization probability P_r [13], the r distribution Y_r is written as

$$Y_r = \int P_r(E_k) w(E_{ion}) dE_{ion} \tag{1}$$

with

$$w(E_{ion}) = \frac{1}{\sqrt{2\pi}\sigma} \exp\left(-\frac{(E_{ion} - E_c)^2}{2\sigma^2}\right) \tag{2}$$

where E_c the most probable ionization energy, σ the standard deviation and E_k the total kinetic

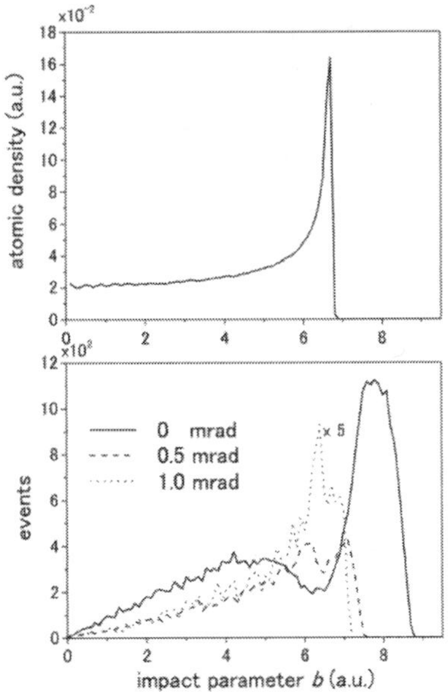

Figure 4. Calculated scattering yields at $\theta = 0, 0.5, 1.0$ mrad plotted as a function of the impact parameter b measured from the center of C_{60}.2MeV C+ on C_{60}. The upper shows number density of carbon atoms in C_{60}.

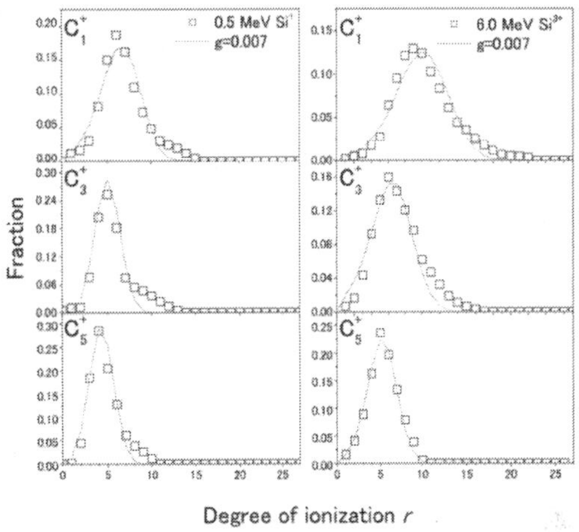

Figure 5. SED model calculations of r distributions correlated with C^+,C^{3+} and C^{5+}obtained for 0.5 MeV Si^+ and 6.0 MeV Si^{3+}single electron loss collisions with C_{60}.

energy of r ejected electrons with the relationship of

$$E_k(r) = E_{ion} - \sum_{i=1}^{r} I_i, \tag{3}$$

with I_i the i-th ionization potential of C_{60}. We applied Eq. (1) to our experimental r-distributions by treating σ and E_c as fitting parameters. Calculated curves are depicted by solid lines in Fig. 5, showing excellently good agreement with experimental distributions. The most probable ionization energies E_c obtained in this way for individual fragment ions are 820eV (C_1^+), 450eV (C_3^+) and 250eV (C_9^+) in case of 6MeV Si^{3+} impact. We stress that E_c can be deduced in our model calculation without knowledge of the kinetic energy carried away by secondary electrons. It is concluded that the SED model works fairly well to account for inelastic energy deposition in complicated collisions involving polyatomic molecules.

3.2. CO fragmentation

Apart from the target of C_{60}, diatomic and polyatomic molecules with line structure provide information about the orientation effect in molecular fragmentation. As a clear demonstration, we show in Fig.6 angular distributions of an ion pair (C^{2+}, O^+) produced from CO bombarded by 6 MeV O^{4+} ions. The data were taken for single electron capture and loss collisions associated with formation of $(CO)^{3+}$ as a predissociation parent ion. Here, the abscissa θ represents the molecular orientation angle between the beam axis and the CO molecular axis. It is well known

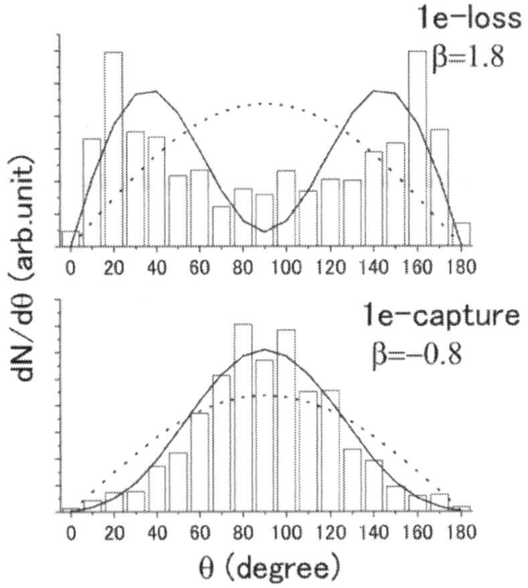

Figure 6. Production cross sections of the ion pair (C^{2+}, O^{+}) as a function of the molecular orientation angle θ measured for 1e-loss (upper) and 1e-capture (lower) collisions. Solid lines are the fitting results with Eq. (3) using anisotropy parameters β as denoted. Dashed lines show $\sin\theta$ for isotropic distributions.6MeV O^{4+} in CO

that the angular distribution in a diatomic molecule is described by the following formula [14]

$$N(\theta) \sim (1 + \beta P_2(\cos\theta))\sin\theta \tag{4}$$

where $P_2(x) = (3x^2 - 1)/2$ and β is the anisotropy parameter. The experimental distributions were fitted to this equation and the results are shown by solid lines. Note that isotropic fragmentation (β =0) is drawn by a dashed line as a reference. Fitting results are reasonably in good agreement with experimental curves. It is obvious that fragmentation associated with capture collisions seems more isotropic while it is highly anisotropic in loss collisions. On the other hand the fragmentation from $(CO)^{2+}$ into (C^{+}, O^{+}) was found to be isotropic in both loss and capture collisions. This implies that the triple ionization of CO occurs preferentially in parallel collisions ($\theta \simeq 0, 180°$) compared to perpendicular collisions. These features can be safely explained as follows. Ionization (electron loss) of O^{4+} with the ionization potential of 114 eV by a neutral CO is restricted to small impact-parameter collisions because a large inelastic energy is required to remove a tightly bound projectile electron. In parallel collisions, a projectile ion can undergo close collisions two times with C and O atoms and increases the electron loss probability compared to perpendicular collisions. By contrast, electron capture by O^{4+} can occur in distant collisions. Obviously the distant collision does not depend greatly on the orientation angle, resulting in isotropic distribution. It is noteworthy to refer to experimental results of Siegmann *et al* obtained for diatomic molecules [14, 15]. In their experiments, however, fragment ion pairs were measured without coincidence with projectile final charge states. They showed that fragmentation from $(CO)^{r+}$ or $(N_2)^{r+}$ becomes anisotropic with increasing r due to close collisions. These results are qualitatively consistent with our results.

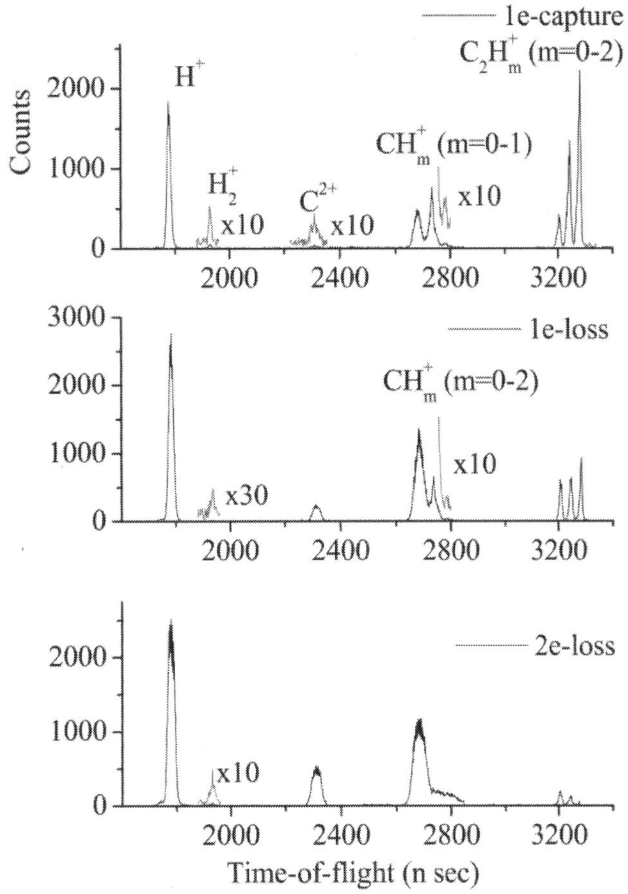

Figure 7. (Color online) TOF spectra in collisions between 2MeV Si^{2+} and C_2H_2

3.3. C_2H_2 fragmentation

Collision induced fragmentation of polyatomic molecular targets opens several new relaxation channels such as emission of fragment molecular ions, bond rearrangement or structure deformation. As for acetylene, there are some experimental investigations using photon impacts [16, 17]. In this work, we investigated acetylene fragmentation in charge changing collisions of 2 MeV Si^{2+} ions. TOF spectra measured in these collisions are shown in Fig. 7. Predominant products are H^+, C^{r+} (r =1,2), CH^+, C_2^+ and $C_2H_m^+$ (m =1,2). In addition, molecular ions formed through bond rearrangement dynamics are also observed weakly; see the peaks of H_2^+ and CH_2^+. It is interesting to point out that the doubly charged parent ion $(C_2H_2)^{2+}$ is not observed in our collision condition, while this peak was observed rather strongly in 1.2 MeV Ar^{8+} collisions [18]. It suggests that, in our lowly charged ion impact, $(C_2H_2)^{2+}$ is formed to unstable electronic or vibrational excited states, resulting in fragmentation. Close inspection of these spectra reveals rather large different profiles depending on the type of charge changing. Namely, the dominant products are $C_2H_m^+$ in single electron capture, while those in two-electron loss collisions are CH_m^+ produced via bond-breaking reactions. This fact reflects different impact parameters effective in capture and loss collisions.

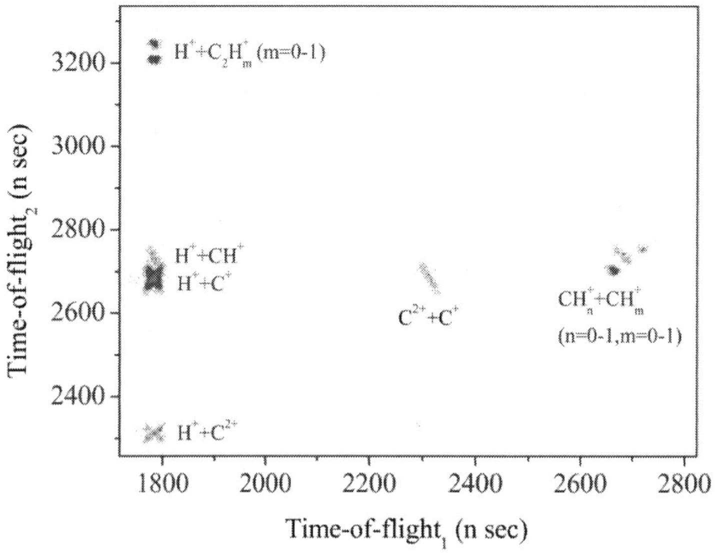

Figure 8. (Color online) TOF 2D spectra of fragment ions obtained in collisions between 2MeV Si^{2+} and C_2H_2

Figure 8 shows two-dimensional correlation map between two fragment ions produced in single collision events measured for 1e-loss collisions of 2MeV Si^{2+} ions. By analyzing 3D momentum data of these fragment ions we obtained kinetic energy release (KER) in various fragmentation channels from acetylene. Figures 9 and 10 present KER spectra in fragmentation channels of $C_2H_2 \longrightarrow (C_2H^+, H^+)$ and (C^+, C^+), respectively. Two peaks at about 3eV and 5 eV in (C_2H^+, H^+) may be assigned as the following fragmentation process [16],

$$Si^{2+} + C_2H_2 \rightarrow \quad C_2H_2^{2+*}(^1\Sigma_g^+) \rightarrow C_2H^+(^3\Pi) + H^+ \quad (KER = 3eV) \tag{5}$$

$$Si^{2+} + C_2H_2 \rightarrow C_2H_2^{2+*}(^1\Pi_u^+) \rightarrow C_2H^+(^1\Delta \text{ or } ^1\Sigma) + H^+ \quad (KER = 5eV) \tag{6}$$

Compared to Fig. 9 the KER spectra in (C^+, C^+) channel are broad and tailing to higher energies and are different considerably from capture to loss collisions. The results indicate that the (C^+, C^+) channel is attained through many intermediate dissociation channels. If the coulomb explosion model is applied to this channel, the KER estimated from $14.4/R$ between two C^+ ions is about 12 eV with $R = 1.21$Åthe interatomic distance between two carbon atoms. Hence, the binary explosion model is ruled out.

4. Summary

In summary, we described briefly our recent experimental data concerning molecular fragmentation of CO, C_2H_2 and C_{60} induced in charge changing collisions of fast heavy ions. It was found that the statistical energy deposition model works fairly well to reproduce experimental charge-state distributions of prefragmented C_{60}^{r+}. In particular, the ionization energy E_{ion} imparted to a molecule can be derived in our model calculations. It is pointed out

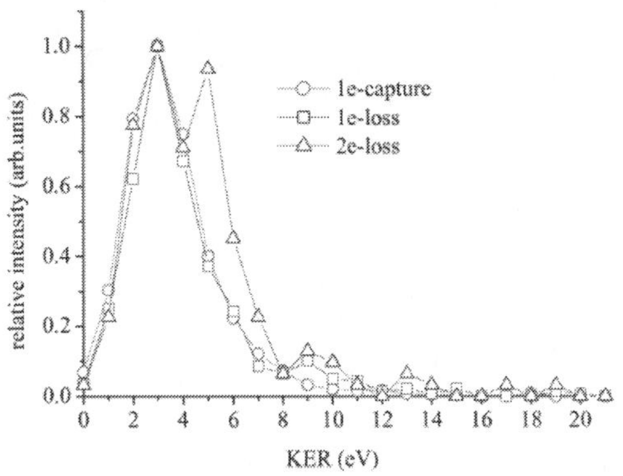

Figure 9. (Color online) KER spectra in a fragmentation channel (C_2H^+,H^+) obtained for 2MeV Si^{2+} collisions with C_2H_2.

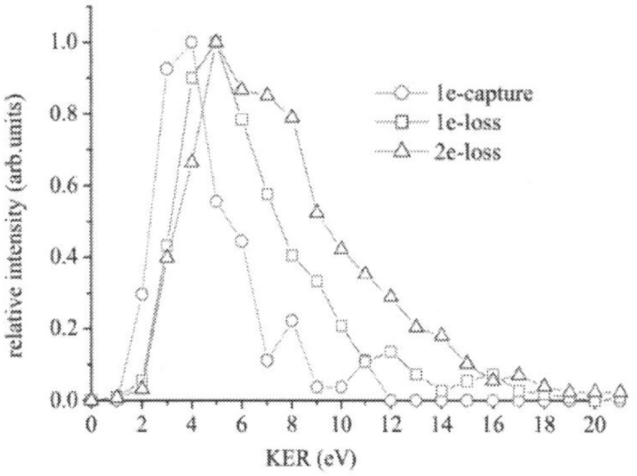

Figure 10. (Color online) The same as in Fig. 8 but for a fragmentation channel (C^+,C^+) .

that, once the amount of E_{ion} is known, the total energy deposition E_d into a target molecule may be deduced.

In collisions with CO and C_2H_2, molecular fragmentation was investigated in relationship with molecular orientation with respect to the incident ion beam, The kinetic energy distribution of fragment ions provides information about transient molecular excited states. To obtain further systematic information about molecular fragmentation, a series of hydrocarbons are investigated in our laboratory.

The work supported by Quantum Science and Engineering Center of Kyoto University.

* Present address: RIKEN (The Institute of Physical and Chemical Research), Wako 351-0198, Japan

Reference

[1] E. E. B. Campbell and E. Rohmund, 2000 *Rep. Prog. Phys.* **63** 1061
[2] J. Ullrich, *et. al.*, 2003*Rep. Prog. Phys.* **66** 1463
[3] T. Kaneyasu, T. Azuma, K. Okuno, 2005 *J. Phys. B: At. Mol. Opt. Phys.* **38** 1341
[4] A. Itoh, H. Tsuchida, T. Majima, and N. Imanishi, 1999 *Phys. Rev. A* **59** 4428
[5] B.L.S Brondsted *et al.*, 2006 *Phys. Rev. Lett.* **97** 133401(2006)
[6] T. Majima, Y. Nakai, T. Mizuno, H. Tsuchida, and A. Itoh, 2006 *Phys. Rev. A* **74** 033201
[7] T. Mizuno, T. Yamada, H. Tsuchida, Y. Nakai, and A. Itoh, 2010 *Phys. Rev. A* **81** 012704
[8] S. Sobottka and M. Williams, 1988 *IEEE Trans. Nucl. Sci.* **35** 348
[9] T. Majima, Y. Nakai, H. Tsuchida and A. Itoh, 2004 *Phys. Rev. A.* **69** 031202.
[10] T. Mizuno, D. Okamoto, T. Majima, Y. Nakai, H. Tsuchida and A. Itoh, 2007 *Phys. Rev. A* **75** 063203
[11] T. Mizuno, H. Tsuchida, T. Majima, Y. Nakai, and A. Itoh, 2008 *Phys. Rev. A* **78** 053202
[12] N. M. Kabachinik, V. N, Kondratyev, Z. Roller-Lutz, and H. O. Lutz, 1997 *Phys. Rev. A* **56** 2848. *ibid.*, 1998 **57** 990
[13] A. Russek, and J. Meli, 1970 *Physica* **46** 222
[14] B. Siegmann et al. 2002 *Phys. Rev. A.* **66** 052701
[15] B. Siegmann, U. Werner, H. Lebius, B. Huber, H. O. Lutz, R. Mann, 2003 *Nucl. Instr. and Meth. B* **205** 629
[16] T. Osipov T *et.al.* 2003 *Phys. Rev. Lett.* **90** 233002
[17] A. Hishikawa, A. Matsuda, M. Fushitani and E. Takahashi, 2007 *Phys. Rev. Lett.* **99** 258302
[18] S. De, J. Rajput, A. Roy, P.N. Ghosh and C. P. Safvan, 2008 *Phys. Rev.A.* **77** 022708

Journal of Physics: Conference Series **288** (2011) 012028 doi:10.1088/1742-6596/288/1/012028

Dynamic correlation in the electron angular distribution in ionization of helium by ion impact

J. M. Monti†, O. A. Fojón†, J. Hanssen‡ and R. D. Rivarola†

† Instituto de Física Rosario (CONICET-UNR) and Facultad de Ciencias Exactas, Ingeniería y Agrimensura, Universidad Nacional de Rosario, Avenida Pellegrini 250, 2000 Rosario, Argentina

‡ Institut de Chimie, Physique et Matériaux, Laboratoire de Physique Moléculaire et des Collisions, Université Paul Verlaine - Metz, 1 Bv. Arago, 57078 Metz Cedex 3, France.

E-mail: rivarola@fceia.unr.edu.ar

Abstract. Single ionization of helium by proton impact is investigated in terms of a four-body distorted wave model. In this approximation both electrons are considered as active, being one of them ionized whereas the other remains in a residual target bound state. The influence of dynamic correlation between electrons is investigated by comparison with a four-body uncorrelated distorted wave model. Double differential cross sections as a function of the emission angle for fixed electron energies and different collision energies are presented.

1. Introduction

The present work deals with single ionization of helium atoms by fast proton impact within a four-body distorted wave model. Originally the three body distorted wave models, like the continuum distorted wave (CDW, [1]) and the continuum distorted wave-eikonal initial state (CDW-EIS, [2]), were developed to investigate ion-atom processes for monoelectronic targets. They were introduced in order to accelerate the convergence of a Born series description. Later an extension of the CDW-EIS description for single ionization of multielectronic targets was made by Fainstein *et al.* [3]. They reduced the multielectronic case to a monoelectronic treatment within a three-body approximation, being the three bodies considered the projectile, the residual target and the *active* electron (the one to be ionized as a consequence of the collision). The other electrons, the *passive* ones, were supposed to remain as frozen in their initial orbitals during the reaction (see [3]). Four-body distorted wave models were also introduced to study single ionization of dielectronic targets [4]. Correct boundary conditions were preserved in both entry and exit channels but only the active electron was considered to be distorted by the projectile. In order to avoid any asymmetry in the description of the target electrons, in a previous work [5], we considered both of them as active electrons in a four-body distorted wave-eikonal initial state formulation (4B-DW-EIS). Both electrons were considered to evolve simultaneously, so that any change in one of them is felt by the other and vice versa, so that they are dynamically correlated. Following reference [5], double differential cross sections for the case of single ionization of He by bare ion impact as a function of the emission angle for fixed ejected electron energies are calculated. The influence of the dynamic electronic correlation is determined by comparison with a four-body dynamically uncorrelated model. This model, called CB-CDW-EIS in [5], treats the

Journal of Physics: Conference Series **288** (2011) 012028 doi:10.1088/1742-6596/288/1/012028

residual target bound electron excitation as in a first Born approximation with correct boundary conditions (Coulomb-Born approximation; CB) and the ionization process as in the CDW-EIS approximation.

2. Theory

Let us consider the impact of a bare ion of nuclear charge Z_P on a dielectronic target of nuclear charge Z_T. We describe the reaction within the straight-line version of the impact parameter approximation, where $\vec{R} = \vec{\rho} + \vec{v}t$, being $\vec{R}$ the internuclear vector, $\vec{v}$ the impact velocity and t the collision time, considering $t = 0$ at the closest distance between the nuclei. The reaction is described from a reference frame fixed to the target nucleus. Both electrons are considered to be active during the collision.

Following a distorted wave formalism, and in order to preserve correct asymptotic conditions (see [5]), the initial and final distorted wavefunctions are chosen as

$$\begin{aligned} \chi_\alpha^+(\vec{x}_1, \vec{x}_2, t) = \varphi_\alpha(\vec{x}_1, \vec{x}_2)\mathcal{L}_\alpha^+(\vec{s}_1, \vec{s}_2)\times \\ \times \exp\left[i\frac{Z_P Z_T}{v}\ln\left(vR - \vec{v}\cdot\vec{R}\right)\right]\times \\ \times \exp\left(-iE_\alpha t\right) \end{aligned} \tag{1}$$

and

$$\begin{aligned} \chi_\beta^-(\vec{x}_1, \vec{x}_2, t) = \frac{1}{\sqrt{2}}(1+\mathcal{P}_{12})\left[\varphi_\beta(\vec{x}_1, \vec{x}_2)\mathcal{L}_\beta^-(\vec{s}_1, \vec{s}_2)\right]\times \\ \times \exp\left[-i\frac{Z_P Z_T}{v}\ln\left(vR + \vec{v}\cdot\vec{R}\right)\right]\times \\ \times \exp\left(-iE_\beta t\right) \end{aligned} \tag{2}$$

with $\vec{x}_j$ and $\vec{s}_j$ denoting the position vector of the j-th electron ($j = 1, 2$) as seen from the target and projectile nuclei, respectively, φ_α is the initial two-electron bound wavefunction with atomic energy E_α and φ_β is a two-electron function with energy E_β describing one of the electrons bound and the other in a continuum state of the target and $\mathcal{P}_{12}$ is the commutation operator.

The two-electron distorting initial $\mathcal{L}_\alpha^+$ and final $\mathcal{L}_\beta^-$ functions are given by

$$\begin{aligned} \mathcal{L}_\alpha^+(\vec{s}_1, \vec{s}_2) = \prod_{j=1,2}\mathcal{E}_\alpha^+(\vec{s}_j) = \\ = \prod_{j=1,2}\exp\left[-i\frac{Z_P}{v}\ln\left(vs_j + \vec{v}\cdot\vec{s}_j\right)\right] \end{aligned} \tag{3}$$

and

$$\begin{aligned} \mathcal{L}_\beta^-(\vec{s}_1, \vec{s}_2) = \mathcal{F}_\beta^-(\vec{s}_1)\mathcal{E}_\beta^-(\vec{s}_2) = \\ = N^*\left(\frac{Z_P}{p}\right)\ {}_1F_1\left[-i\frac{Z_P}{p}, 1, -i(ps_1 + \vec{p}\cdot\vec{s}_1)\right]\times \\ \times \exp\left[i\frac{Z_P}{v}\ln\left(vs_2 - \vec{v}\cdot\vec{s}_2\right)\right] \end{aligned} \tag{4}$$

where $\vec{p}$ is the linear momentum of the ionized electron with respect to the projectile, ${}_1F_1$ is the confluent hypergeometric function and N its normalization factor.

For the He case the initial target bound wavefunction was considered within a five-zeta

Roothaan-Hartree-Fock representation (see [6]) in which it is given by the product of one-electron five-functions $\phi_\alpha(\vec{x})$

$$\varphi_\alpha(\vec{x}_1, \vec{x}_2) = \phi_\alpha(\vec{x}_1)\phi_\alpha(\vec{x}_2) \tag{5}$$

The two-electron final wavefunction φ_β is chosen as

$$\begin{aligned}\varphi_\beta(\vec{x}_1, \vec{x}_2) &= \phi_\beta^c(\vec{x}_1)\phi_\beta(\vec{x}_2) = \\ &= \phi_\beta(\vec{x}_2)\frac{\exp(i\vec{k}\cdot\vec{x}_1)}{(2\pi)^{3/2}} \times \\ &\times N^*\left(\frac{Z_T - 1}{k}\right) \times \\ &\times\, {}_1F_1\left[-i\frac{(Z_T-1)}{k}, 1, -i(kx_1 + \vec{k}\cdot\vec{x}_1)\right]\end{aligned} \tag{6}$$

where $\vec{k}$ is the momentum of the ejected electron seen from the target nucleus. In (6), ϕ_β represents a hydrogenic state corresponding to the residual target. The continuum wavefunction in (6) corresponds to the interaction between the ionized electron and the residual target at large asymptotic separation. We have verified [5] that a possible ionization path in which the projectile impacts on the electron that remains bound to the target while the other one is ejected through a dynamical shake-off mechanism gives negligible contributions to the transition amplitude by comparison with the mechanism of direct interaction of the projectile with the emitted electron. Thus, for the 4B-DW-EIS approximation the prior version of the scattering amplitude results

$$\mathcal{A}_{\alpha\beta}^-(\vec{\rho}) = i\,(\rho v)^{2iZ_PZ_T/v}\sqrt{2}\int_{-\infty}^{+\infty} dt\left[S_{\alpha\beta}^{bound}(\vec{\rho}, t)\, a_{\alpha\beta}^{ion,-}(\vec{\rho}, t)\right], \tag{7}$$

where $S_{\alpha\beta}^{bound}(\vec{\rho}, t)$ is a one-electron time dependent overlap integral corresponding to a final bound state, and $a_{\alpha\beta}^{ion,-}(\vec{\rho}, t)$ a one-electron time dependent scattering amplitude corresponding to electron ionization (see [5] for further details). Thus, the excitation channel is populated through a dynamical shake-off mechanism.

A four-body dynamically uncorrelated model can be obtained by a further approximation to the 4B-DW-EIS one using the distortions

$$\mathcal{L}_\alpha^+(\vec{s}_1, \vec{R}) = \mathcal{E}_\alpha^+(\vec{s}_1)\exp\left[-i\frac{Z_P}{v}\ln(vR - \vec{v}\cdot\vec{R})\right] \tag{8}$$

and

$$\mathcal{L}_\beta^-(\vec{s}_1, \vec{R}) = \mathcal{F}_\beta^-(\vec{s}_1)\exp\left[i\frac{Z_P}{v}\ln(vR + \vec{v}\cdot\vec{R})\right]. \tag{9}$$

This leads to a transition amplitude given by

$$\mathcal{A}_{\alpha\beta}^-(\vec{\rho}) = i\,(\rho v)^{2iZ_P(Z_T-1)/v}\sqrt{2}\; S_{\alpha\beta}^{bound}(\vec{\rho})\mathcal{A}_{\alpha\beta}^{ion,-}(\vec{\rho}) \tag{10}$$

where now $S_{\alpha\beta}^{bound}(\vec{\rho})$ is the time-independent overlap integral between the initial and final bound states, and $\mathcal{A}_{\alpha\beta}^{ion,-}(\vec{\rho})$ is a one-electron transition amplitude given by

$$\mathcal{A}_{\alpha\beta}^{ion,-}(\vec{\rho}) = \int_{-\infty}^{+\infty} dt\; a_{\alpha\beta}^{ion,-}(\vec{\rho}, t). \tag{11}$$

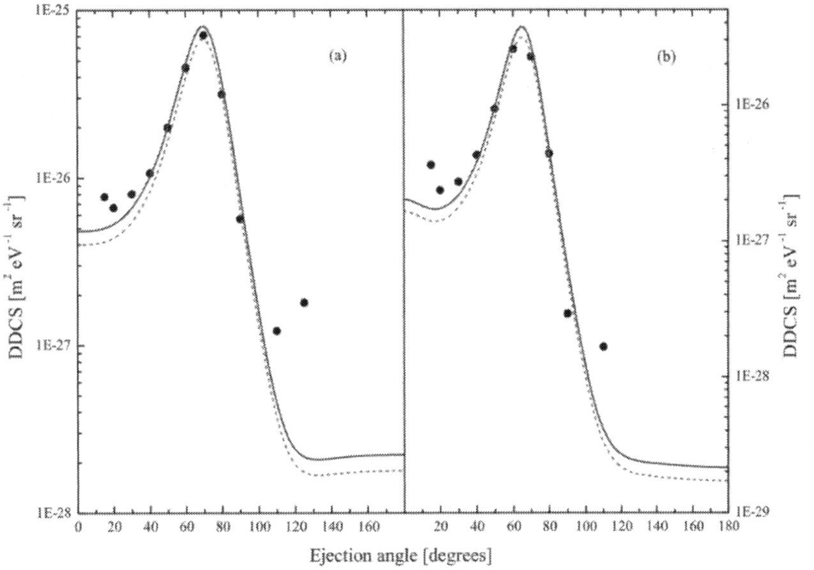

Figure 1. Double differential cross section for single ionization of He by 1.5 MeV proton impact as a function of the emission angle for fixed electron energy of 300 eV (figure a) and 500 eV (figure b). Theory: ——, 4B-DW-EIS calculations; - - - -, CB-CDW-EIS calculations [5]. Experiments: ●, Toburen *et al.* extracted from [7].

3. Results and discussions

Double differential cross sections (DDCS) for single ionization of He calculated with the 4B-DW-EIS and CB-CDW-EIS ([5]) models are shown in figures 1 and 2 for the cases of 1.5 MeV and 4.2 MeV proton impact, respectively. Only the final $He^+(1s)$ state is considered in the calculations which is the dominant channel for the ionization reaction studied. A general good agreement is found between both theoretical models and existing experimental data [7].

4B-DW-EIS results are normalized according to the procedure given in our previous work [5]. Thus, as there discussed, the difference between both theoretical models will give a lower bound to the contribution of the dynamic electron correlation.

From figures 1 and 2, the influence of dynamic correlation appears to be more important at forward and backward emission angles. In the binary peak region, where the reaction is dominated by the projectile-electron interaction the dynamic correlation seems to play a weaker role. It can be seen that, for the cases presented, the correction introduced by this interaction can be at least 20% for the case of 1.5 MeV proton impact whereas for the 4.2 MeV a correction at least as high as 50% is found.

To the light of these results we can conclude that for this system the dynamic electron correlation may play an important role on the doubly differential angular spectra for single ionization. However, a further systematic study of its importance as a function of the impact and ejected electron energies and on the projectile charge state appears as necessary. This is matter of our present research.

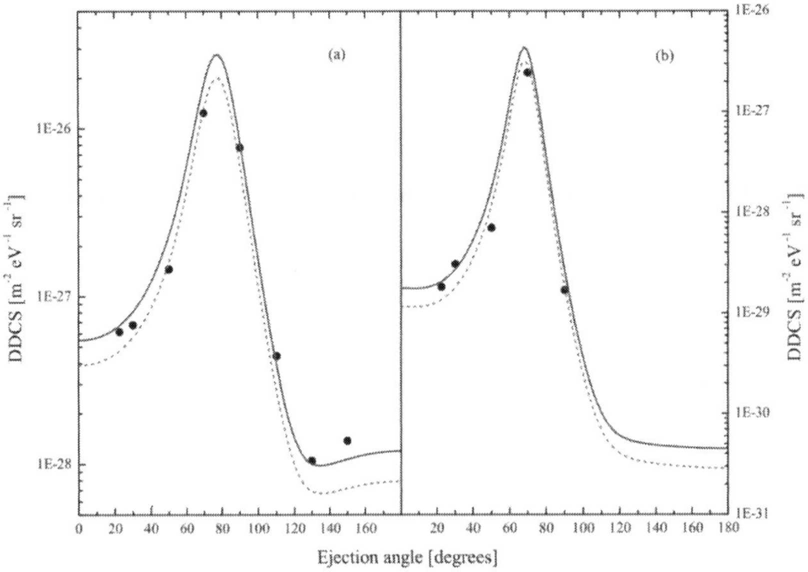

Figure 2. Double differential cross section for single ionization of He by 4.2 MeV proton impact as a function of the emission angle for fixed electron energy of 341 eV (figure a) and 1166 eV (figure b). Theory: ——, 4B-DW-EIS calculations; - - - -, CB-CDW-EIS calculations [5]. Experiments: ●, Stolterfoht *et al.* extracted from [7].

4. Conclusions

A four-body distorted wave model has been developed and applied to the case of single ionization of He by high energy proton impact. Double differential cross sections have been calculated as a function of the emission angle for a fixed energy of the ionized electron. The comparison of the 4B-DW-EIS results with the ones from CB-CDW-EIS shows that the dynamic electron correlation can give an important contribution on the DDCS. This effect is shown to be more important at forward and backward electron emission.

Acknowledgments

The authors thanks the Agencia Nacional de Promoción Científica y Tecnológica and the Consejo Nacional de Investigaciones Científicas y Técnicas, both from República Argentina, for financial support through the projects PICT N° 1912 and PIP N° 1026/10, respectively.

References

[1] Belkić Dž 1978 *J. Phys. B: At. Mol. Phys.* **11** 3529.
[2] Crothers D S F and McCann J F 1983 *J. Phys. B: At. Mol. Phys.* **16** 3229.
[3] Fainstein P D, Ponce V H and Rivarola R D 1988 *J. Phys. B: At. Mol. Opt. Phys.* **21** 287.
[4] Belkić Dž, *J. Phys. B: At. Mol. Opt. Phys.* **30**, 1731 (1997); *Nucl. Instrum. Methods B* **124**, 365 (1997)
[5] J. M. Monti, O. A. Fojón, J. Hanssen and R. D. Rivarola *J. Phys. B: At. Mol. Opt. Phys.* **42**, 195201 (2009)
[6] Clementi E and Roetti C 1974 *Atomic Data and Nuclear Data Tables* **14** 177.
[7] Rudd M E, Toburen L H and Stolterfoht N 1976 *Atomic Data and Nuclear Data Tables* **18** 413.

AUTHOR INDEX

Al-Hagan, O....49
Artamonov, O M....99
Avaldi, L....40, 142
Bellm, S M....49, 94
Berakdar, J....40
Berrah, N....133
Blanco, F....94
Bolognesi, P....40, 142
Bray, I....10
Camilloni, R....40
Cappello, C Dal....22
Casagrande, E M Staicu....16
Champion, C....22
Chen, Zhangjin....112
Chohda, M....156
Colgan, J....1, 10
Colyer, C J....49, 94
Cvejanovic, D....40, 99
De Sanctis, M L....68
Dubois, R D....75
Ehara, M....174
Fang, L....133
Fernández, J....185
Fojón, O A....34, 68, 185, 216
Fukuda, R....174
Fursa, D V....10
Galassi, M E....197
Garcia, G....94
Gejo, T....156, 167
Guagliardo, P....99
Hanne, G F....49
Hanssen, J....216
Harries, J R....167
Hashimoto, Shogo....107
Higuchi, I....156
Hino, Yuta....87
Hirayama, Takato....81
Hoener, M....133
Honma, K....156, 167
Horikawa, T....174
Hoshino, M....174
Huang, C....149
Itoh, A....207
Jinno, Satoshi....81
Kada, I....22
Kadyrov, A S....10
Kakiuchi, Takuhiro....107
Katayanagi, H....149
King, G C....142
Kouzakov, Konstantin A....63
Kubotera, Hayato....81
Kuniwake, M....167
Lahmam-Bennani, A....16
Li, C....16
Liang, Yaqiu....112
Lin, C D....112
Lohmann, B....49, 94
Madison, D....49
Majima, T....207
Mansouri, A....22
Martín, F....185, 197
Mase, Kazuhiko....107
Mihelic, A....142
Mitsuke, K....149
Mizuno, T....207
Moise, A....142
Monti, J M....216
Nagaoka, S....156
Nakashima, T....149
Nakatsuji, H....174
Ning, C G....49
Nogami, Keisuke....87
Okada, K....149
O'Keeffe, P....142
Oura, M....167
Ovcharenko, E....142
Palacios, A....185
Patidar, Vinod....55
Pavlyukh, Y....40
Pindzola, M S....1, 10
Politis, M F....68
Popov, Yuri V....63
Pravica, L....40, 99
Purohit, G....55
Richter, R....142
Rivarola, R D....34, 197, 216
Sakai, M....149
Sakai, Yasuhiro....87
Samarin, S N....99
Sekitsuka, Tatsuya....81
Sergeant, A D....99
Stelbovics, A T....10
Stia, C R....34, 68
Sud, K K....55
Suemitsu, A....149
Suvorov, A A....99
Suzuki, I H....156
Tabayashi, K....149, 156
Tachibana, Takayuki....81
Tachino, C A....197
Takahashi, Masahiko....63
Takahashi, O....156
Tamenori, Y....156
Tanaka, H....174
Tanaka, Masatoshi....107
Tanaka, T....174
Tong, X M....128
Toshima, N....128
Tsutsumi, Y....156
Ueda, K....174

AUTHOR INDEX

Vuilleumier, R 68
Wilkie, P 99
Williams, J F 99
Xu, Junliang 112
Yagi, H 149
Yamamoto, Karin 87
Yamazaki, Takahiko 107

Institute of Physics
Dirac House, Temple Back
Bristol BS1 6BE UK